*Elementary Applied Topology*, edition 1.0



ISBN: 978-1-5028-8085-7

Robert Ghrist is the *Andrea Mitchell PIK University Professor*
of Mathematics and Electrical & Systems Engineering
at the University of Pennsylvania.

# Contents

**Preface** 3

**1 Manifolds** 9
1.1 Manifolds . . . 10
1.2 Configuration spaces of linkages . . . 11
1.3 Derivatives . . . 13
1.4 Vector fields . . . 14
1.5 Braids and robot motion planning . . . 15
1.6 Transversality . . . 17
1.7 Signals of opportunity . . . 20
1.8 Stratified spaces . . . 22

**2 Complexes** 25
2.1 Simplicial and cell complexes . . . 26
2.2 Vietoris-Rips complexes and point clouds . . . 28
2.3 Witness complexes and landmarks . . . 29
2.4 Flag complexes and networks . . . 29
2.5 Čech complexes and random samplings . . . 30
2.6 Nerves and neurons . . . 31
2.7 Phylogenetic trees and links . . . 32
2.8 Strategy complexes and uncertainty . . . 34
2.9 Decision tasks and consensus . . . 35
2.10 Discretized graph configuration spaces . . . 37
2.11 State complexes and reconfiguration . . . 39

**3 Euler Characteristic** 43
3.1 Counting . . . 44
3.2 Curvature . . . 45
3.3 Nonvanishing vector fields . . . 46
3.4 Fixed point index . . . 47
3.5 Tame topology . . . 48
3.6 Euler calculus . . . 50
3.7 Target enumeration . . . 51
3.8 A Fubini Theorem . . . 52

3.9 Euler integral transforms . . . . . . . . . . . . . . . . . . . . 54
3.10 Intrinsic volumes . . . . . . . . . . . . . . . . . . . . 56
3.11 Gaussian random fields . . . . . . . . . . . . . . . . . . . . 58

**4 Homology 61**
4.1 Simplicial and cellular homology . . . . . . . . . . . . . . . . . . . . 62
4.2 Homology examples . . . . . . . . . . . . . . . . . . . . 64
4.3 Coefficients . . . . . . . . . . . . . . . . . . . . 65
4.4 Singular homology . . . . . . . . . . . . . . . . . . . . 68
4.5 Reduced homology . . . . . . . . . . . . . . . . . . . . 69
4.6 Čech homology . . . . . . . . . . . . . . . . . . . . 70
4.7 Relative homology . . . . . . . . . . . . . . . . . . . . 70
4.8 Local homology . . . . . . . . . . . . . . . . . . . . 72
4.9 Homology of a relation . . . . . . . . . . . . . . . . . . . . 72
4.10 Functoriality . . . . . . . . . . . . . . . . . . . . 74
4.11 Inverse kinematics . . . . . . . . . . . . . . . . . . . . 76
4.12 Winding number and degree . . . . . . . . . . . . . . . . . . . . 77
4.13 Fixed points and prices . . . . . . . . . . . . . . . . . . . . 79

**5 Sequences 83**
5.1 Homotopy invariance . . . . . . . . . . . . . . . . . . . . 84
5.2 Exact sequences . . . . . . . . . . . . . . . . . . . . 85
5.3 Pairs and Mayer-Vietoris . . . . . . . . . . . . . . . . . . . . 87
5.4 Equivalence of homology theories . . . . . . . . . . . . . . . . . . . . 89
5.5 Cellular homology, redux . . . . . . . . . . . . . . . . . . . . 90
5.6 Coverage in sensor networks . . . . . . . . . . . . . . . . . . . . 92
5.7 Degree and computation . . . . . . . . . . . . . . . . . . . . 94
5.8 Borsuk-Ulam theorems . . . . . . . . . . . . . . . . . . . . 97
5.9 Euler characteristic . . . . . . . . . . . . . . . . . . . . 98
5.10 Lefschetz index . . . . . . . . . . . . . . . . . . . . 99
5.11 Nash equilibria . . . . . . . . . . . . . . . . . . . . 100
5.12 The game of Hex . . . . . . . . . . . . . . . . . . . . 102
5.13 Barcodes and persistent homology . . . . . . . . . . . . . . . . . . . . 104
5.14 The space of natural images . . . . . . . . . . . . . . . . . . . . 106
5.15 Zigzag persistence . . . . . . . . . . . . . . . . . . . . 108

**6 Cohomology 111**
6.1 Duals . . . . . . . . . . . . . . . . . . . . 112
6.2 Cochain complexes . . . . . . . . . . . . . . . . . . . . 113
6.3 Cohomology . . . . . . . . . . . . . . . . . . . . 115
6.4 Poincaré duality . . . . . . . . . . . . . . . . . . . . 117
6.5 Alexander duality . . . . . . . . . . . . . . . . . . . . 118
6.6 Helly's Theorem . . . . . . . . . . . . . . . . . . . . 119
6.7 Numerical Euler integration . . . . . . . . . . . . . . . . . . . . 120
6.8 Forms and Calculus . . . . . . . . . . . . . . . . . . . . 121

6.9 De Rham cohomology . . . 124
6.10 Cup products . . . 125
6.11 Currents . . . 126
6.12 Laplacians and Hodge Theory . . . 129
6.13 Circular coordinates in data sets . . . 131

**7 Morse Theory 135**
7.1 Critical points . . . 136
7.2 Excursion sets and persistence . . . 137
7.3 Morse homology . . . 138
7.4 Definable Euler integration . . . 140
7.5 Stratified Morse theory . . . 141
7.6 Conley index . . . 143
7.7 Lefschetz index, redux . . . 147
7.8 Discrete Morse theory . . . 149
7.9 LS category . . . 151
7.10 Unimodal decomposition in statistics . . . 153

**8 Homotopy 157**
8.1 Group fundamentals . . . 158
8.2 Covering spaces . . . 160
8.3 Knot theory . . . 163
8.4 Higher homotopy groups . . . 166
8.5 Biaxial nematic liquid crystals . . . 168
8.6 Homology and homotopy . . . 169
8.7 Topological social choice . . . 170
8.8 Bundles . . . 171
8.9 Topological complexity of path planning . . . 173
8.10 Fibrations . . . 175
8.11 Homotopy type theory . . . 176

**9 Sheaves 179**
9.1 Cellular sheaves . . . 180
9.2 Examples of cellular sheaves . . . 181
9.3 Cellular sheaf cohomology . . . 184
9.4 Flow sheaves and obstructions . . . 186
9.5 Information flows and network coding . . . 188
9.6 From cellular to topological . . . 189
9.7 Operations on sheaves . . . 193
9.8 Sampling and reconstruction . . . 195
9.9 Euler integration, redux . . . 197
9.10 Cosheaves . . . 198
9.11 Bézier curves and splines . . . 200
9.12 Barcodes, redux . . . 202

**10 Categorification** **207**
10.1 Categories . . . . . . . . . . . . . . . . . . . . . . . . . . . . 208
10.2 Morphisms . . . . . . . . . . . . . . . . . . . . . . . . . . . . 211
10.3 Functors . . . . . . . . . . . . . . . . . . . . . . . . . . . . . 213
10.4 Clustering functors . . . . . . . . . . . . . . . . . . . . . . . . 215
10.5 Natural transformations . . . . . . . . . . . . . . . . . . . . . 217
10.6 Interleaving and stability in persistence . . . . . . . . . . . . . 219
10.7 Limits . . . . . . . . . . . . . . . . . . . . . . . . . . . . . . . 220
10.8 Colimits . . . . . . . . . . . . . . . . . . . . . . . . . . . . . 223
10.9 Sheaves, redux . . . . . . . . . . . . . . . . . . . . . . . . . . 224
10.10 The genius of categorification . . . . . . . . . . . . . . . . . . 226
10.11 "Bring out number" . . . . . . . . . . . . . . . . . . . . . . . 229

**A Background** **237**
A.1 On point-set topology . . . . . . . . . . . . . . . . . . . . . . . 238
A.2 On linear and abstract algebra . . . . . . . . . . . . . . . . . . 239

**Bibliography** **245**

# Preface

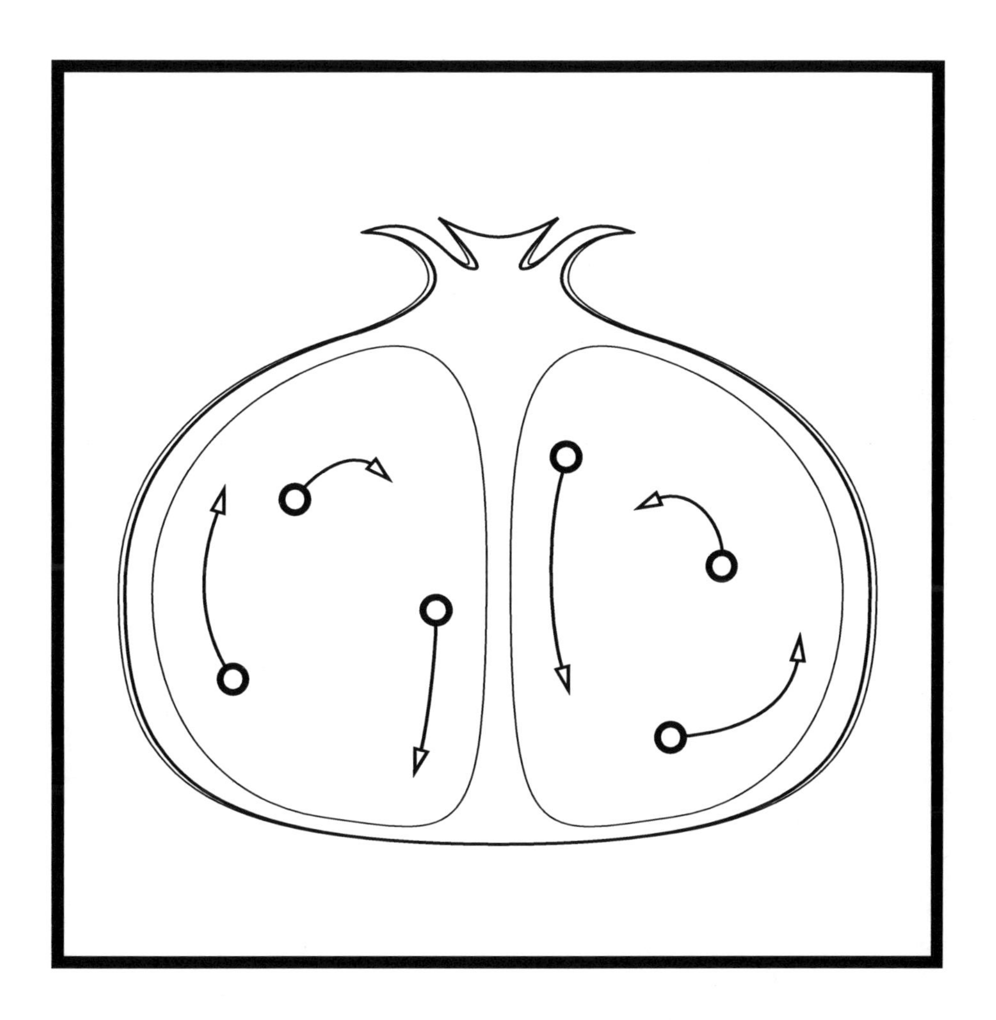

## What topology is

Spaces and maps between spaces generate that branch of Mathematics known as *Topology* – the natural evolution of the notions of proximity and continuity. Nested systems of open neighborhoods communicate nearness without necessitating a metric distance. The collection of all open sets in a space is (confusingly) called its **topology**. This thin notion of closeness suffices to define continuity, convergence, and connectivity familiar to students of calculus. A **map** between two spaces carries the implication of continuity.

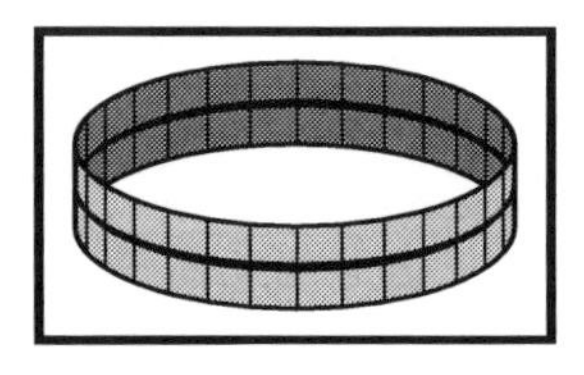

Topology (the *subject*) explores deformations of maps and spaces, and the basic equivalence relations of topology emphasize qualitative features. Certainly, a **homeomorphism** (a bijective map with continuous inverse) counts as a topological isomorphism. A more general equivalence emerges at the level of maps. A **homotopy** of maps $f_0 \simeq f_1$ from $X$ to $Y$ is a continuous 1-parameter family of maps $f_t \colon X \to Y$ interpolating $f_0$ and $f_1$. **Homotopy equivalent** (or *homotopic*) spaces are those $X \simeq Y$ having maps $g \colon X \to Y$ and $h \colon Y \to X$ with $h \circ g \simeq \mathrm{Id}_X$ and $g \circ h \simeq \mathrm{Id}_Y$, where Id denotes an identity map. The simplest (nonempty) space is one which is **contractible**, *i.e.*, homotopic to a single point.

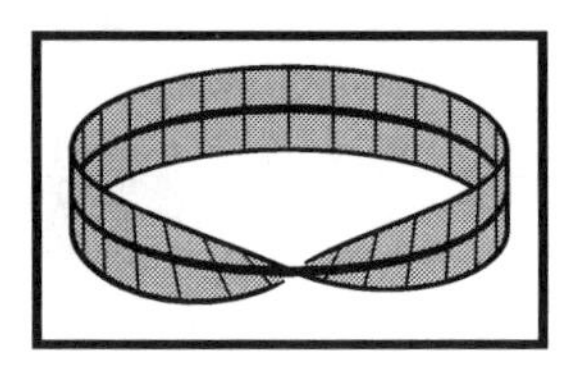

One example of homotopy equivalence is a **deformation retraction** of $X$ to a subspace $A \subset X$. This is defined by a map $r \colon X \to A$ such that the inclusion map $\iota \colon A \hookrightarrow X$ satisfies $r \circ \iota = \mathrm{Id}_A$ (that is, $r$ is a **retraction**), and $\iota \circ r \simeq \mathrm{Id}_X$ (the deformation). The homotopic simplification of $X$ to $A$ in a deformation retraction illustrates nicely the difference between homotopy and homeomorphism. The reader to whom these concepts are new should take some time to play with a few examples. Chapter 0 of the excellent text of Hatcher [176] is recommended.

Topology does its work by means of distinguishing spaces and maps to various degrees of resolution. A **topological invariant** of spaces is an assignment of some (usually algebraic) datum to spaces which respects the equivalence relation of homotopy: homotopic spaces are sent to the same invariant. Counting the number of connected components of a space is a simple topological invariant.

## What topology can do

Topology was built to distinguish qualitative features of spaces and mappings. It is good for, *inter alia*:

1. **Characterization:** Topological properties encapsulate qualitative signatures. For example, the genus of a surface, or the number of connected components of an object, give global characteristics important to classification.
2. **Continuation:** Topological features are robust. The number of components or holes is not something that should change with a small error in measurement. This is vital to applications in scientific disciplines, where data is never not noisy.

3. **Integration:** Topology is the premiere tool for converting local data into global properties. As such, it is rife with principles and tools (Mayer-Vietoris, Excision, spectral sequences, sheaves) for integrating from local to global.
4. **Obstruction:** Topology often provides tools for answering feasibility of certain problems, even when the answers to the problems themselves are hard to compute. These characteristics, classes, degrees, indices, or obstructions take the form of algebraic-topological entities.

## What topology cannot do

Topology is fickle. There is no recourse to tweaking epsilons should desiderata fail to be found. If the reader is a scientist or applied mathematician hoping that topological tools are a quick fix, take this text with caution. The reward of reading this book with care may be limited to the realization of new questions as opposed to new answers. It is not uncommon that a new mathematical tool contributes to applications not by answering a pressing question-of-the-day but by revealing a different (and perhaps more significant) underlying principle.

## What this text is

This text is a quick tour of applied topology, with just enough detail to motivate further study elsewhere. The intent is breadth in ideas, tools, perspectives, and applications; this precludes depth, both in the mathematics and in its applications. The subject of *applied topology* is in its infancy, and it seems certain that a more detailed treatment of the examples given here would appear quaint in less than a decade. The best approach, perhaps, is to make the text intentionally shallow, in the hopes that it will lure the unsuspecting reader to greater depths and prepare for the field as it will be. The author would have called this text *"Cartoons in Applied Topology"* were it not for the resulting confusion.

The chapters are organized according to mathematical topic, rather than according to application domain. This raises an interesting philosophical question about the nature of applied mathematics: is it how different branches of mathematics embed in the physical world, or is it how different applications implicate and are aided by mathematics? The organization of the text reflects a firmly-held belief: *applied mathematics concerns the incarnation of mathematical objects and structures.*

The text begins with an informal introduction to spaces, emphasizing examples and avoiding the set-theoretic technicalities which, though not unnecessary, may overly discourage the interested scientist. The goal is to get to applications as quickly as possible. This reflects the author's learning of the subject of topology: despite the best efforts of brilliant topologists (including Profs. Dranishnikov, Hatcher, Kahn, Krstić, and Vogtmann), the author never learned much of anything in the subject without first finding some physical manifestation of the principle, no matter how cartoonish. This book represents a partial collection of such cartoons.

This is not a mathematics text in the classical sense: some theorems are in a stripped-down version, and proofs are usually skipped, for the sake of making the exposition quick and painless. The reader should not conclude that the subject is quickly or painlessly learned. This text, properly used, is the impetus for future work: hard, slow, and fruitful.

This is not a text in computational topology: the reader may look to several excellent sources [104, 186] for the problem of algorithmic complexity of the topological objects explored in this text. The questions of *"What is it good for?"* and *"How do I compute that?"* are neither independent nor inseparable.

Experts may be exasperated with this text, for many reasons. The text is meant not for experts but for beginners, to point them in the direction of better things to come.

## How to use this text

With the advent of *Wikipedia*, *MathOverflow*, and other searchable resources, the need for comprehensive reference texts has perhaps diminished and may continue to do so. To some extent, the demand for the classical definition-theorem-proof text is also somewhat lessened, since one can look up a standard proof on-demand. What has not been eliminated is the need for a story with drama and characters. Overarching narratives are not easily modularized; connections and applications between areas require a global view.

This text attempts to tell a story. Even excised from applications, this story is unorthodox, in content and in tone. Manifolds are quickly marched past a chorus of cell complexes. The Euler characteristic is given stage time far out of measure with its more discerning invariant cousins, homology and, subsequently, cohomology. Classical Morse theory is glossed, interrupted by the ghost of stratified Morse theory, colliding with the Conley index. The shock of introducing the fundamental group after homology and cohomology is surpassed by the scandal of a stripped sheaf theory, preempting the categorical language that would have made for a simpler-seeming entrance.

Perhaps this text will be best used if simply read, for pleasure. It may also serve as a basis text for a graduate-level course in applied topology for mathematical scientists, in which case the lack of a formal theorem-proof delivery seems no impediment. If used as the text for a course in applied topology for mathematicians, this book should provide structure and lots of examples: the instructor for such a course can add details and proofs of the classical material to taste. It is hoped that this book will also make a good accompaniment to a principal text in a traditional algebraic topology course. Those students who struggle in this subject may find some motivation to persevere here, and even those students not interested in applications may find the story entertaining.

A good text should have numerous exercises. This text does not (save for some cryptic figures), for good reasons. First, exercises should be tested and refined via years of teaching from the text. Applied topology is too new a subject, and this author's teaching vocation is, at present, calculus. Second, the diversity of the audience for this text prompts a partitioning of the exercises into those meant for applied mathematical scientists and those meant for mathematicians in a more classical topology course.

Such an array of exercises will require experimentation with level of rigor demanded: it is best not to conduct that experiment in print. The author *will* do this experiment on-line: for the present, this will take the form of an evolving list of exercises linked to the web site for this text. The reader is encouraged to use and comment on these exercises.

## Acknowledgements

A debt of thanks is due to the organizers and attendees of the 2009 meeting at Cleveland State University, Peter Bubenik and John Oprea above all. This text was inspired by that meeting and reframed over the years, especially during various visits to the IMA in Minneapolis.

The young field of Applied Topology has numerous practitioners, and their consistent support for the author are worth more than words can repay. Of particular note is Benjamin Mann, the *primum mobile* of the field. The vision and personal guidance of (in chronological order) Henry Wente, Phil Holmes, Bob Williams, Konstantin Mischaikow, Dan Koditschek, and Gunnar Carlsson have been invaluable. The author has had the pleasure of collaborating with and learning from Yuliy Baryshnikov, Vin de Silva, John Etnyre, Yasu Hiraoka, Sanjeevi Krishnan, David Lipsky, Vidit Nanda, Michael Robinson, Rob Vandervorst, and others during this last decade of development in the field. Thank you for your patience.

A special thanks is due to those few who helped find the many mistakes present in rough drafts, especially Iris Yoon and Vidit Nanda, the first persons to read the text carefully.

The work behind and writing of this text was made possible in part through the support of several United States agencies: AFOSR, DARPA, NSF, ONR, and OSD. Without their support, this book would have been thinner and more quickly written.

All figures were drawn by the author using *Adobe Illustrator*.

Chapter 1

# Spaces: Manifolds

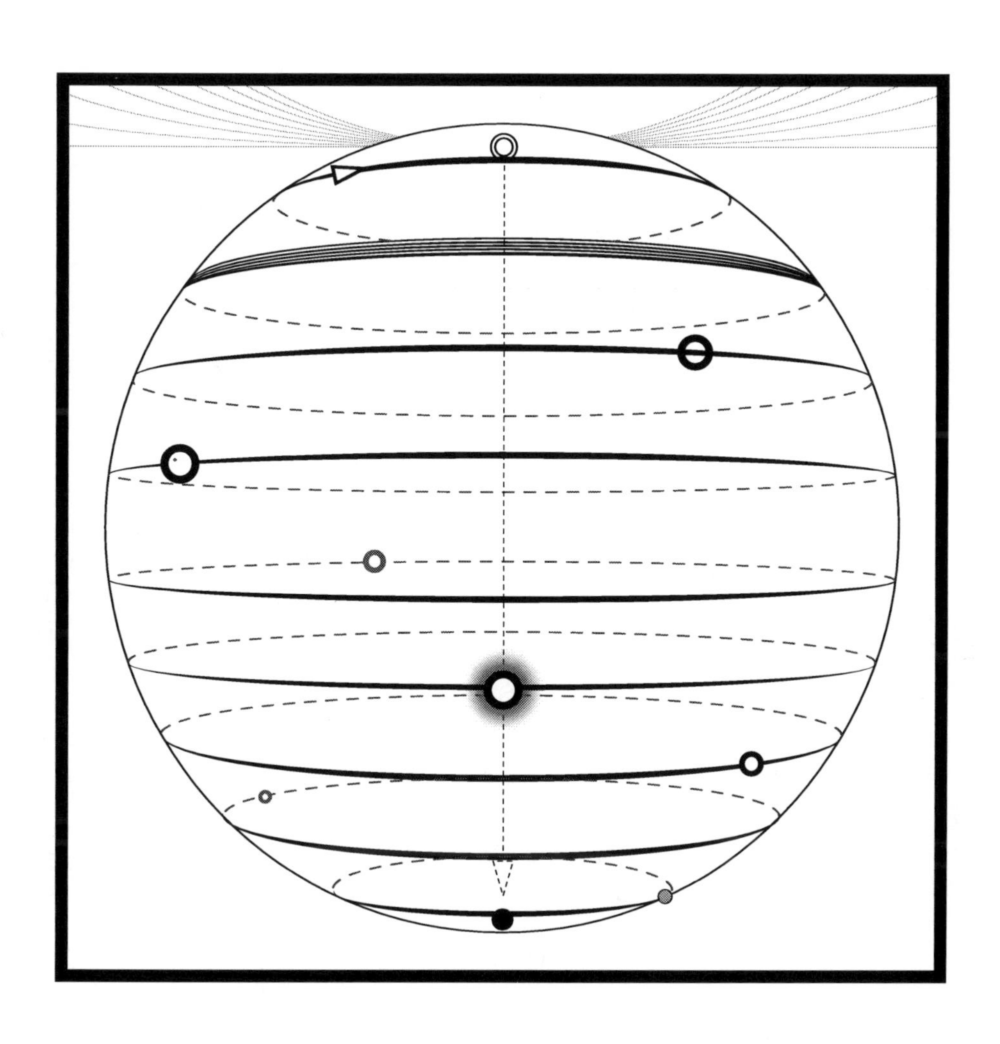

Manifolds are the extension of domains familiar from calculus – curves and surfaces – to higher-dimensional settings. As the most intuitive and initially useful topological spaces, these smooth domains provide a first glimpse of the technicalities implicit in passing from local to global.

## 1.1 Manifolds

A topological $n$-**manifold** is a space[1] $M$ locally homeomorphic to $\mathbb{R}^n$. That is, there is a cover $\mathcal{U} = \{U_\alpha\}$ of $M$ by open sets along with maps $\phi_\alpha \colon U_\alpha \to \mathbb{R}^n$ that are continuous bijections onto their images with continuous inverses. In order to do differential calculus, one needs a smoothing of a manifold. This consists of insisting that the maps

$$\phi_\beta \circ \phi_\alpha^{-1} \colon \phi_\alpha(U_\alpha \cap U_\beta) \to \phi_\beta(U_\alpha \cap U_\beta)$$

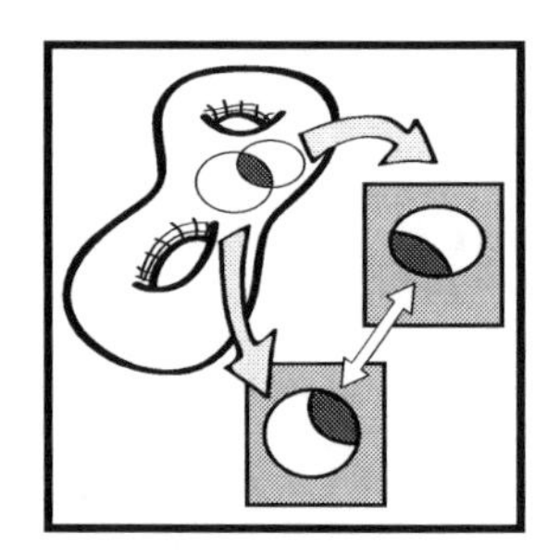

are smooth (infinitely differentiable, or $C^\infty$) whenever $U_\alpha \cap U_\beta \neq \varnothing$. The pairs $(U_\alpha, \phi_\alpha)$ are called **charts**; they generate a maximal **atlas** of charts which specifies a smooth structure on $M$. Charts and atlases are rarely explicitly constructed, and, if so, are often immediately ignored. The standard tools of multivariable calculus – the Inverse and Implicit Function Theorems – lift to manifolds and allow for a simple means of producing interesting examples.

Smooth curves are 1-manifolds, easily classified. Any connected curve is **diffeomorphic** (smoothly homeomorphic with smooth inverse) to either $\mathbb{R}$ or to the circle $\mathbb{S}^1$; thus, compactness suffices to distinguish the two. The story for 2-manifolds – surfaces – introduces two more parameters. Compact surfaces can be orientable or non-orientable, and the existence of holes or handles is captured in the invariant called **genus**.

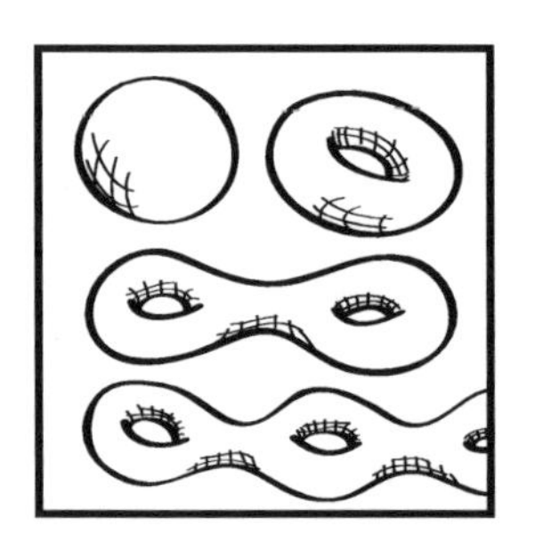

The **sphere** $\mathbb{S}^2$ is the orientable surface of genus zero; the **torus** $\mathbb{T}^2$ the orientable surface of genus one; their nonorientable counterparts are the **projective plane** $\mathbb{P}^2$ and the **Klein bottle** $K^2$ respectively. The Classification Theorem for Surfaces states that any compact surface is diffeomorphic to the orientable or non-orientable surface of some fixed genus $g \geq 0$. The spatial universe is, seemingly, a 3-manifold[2]. The classification of 3-manifolds is a delightfully convoluted story [290], with recent, spectacular progress [233] that perches this dimension between the simple (2) and the wholly bizarre (4).

[1] All manifolds in this text should be assumed Hausdorff and paracompact. The reader for whom these terms are unfamiliar is encouraged to ignore them for the time being.

[2] This is a simplification, ignoring whatever complexities black holes, strings, and other exotica produce.

**Example 1.1 (Spheres)** ⊙

The $n$-sphere, $\mathbb{S}^n$, is the set of points in Euclidean $\mathbb{R}^{n+1}$ unit distance to the origin. The $n$-sphere is an $n$-dimensional manifold.

The 0-dimensional sphere $\mathbb{S}^0$ is disconnected – it is the disjoint union of two points. For $n > 0$, $\mathbb{S}^n$ is a connected manifold diffeomorphic to the compactification of $\mathbb{R}^n$ as follows. Consider the quotient space obtained from $\mathbb{R}^n \sqcup \star$, where $\star$ is an abstract point whose neighborhoods consist of $\star$ union the points in $\mathbb{R}^n$ sufficiently far from the origin. This abstract space is diffeomorphic to $\mathbb{S}^n$ via a diffeomorphism that sends the origin and $\star$ to the south and north *poles* of the sphere $\mathbb{S}^n$ respectively. ⊙

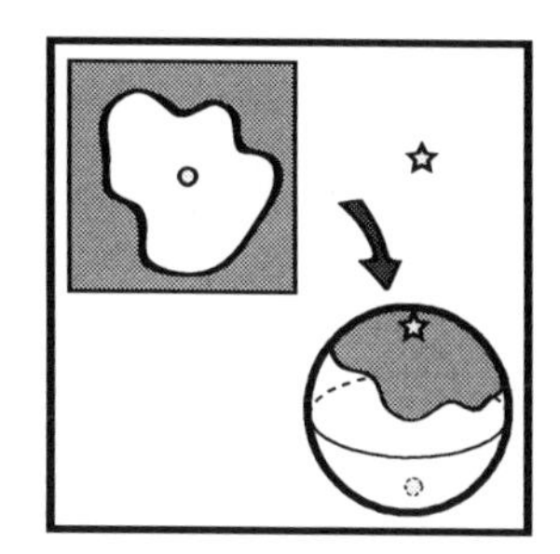

**Example 1.2 (Projective spaces)** ⊙

The **real projective space**, $\mathbb{P}^n$, is defined as the space of all 1-dimensional linear subspaces of $\mathbb{R}^{n+1}$, with the topology that says neighborhoods of a point in $\mathbb{P}^n$ are generated by small open cones about the associated line. That $\mathbb{P}^n$ is an $n$-manifold for all $n$ is easily shown (but should be contemplated until it appears obvious). Projective 1-space, $\mathbb{P}^1$, is diffeomorphic to $\mathbb{S}^1$. The projective plane, $\mathbb{P}^2$, is a non-orientable surface diffeomorphic to the following quotient spaces:

1. Identify opposite sides of a square with edge orientations reversed;
2. Identify antipodal points on the boundary of the closed unit ball $B \subset \mathbb{R}^2$;
3. Identify antipodal points on the 2-sphere $\mathbb{S}^2$.

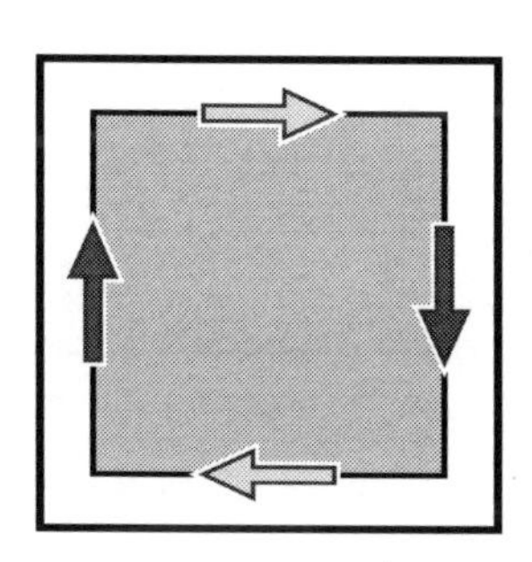

For any $n$, $\mathbb{P}^n$ is diffeomorphic to the quotient $\mathbb{S}^n/a$, where $a\colon \mathbb{S}^n \to \mathbb{S}^n$ is the **antipodal** map $a(x) = -x$. The space $\mathbb{P}^3$ is diffeomorphic to the space of rotation matrices, $SO_3$, the group of real 3-by-3 orthogonal matrices with determinant 1. Among the many possible extensions of projective spaces, the **Grassmannian** spaces arise in numerous contexts. The Grassmannian $\mathbb{G}^n_k$ is defined as the space of all $k$-dimensional subspaces of $\mathbb{R}^n$, with topology induced in like manner to $\mathbb{P}^n$. The Grassmannian is a manifold that specializes to $\mathbb{P}^n = \mathbb{G}^{n+1}_1$. ⊙

## 1.2 Configuration spaces of linkages

Applications of manifolds and differential topology are ubiquitous in rational mechanics, Hamiltonian dynamics, and mathematical physics and are well-covered in standard texts [1, 15, 217, 231]. A simple application of (topological) manifolds to robotics falls under a different aegis. Consider a planar mechanical linkage consisting of several flat, rigid rods joined at their ends by pins that permit free rotation in the plane. One can use out-of-plane height (or mathematical license) to assert that interior intersections of rods are ignorable. The **configuration space** of the linkage is a topological space that assigns a point to each configuration of the linkage – a relative positioning of the rods up to equivalence generated by rotations and translations in the plane – and which assigns neighborhoods in the obvious manner. A neighborhood of a configuration is all configurations obtainable via a small perturbation of the mechanical linkage. The configuration space of a planar linkage is almost always a manifold, the dimension of which conveys the number of mechanical degrees of freedom of the device.

**Example 1.3 (Crank-rocker)** ⊚

The canonical example of a simple useful linkage is the **Grashof 4-bar**, or **crank-rocker** linkage, used extensively in mechanical components. Four rods of lengths $\{L_i\}_1^4$ are linked in a cyclic chain. When one rod is anchored, the system is seen to have one mechanical degree of freedom. The configuration space is thus one-dimensional and almost always a manifold. If one has a single short rod, then this rod can be rotated completely about its anchor, causing the opposing rod to rock back-and-forth.

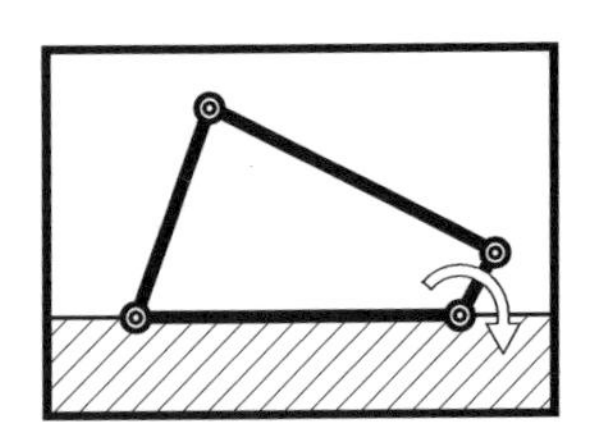

This linkage is used to transform spinning motion (from, say, a motor) into rocking motion (as in a windshield wiper). The configuration space of such a linkage is $\mathbb{S}^1 \sqcup \mathbb{S}^1$, the **coproduct** or **disjoint union** of two circles. The second circle comes from taking the mirror image of the linkage along the axis of its fixed rod in the plane and repeating the circular motion there: this forms an entirely separate circle's worth of configuration states. ⊚

Many other familiar manifolds are realized as configuration spaces of planar linkages (with judicious use of the third dimension to mitigate bar crossings). The undergraduate (!) thesis of Walker [299] has many examples of orientable 2-manifolds as configuration spaces of planar linkages. A simple 5-bar linkage has configuration space which can be a closed, orientable surface of genus $g$ ranging between 0 and 4, depending on the lengths of the edges. The reader is encouraged to try building a linkage whose configuration space yields an interesting 3-manifold. The realization question this exercise prompts has a definitive answer (albeit with a convoluted attribution and history):

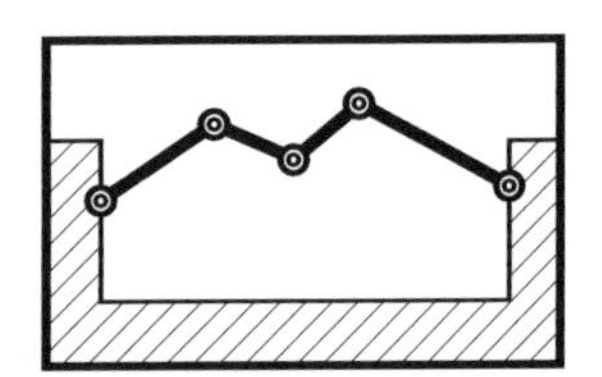

**Theorem 1.4 ([189]).** *Any smooth compact manifold is diffeomorphic to the configuration space of some planar linkage.*

This remarkable result provides great consolation to students whose ability to conceptualize geometric dimensions greater than three is limited: one can sense all the complexities of manifolds *by hand* via kinematics. The reader is encouraged to build a few configurable linkages and to determine the dimensions of the resulting configuration spaces.

**Example 1.5 (Robot arms)** ⊚

A robot arm is a special kind of mechanical linkage in which joints are sequentially attached by rigid rods. One end of the arm is fixed (mounted to the floor) and the other is free (usually ending in a manipulator for manufacturing, grasping, pick-and-place, etc.).

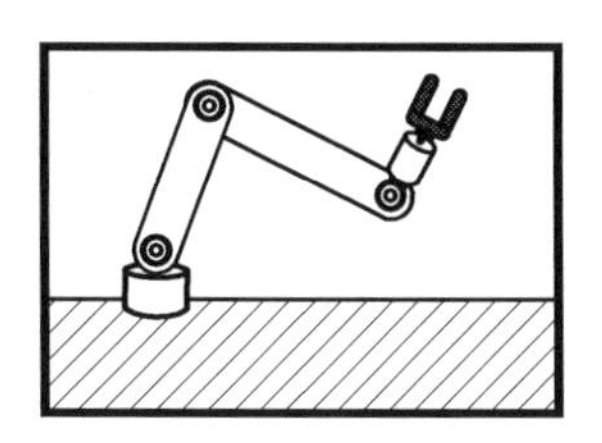

Among the most commonly available joints are pin joints (*cf.* an elbow) and rotor joints (*cf.* rotation of a forearm), each with configuration space $\mathbb{S}^1$. Ignoring the (nontrivial!) potential for collision, the configuration space of such an arm in $\mathbb{R}^3$ has the topology of the $n$-torus, $\mathbb{T}^n := \prod_1^n \mathbb{S}^1$, the **cartesian product** of $n$ circles, where $n$ is the number of rotational or pin joints. There are natural maps associated with this configuration space, including the map to $\mathbb{R}^3$ which records the location of the end of the arm, or the map to $SO_3$ that records the orientation (but not the position) of an asymmetric part grasped by the end manipulator. ⊚

## 1.3 Derivatives

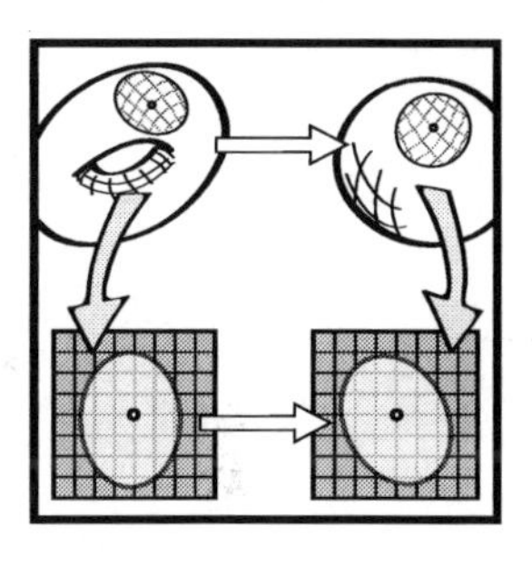

Derivatives, vector fields, gradients, and more are familiar constructs of calculus that extend to arbitrary manifolds by means of localization. Differentiability is a prototypical example. A map between manifolds $f: M \to N$ is **differentiable** if pushing it down via charts yields a differentiable map.

Specifically, whenever $f$ takes $p \in U_\alpha \subset M$ to $f(p) \in V_\beta \subset N$, one has $\psi_\beta \circ f \circ \phi_\alpha^{-1}$ a smooth map from a subset of $\mathbb{R}^m$ to a subset of $\mathbb{R}^n$. The derivative of $f$ at $p$ is therefore defined as the derivative of $\psi_\beta \circ f \circ \phi_\alpha^{-1}$, and one must check that the choice of chart does not affect the result.

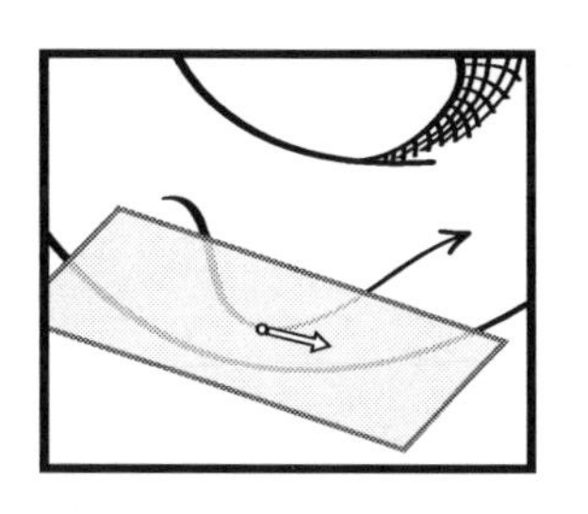

It suffices to use charts and coordinates to understand derivatives, but it is not satisfying. A deeper inquiry leads to a significant construct in differential topology. The **tangent space** to a manifold $M$ at a point $p \in M$, $T_pM$, is a vector space of tangent directions to $M$ at $p$, where the origin $0 \in T_pM$ is abstractly identified with $p$ itself.

This notion is the first point of departure from the calculus mindset – in elementary calculus classes there is a general confusion between tangent vectors (*e.g.*, from vector fields) and points in the space itself. It is tempting to illustrate the tangent space as a vector space of dimension dim $M$ that is *tangent* to the manifold, but this pictorial representation is dangerously ill-defined – in what larger space does this tangent space reside? Do different tangent spaces intersect? There are several ways to correctly define a tangent space. The most intuitive uses smooth curves. Define $T_pM$ to be the vector space of equivalence classes of differentiable curves $\gamma: \mathbb{R} \to M$

where $\gamma(0) = p$. Two such curves, $\gamma$ and $\tilde{\gamma}$ are equivalent if and only if $\gamma'(0) = \tilde{\gamma}'(0)$ in some (and hence any) chart. An element of $T_pM$ is of the form $\xi = [\gamma'(0)]$, where $[\cdot]$ denotes the equivalence relation. The vector space structure is inherited from that of the chart in $\mathbb{R}^n$. A tangent vector coincides with the intuition from calculus in the case of $M = \mathbb{R}^n$. Invariance with respect to charts implies that the derivative of $f\colon M \to N$ at $p \in M$ is realizable as a linear transformation $Df_p\colon T_pM \to T_{f(p)}N$. In any particular chart, a basis of tangent vectors may be chosen to realize $Df_p$ as the Jacobian matrix of partial derivatives at $p$.

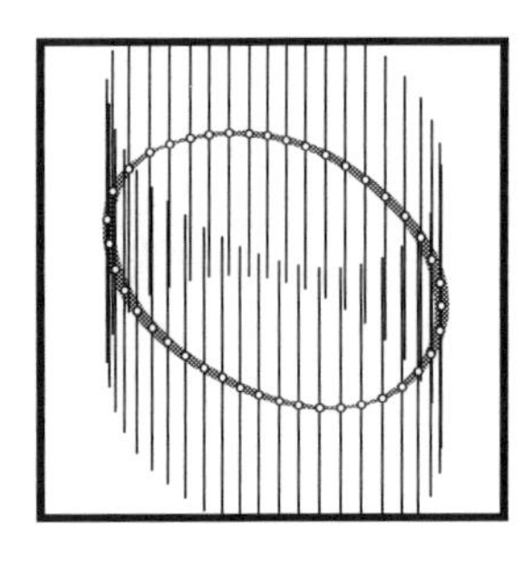

The next step is crucial: one glues the disjoint union of all tangent spaces $T_pM$, $p \in M$, into a single space $T_*M$ called the **tangent bundle** of $M$. An element of $T_*M$ is of the form $(p, V)$, where $V \in T_pM$. The natural topology on $T_*M$ is one for which a neighborhood of $(p, V)$ is a product of a neighborhood of $V$ in $T_pM$ with a neighborhood of $p$ in $M$. In this topology, $T_*M$ is a smooth manifold of dimension equal to $2\dim M$. For example, the tangent bundle of a circle is diffeomorphic to $\mathbb{S}^1 \times \mathbb{R}^1$. However, it is *not* the case that $T_*\mathbb{S}^2 \cong \mathbb{S}^2 \times \mathbb{R}^2$. That this is so is not so obvious.

## 1.4 Vector fields

The formalism of tangent spaces and tangent bundles simplifies the transition of calculus-based ideas to arbitrary manifolds; a ready and recurring example is the topology and dynamics of vector fields. A **vector field** on $M$ is a choice of tangent vectors $V(p) \in T_p(M)$ which is continuous in $p$. Specifically, $V\colon M \to T_*M$ is a map satisfying $\pi \circ V = \mathrm{Id}_M$, for $\pi\colon T_*M \to M$ the projection map taking a tangent vector at $p$ to $p$ itself. Such a map $V$ is called a **section** of $T_*M$.

As with sufficiently smooth differential equations on $\mathbb{R}^n$, vector fields can be integrated to yield a flow. Given $V$ a vector field on $M$, the **flow** associated to $V$ is the family of diffeomorphisms $\varphi_t\colon M \to M$ satisfying:

1. $\varphi_0(x) = x$ for all $x \in M$;
2. $\varphi_{s+t}(x) = \varphi_t(\varphi_s(x))$ for all $x \in M$ and $s, t \in \mathbb{R}$;
3. $\frac{d}{dt}\varphi_t(x) = V(x)$.

One thinks of $\varphi_t(x)$ as determining the location of a particle starting at $x$ and moving via the velocity field $V$ for $t$ units of time. For $M$ non-compact or $V$ insufficiently smooth, one must worry about existence and uniqueness of solutions: such questions are not considered in this text. Smooth vector fields on compact manifolds yield smooth flows whose dynamics links topology and differential equations.

**Example 1.6 (Equilibria)** ◎

The primal objects of inquiry in dynamics are the equilibrium solutions: a vector field is said to have a **fixed point** or **equilibrium** at $p$ if $V(p) = 0$. An isolated fixed point may have several qualitatively distinct features based on stability. The **stable**

**manifold** of a fixed point $p$ is the set

$$W^s(p) := \{x \in M : \lim_{t \to \infty} \varphi_t(x) = p\}. \tag{1.1}$$

For "typical" fixed points of a "typical" vector field, $W^s(p)$ is in fact a manifold, as is the **unstable manifold**, $W^u(p)$, defined by taking the limit $t \to -\infty$ in (1.1) above [258]. A *sink* is a fixed point $p$ whose stable manifold contains an open neighborhood of $p$; such equilibria are fundamentally stable solutions. A *source* is a $p$ whose unstable manifold contains an open neighborhood of $p$; such equilibria are fundamentally unstable. A *saddle* equilibrium satisfies dim $W^s(p) > 0$ and dim $W^u(p) > 0$; such solutions are balanced between stable and unstable behavior. ⊙

**Example 1.7 (Periodic orbits)** ⊙

If a continuous-time dynamical system – a flow – is imagined to be supported by the *skeleton* of its equilibria, then the analogous *circulatory system* would be comprised of the periodic orbits.

A **periodic orbit** of a flow is an orbit $\{\varphi_t(x)\}_{t\in\mathbb{R}}$ satisfying $\varphi_{t+T}(x) = \varphi(x)$ for some fixed $T > 0$ and all $t \in \mathbb{R}$. The minimal such $T > 0$ is the **period** of the orbit. Periodic orbits are submanifolds diffeomorphic to $\mathbb{S}^1$. One may classify periodic orbits as being stable, unstable, saddle-type, or degenerate. The existence of periodic orbits – in contrast to that of equilibria – is a computationally devilish problem. On $\mathbb{S}^3$, it is possible to find smooth, fixed-point-free vector fields whose set of periodic orbits is all of $\mathbb{S}^3$ or empty: see Example 8.11. ⊙

The dynamics of vector fields goes well beyond equilibria and periodic orbits (see, *e.g.*, [167, 258]); however, for typical systems, the skeleton of periodic orbits and equilibria, together with the musculature of their stable and unstable manifolds, give the basic frame for reasoning about the body of behavior.

## 1.5 Braids and robot motion planning

A different class of configuration spaces is inspired by applications in multi-agent robotics. Consider an automated factory equipped with mobile robots. A common goal is to place several such robots in motion simultaneously, controlled by an algorithm that either guides the robots from initial positions to goal positions (in a warehousing application), or executes a cyclic pattern (in manufacturing applications). These robots are costly and cannot tolerate collisions. As a first step at modeling such a system, assume the location of each robot is a point in a space $X$ (typically a domain in $\mathbb{R}^2$ or $\mathbb{R}^3$). The **configuration space** of $n$ distinct labeled points on $X$, denoted $\mathcal{C}^n(X)$, is the space

$$\mathcal{C}^n(X) := \left(\prod_1^n X\right) - \Delta \quad ; \quad \Delta := \{(x_k)_1^n : x_i = x_j \text{ for some } i \neq j\}. \tag{1.2}$$

The set $\Delta$, the **pairwise diagonal**, represents those configurations of $n$ points in $X$ which experience a collision – this is the set of illegal configurations for the robots. Of course, robots are not point-like, and near-collisions are unacceptable. From the point of view of topology, however, removing a sufficiently small neighborhood of $\Delta$ gives a space equivalent to $\mathcal{C}^n(X)$, and the configuration space $\mathcal{C}^n(\mathbb{R}^2)$ forms an acceptable model for robot motion planning on an unobstructed floor.

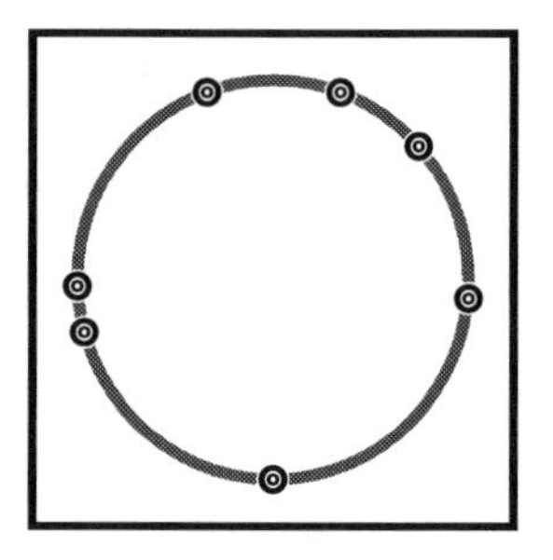

There are applications for which labeling the points is important, with warehousing, in which robots move specific packages, being a prime example. However, in some settings, such as mobile security cameras in a building, anonymity is not detrimental – any camera will do. The **unlabeled configuration space**, denoted $\mathcal{UC}^n(X)$, is defined to be the quotient of $\mathcal{C}^n(X)$ by the natural action of the symmetric group $S_n$ which permutes the ordered points in $X$:

$$\mathcal{UC}^n(X) := \mathcal{C}^n(X)/S_n.$$

This space is given the **quotient topology**: a subset in $\mathcal{UC}^n(X)$ is open if and only if the union of preimages in $\mathcal{C}(X)$ is open. Configuration spaces of points on a manifold $M$ are all (non-compact) manifolds of dimension dim $\mathcal{C}^n(M) = n$ dim $M$. The space $\mathcal{C}^n(\mathbb{S}^1)$ is homeomorphic to $(n-1)!$ disjoint copies of $\mathbb{S}^1 \times \mathbb{R}^{n-1}$, while $\mathcal{UC}^n(\mathbb{S}^1)$ is a connected space.

**Example 1.8 (Braids)** 

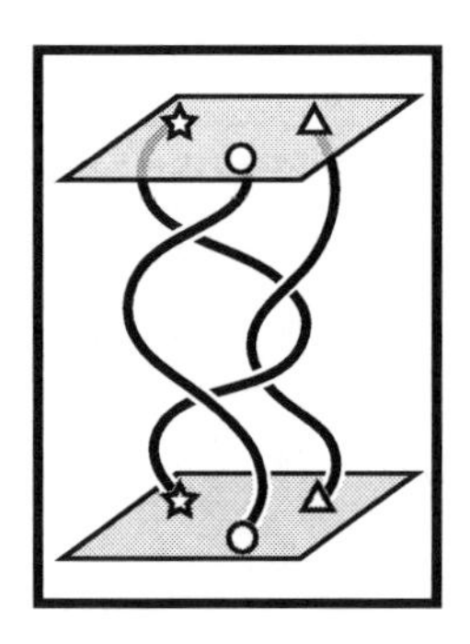

The configuration space of points on $\mathbb{R}^2$ is relevant to mobile robot motion planning (*e.g.*, on a factory floor); it is also among the most topologically interesting configuration spaces [38]. Consider the case of $n$ robots which begin and end at fixed configurations, tracing out non-colliding routes in between. This complex motion corresponds to a path in $\mathcal{C}^n(\mathbb{R}^2)$, or, perhaps, $\mathcal{UC}^n(\mathbb{R}^2)$, if the robots are not labeled. (If the path has the same beginning and ending configuration, it is a loop, an image of $\mathbb{S}^1$ in the configuration space.) How many different ways are there for the robots to wind about one another *en route* from their starting to ending locations? The space-time graph of a path in configuration space is a **braid**, a weaving of strands encapsulating positions. A deformation in the motion plan equals a homotopy of the path (fixing the endpoints), which itself corresponds to moving the braid strands in such a way that they cannot intersect. From this, and a few sketches, the reader reasons correctly that there are infinitely many fundamentally inequivalent motion plans between starting and ending configurations. See §8.3 for a more algebraic description. ⊚

**Example 1.9 (Navigation fields)** ⊚

Configuration spaces in robotics are widespread, both as models of complex articulated agents or automated guided vehicles. One of many useful techniques for controlling robotic systems to execute behaviors and avoid collisions is to place a vector field on the configuration space and use the resulting flow to guide systems towards a goal configuration. The task: given a configuration space $X$ and a specified goal point or loop $G \subset X$, construct an explicit vector field on $X$ having $G$ as a global **attractor**, so that (almost) all initial conditions will converge to and remain near the goal. Programming such a vector field is a challenge, usually requiring explicit coordinate systems. Additional features are also desirable: near the collision set (*e.g.*, $\Delta$ from Equation (1.2)), the vector field should be pointed away from the collision set, so that $\Delta$ is a **repeller**.

For example, consider the problem of mobilizing a pair of robots on a circular track so as to patrol the domain while remaining as far apart as possible. This can be done by means of a vector field on $\mathcal{C}^2(\mathbb{S}^1)$. Coordinatize $\mathcal{C}^2(\mathbb{S}^1) \subset \mathbb{T}^2$ as $\{(\theta_1, \theta_2)\,;\, \theta_1 \neq \theta_2\}$. The reader may check that the following vector field has $\{\theta_1 = \theta_2 \pm \pi\}$ as an attracting periodic solution and $\Delta = \{\theta_1 = \theta_2\}$ as a repelling set:

$$\dot{\theta}_1 = 1 + \sin(\theta_1 - \theta_2) \quad ; \quad \dot{\theta}_2 = 1 + \sin(\theta_2 - \theta_1).$$

All initial conditions evolve to this desired state of circulation on $\mathbb{S}^1$. This is an excellent model not only for motion on a circular track, but also for an alternating gait in legged locomotion, where each leg position/momentum is represented by $\mathbb{S}^1$.

One reason for specifying a vector field on the entire configuration space (instead of simply dictating a path from initial to goal states) is so that if the robot experiences an unexpected failure in executing a motion, the vector field automatically corrects for the failure. Enlarging the problem from *path-planning* to *field-planning* returns stability and robustness [248]. This technique has been successfully applied in visual servoing [75], in robots that juggle [250], in automated guided vehicles [149], and in hopping [195] and insectoid [177] robot locomotion. This approach to robot motion planning makes extensive use of differential equations and dynamics on manifold configuration spaces. ⊚

## 1.6 Transversality

Genericity is often invoked in applications, but seldom explained in detail. Intuition is an acceptable starting point: consider the following examples of generic features of smooth manifolds and mappings:

1. Two intersecting curves in $\mathbb{R}^2$ generically intersect in a discrete set of points.
2. Three curves in $\mathbb{R}^2$ generically do not have a point of mutual intersection.
3. Two curves in $\mathbb{R}^n$ generically do not intersect for $n > 2$.
4. Two intersecting surfaces in $\mathbb{R}^3$ generically intersect along curves.
5. A real square matrix $A$ is generically invertible.
6. Critical points of a $\mathbb{R}$-valued function on a manifold are generically discrete.
7. The roots of a polynomial are generically non-repeating.
8. The fixed points of a vector field are generically discrete.
9. The configuration space of a planar linkage is generically a manifold.
10. A generic map of a surface into $\mathbb{R}^5$ is injective.

Some of these seem obviously true; others less obviously so. All are provably true with precise meaning using the theory of transversality.
Two submanifolds $V, W$ in $M$ are transverse, written $V \pitchfork W$, if and only if,

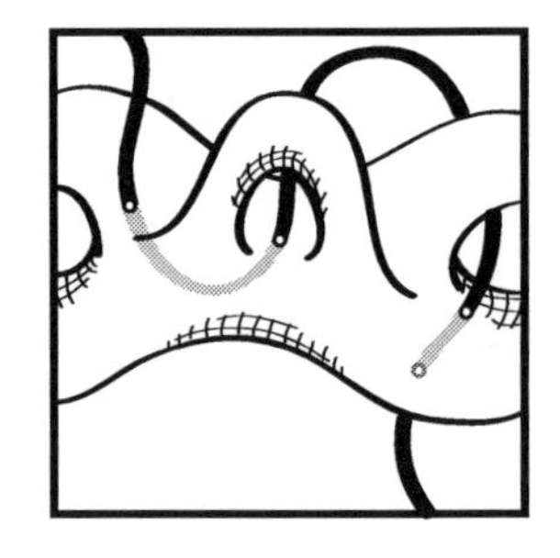

$$T_pV + T_pW = T_pM \quad \forall\, p \in V \cap W. \tag{1.3}$$

Otherwise said, at an intersection, the tangent spaces to $V$ and $W$ span that of $M$. For example, two planes in $\mathbb{R}^3$ are transverse if and only if they are not identical. Note that the absence of intersection is automatically transverse. A central theme of topology is the lifting of concepts from spaces to maps between spaces. The notion of transverse maps is the first of many examples of this principle. Two smooth maps $f\colon V \to M$ and $g\colon W \to M$ are transverse, written $f \pitchfork g$, if and only if:

$$Df_v(T_vV) + Dg_w(T_wW) = T_pM \quad \forall\, f(v) = g(w) = p. \tag{1.4}$$

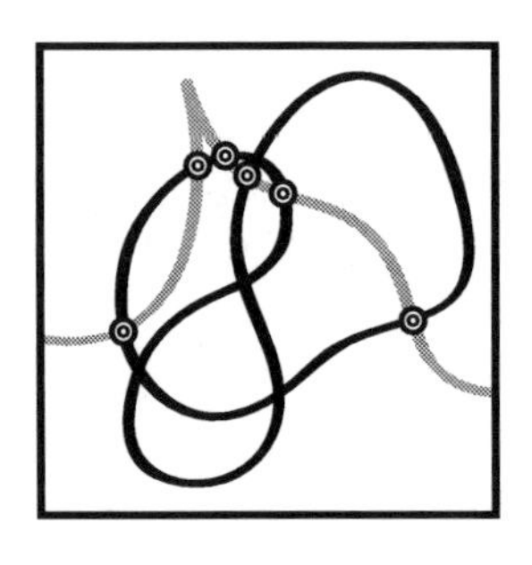

This means that the degrees of freedom in the intersection of images of $f$ and $g$ span the full degrees of freedom in $M$. Note that submanifolds $V, W \subset M$ are transverse if and only if the inclusion maps $\iota_V\colon V \to M$ and $\iota_W\colon W \to M$ are transverse as maps. Likewise, a map $f$ is transverse to a submanifold $W \subset M$ if and only if $f \pitchfork \iota_W$. This map-centric definition does not constrain the images of maps to be submanifolds. This is one hint that differential tools are efficacious in the management of singular behavior. A point $q \in N$ is a **regular value** of $f\colon M \to N$ if $f \pitchfork \{q\}$. This is equivalent to the statement that, for each $p \in f^{-1}(q)$, the derivative is a surjection – the matrix of $Df_p$ is of rank at least $n = \dim N$.

One benefit of transversality is that localized linear-algebraic results can be pulled back to global results. The following local result from linear algebra about dimensions of intersecting subspaces of a vector space is crucial:

**Lemma 1.10 (Rank-Nullity Theorem).** *For a linear transformation $T\colon U \to V$ between finite-dimensional vector spaces,*

$$\dim \ker T - \dim \operatorname{coker} T = \dim U - \dim V. \tag{1.5}$$

By applying Lemma 1.10 pointwise along the inverse image of transverse maps, one obtains a fundamental useful theorem:

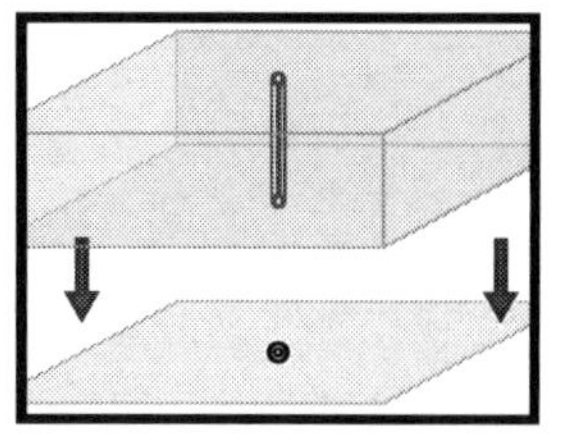

**Theorem 1.11 (Preimage Theorem).** *Consider a differentiable map $f\colon M \to N$ between smooth manifolds. If $f \pitchfork W$ for $W \subset N$ a submanifold, then $f^{-1}(W)$ is a submanifold of $M$ of dimension*

$$\dim f^{-1}(W) = \dim M - \dim N + \dim W. \tag{1.6}$$

This provides an effective means of constructing manifolds without the need for explicit charts: it is often used in the context of a regular value of a map.

1. The sphere $\mathbb{S}^n$ is the inverse image of 1 under $f\colon \mathbb{R}^{n+1} \to \mathbb{R}$ given by $f(x) = \|x\|$. It is a manifold of dimension $n = (n+1) - 1 + 0$.
2. The torus $\mathbb{T}^n$ is the inverse image $f^{-1}(1, \ldots, 1)$ of the map $f\colon \mathbb{C}^n \to \mathbb{R}^n$ given by $f(z) = (\|z_1\|, \ldots, \|z_n\|)$. Its dimension is $n = 2n - n + 0$.
3. The matrix group $O_n$ of rigid rotations of $\mathbb{R}^n$ (both orientation preserving and reversing) can be realized as the inverse image $f^{-1}(\mathrm{Id})$ of the identity under the map from $n$-by-$n$ real matrices to symmetric $n$-by-$n$ real matrices given by $f(A) = AA^T$. The dimension of $O_n$ is $n^2 - \frac{1}{2}n(n+1) + 0 = \frac{1}{2}n(n-1)$.
4. The determinant map restricted to $O_n$ is in fact a smooth map to the 0-manifold $\mathbb{S}^0 = \{\pm 1\}$. As such, the special orthogonal group $SO_n$, as the inverse image of $+1$ under this restricted det, is a manifold of the same dimension as $O_n$; thus, $O_n$ is a disjoint union of two copies of $SO_n$.

In the above examples, the transversality condition is checked by showing that the mapping $f$ has a derivative of full rank at the appropriate (regular) value. Such regularity seems to fail rarely, for special *singular* values. This intuition is the driving force behind the Transversality Theorem. A subset of a topological space is said to be **residual** if it contains a countable intersection of open, dense subsets. A property dependent upon a parameter $\lambda \in \Lambda$ is said to be a **generic** property if it holds for $\lambda$ in a residual subset of $\Lambda$ – even when that subspace is not explicitly given. For reasonable (*e.g.*, **Baire**) spaces, residual sets are dense, and hence form a decent notion of topological typicality.

**Theorem 1.12 (Transversality Theorem).** *For $M$ and $N$ smooth manifolds and $W \subset N$ a submanifold, the set of smooth maps $f\colon M \to N$ with $f \pitchfork W$ is residual in $C^\infty(M, N)$, the space of all smooth maps from $M$ to $N$. If $W$ is closed, then this set of transverse maps is both open and dense.*

The proof of this theorem relies heavily on **Sard's Lemma**: the regular values of a sufficiently smooth map between manifolds are generic in the codomain.

**Example 1.13 (Fixed points of a vector field)** ◎

The fixed point set of a differentiable vector field on a compact manifold $M$ is generically finite, thanks to transversality.

Recall from §1.4 that a smooth vector field is a section, or smooth map $V: M \to T_*M$ with $\pi(V(p)) = p$ for all $p \in M$. The **zero-section** $Z \subset T_*M$ is the set $\{(p, 0)\}$ of all zero vectors. The fixed point set of $V$ is therefore the preimage $V^{-1}(Z)$. For a generic perturbation of $V$, this set is a submanifold of dimension

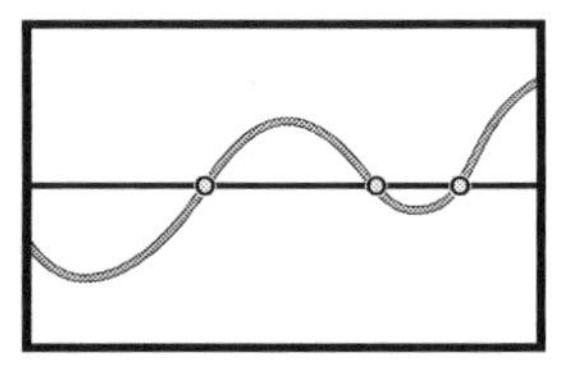

$$\dim \operatorname{Fix}(V) = \dim M + \dim M - \dim T_*M = 0.$$

A 0-dimensional compact submanifold is a finite point set. With more careful analysis of the meaning of transversality, it can be shown that the type of fixed point is also constrained: on 2-d surface, only source, sinks, and saddles are generic fixed points. ◎

**Example 1.14 (Beacon alignment)** ◎

Consider three people walking along generic smooth paths in the plane $\mathbb{R}^2$. How often are their positions collinear? This is relevant to robot navigation via beacon triangulation.

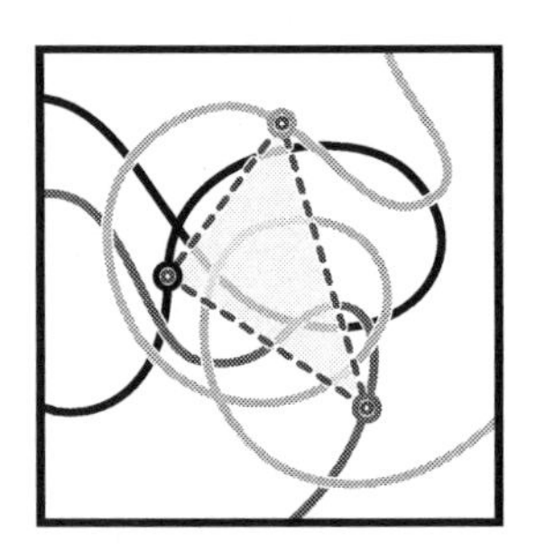

One considers the map $f: \mathcal{C}^3(\mathbb{R}^2) \to \mathbb{R}$ which computes the (signed) area of the triangle spanned by the three locations at an instant of time:

$$f(v_1, v_2, v_3) = \det [v_2 - v_1; v_3 - v_1].$$

This map has zero as a regular value, and Theorem 1.11 implies that the set of collinear configurations is a submanifold $W = f^{-1}(0)$ of $\mathcal{C}^3(\mathbb{R}^2)$. A set of three paths is a generic map from $\mathbb{R}$ (time) into $\mathcal{C}^3(\mathbb{R}^2)$. It follows from Theorems 1.12 and 1.11 that, generically, one expects collinearity at a discrete set of times, since

$$\dim \mathbb{R}^1 + \dim W - \dim \mathcal{C}^3(\mathbb{R}^2) = 1 + 5 - 6 = 0.$$

This, moreover, implies a stability in the phenomenon of collinearity: at such an alignment, a generic perturbation of the paths perturbs where the alignment occurs, but does not remove it. ◎

## 1.7 Signals of opportunity

Applications of transversality are alike: (1) set up the correct maps/spaces; (2) invoke transversality; (3) count dimensions. This has some simple consequences, as in computing the generic intersection of curves and surfaces in $\mathbb{R}^3$. Other consequences are not so obvious. The following theorem states that any continuous map of a source manifold into a target manifold has, after generic perturbation, a submanifold as its image when the dimension of the target is large enough. How large? The critical bound comes, as it must, from the proper transversality criterion and a dimension count:

**Theorem 1.15 (Whitney Embedding Theorem).** *Any continuous function $f\colon M \to N$ between smooth manifolds is generically perturbed to a smooth injection when* $\dim N > 2\dim M$.

**Proof.** The configuration space $\mathcal{C}^2M = M \times M - \Delta_M$ of two distinct points on $M$ is a manifold of dimension $2\dim M$. The graph of $f$ induces a map

$$\mathcal{C}^2 f : \mathcal{C}^2 M \to \mathcal{C}^2 M \times N \times N \quad : \quad (x, y) \mapsto (x, y, f(x), f(y)).$$

The key step is this: observe that the set of points on which $f$ is non-injective is $(\mathcal{C}^2 f)^{-1}(\mathcal{C}^2 M \times \Delta_N)$, where $\Delta_N \subset N \times N$ is the diagonal. According to the appropriate transversality theorem (specifically, the **multi-jet** transversality theorem [161, Thm. 4.13]), generic perturbations of $f$ induce generic perturbations of $\mathcal{C}^2 f$. From Theorems 1.11, 1.12, and the hypothesis that $\dim N > 2\dim M$, the generic dimension of the non-injective set of $f$ is:

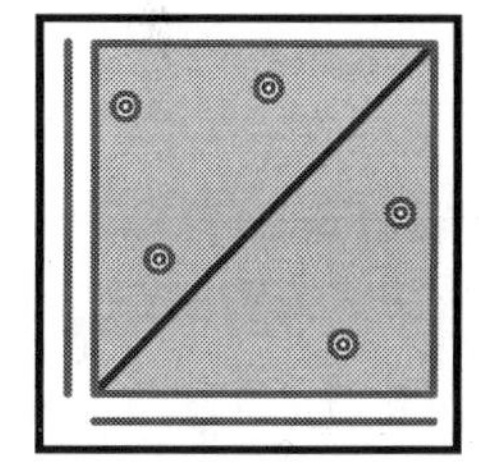

$$\dim \mathcal{C}^2 M - \dim(\mathcal{C}^2 M \times N \times N) + \dim(\mathcal{C}^2 M \times \Delta_N) = 2\dim M - \dim N < 0.$$

Thus, there are no self-intersections, and the result is a smooth submanifold. ⊙

A simple application of Whitney's theorem [257] informs a problem of localization via signals. Consider a scenario in which one wants to determine location in an unknown environment. Certainly, a GPS device would suffice. Such a device works by receiving signals from multiple satellite transmitters and utilizing known timing data about the transmitters to determine location, within the tolerances of the signal reception. This sophisticated system requires many independent components to operate, including geosynchronous satellites, synchronized clocks, and more, all with nontrivial power constraints. Though wonderfully useful, GPS devices are not universally available: they do not operate underwater or indoors; they are unreliable in *urban canyons*; the need for satellites and synchronized clocks can limit availability.

There is no reason, however, why other signals could not be used. In fact, passive signals – arising naturally from TV transmissions, radio, even ionospheric waves induced by lightning strikes [99] – provide easily measured pulses with which to attempt to reconstruct location. There is a small but fascinating literature on the use of such **signals of opportunity** to solve localization and mapping problems.

The following is a simple mathematical model for localization via signals of opportunity. Consider a connected open domain $\mathcal{D} \subset \mathbb{R}^k$ which is a $k$-manifold. Assume there exist $N$ transmitters of fixed location which asynchronously emit pulses whose times of arrival can be measured by a receiver at any location in $\mathcal{D}$. Given a receiver located at an unknown point of $\mathcal{D}$, do the received TOA (time of arrival) signals uniquely localize the receiver? Consider the signal profile mapping $T: \mathcal{D} \to \mathbb{R}^N$ which records the TOA of the (received, identified, and ordered) transmitter pulses. If one assumes that the generic placement of transmitters associated with this system provides a generic perturbation to $T$ within $C^\infty(\mathcal{D}, \mathbb{R}^N)$, then the resulting perturbation embeds the domain $\mathcal{D}$ smoothly for $N > 2k$. Thus, the mapping $T$ is generically injective, implying unique *channel response* and the feasibility of localization in $\mathcal{D}$ via TOA. For example, this implies that a receiver can be localized to a unique position in a planar domain $\mathcal{D}$ using only a sequence of *five* or more transmission signals, globally readable from generically-placed transmitters. Note: it is not assumed that the signals move along round waves, nor does the receiver compute any distances-to-transmitters.

This result is greatly generalizable to a robust *signals embedding theorem* [257]. First, one may modify the codomain to record different signal inputs. For example, using TDOA (time difference of arrival) merely reduces the dimension of the signal codomain by one and preserves injectivity for sufficiently many pulses. Second, one may quotient out the signal codomain by the action of the symmetric group $S_N$ to model inability to identify target sources. This does not change the dimension of the signals codomain: the system has the same number of degrees of freedom. The only change is that the codomain has certain well-mannered singularities inherited from the action of $S_N$. It follows from transversality that *any* reasonable signal space of sufficient dimension preserves the ability to localize based on knowledge of the image of $\mathcal{D}$ under $T$. Transversality and dimension-counting provide a critical bound on the number of signals needed to disambiguate position, independent of the types of signals used.

## 1.8 Stratified spaces

The application of Whitney's Theorem to signal localization in the previous subsection is questionable in practice. Signals do not propagate unendingly, and the physical realities of signal reflection/echo, multi-bounce, and diffraction conspire to make manifold theory suboptimal in this setting. The addition of signal noise further frustrates a differential-topological approach. Finally, the assumption that $\mathcal{D}$ is a manifold is a poor one. In realistic settings, the domain has a boundary: signals are bouncing off of walls, building exteriors, and other structures that, at best, are piecewise-manifolds.

One approach to this last difficulty is to enlarge the class of manifolds. An $n$-**manifold with boundary** is a space locally homeomorphic to *either* $\mathbb{R}^n$ or $\mathbb{R}^{n-1} \times [0, \infty)$, with the usual compatibility conditions required for a smoothing. The **boundary** of $\mathcal{D}$, $\partial\mathcal{D}$, is therefore a manifold of dimension $n-1$. Many of the tools and theorems of this chapter (*e.g.*, tangent spaces and transversality) apply with minor modifications to manifolds with boundary. Yet this is not enough in practice: further generalization is needed. An $n$-**manifold with corners** is a space, each point of which has a neighborhood locally homeomorphic to

$$\{x \in \mathbb{R}^n : x_i \geq 0, i = 1, \ldots, m\},$$

for some $0 \leq m \leq n$, where $m$ may vary from point to point. A true manifold has $m = 0$ everywhere; a manifold with boundary has $m \leq 1$. The analogues of smoothings, derivatives, tangent spaces, and other constructs are not difficult to generate. The boundary of a manifold with corners no longer has the structure of a smooth manifold, as, *e.g.*, is clear in the case of a cube or other platonic solid. Note however, that such a boundary is assembled from manifolds of various dimensions, suitably glued together. Such piecewise-manifolds are common in applications. Consider the solution to a polynomial equation $p(x) = 0$, for $x \in \mathbb{R}^n$. An application of transversality theory shows that the solution set is, for generic choices of coefficients of $p$, a manifold of dimension $n - 1$; however, nature does not always deal out such conveniences. Innumerable applications call for the solution to a specific polynomial equation. The null set of a polynomial, even when not a true manifold, can nevertheless be decomposed into manifolds of various dimensions, glued together in a particular manner.

There is a hierarchy of such **stratified spaces** which deviate from the smooth regularity of a manifold. An intuitive definition of a stratified space is a space $X$, along with a finite partition $X = \cup_i X_i$, such that each $X_i$ is a manifold. Precise control over how these manifolds are pieced together is needed but is too intricate for this introduction: let the reader think of a stratified space as a piecewise-manifold.

Typical examples of stratified spaces include singular solutions to polynomial or real-analytic equations. More physical examples are readily generated. Recall the setting of planar linkage configuration spaces. The 4-bar mechanism gives a 1-d manifold, *except* when the lengths satisfy $L_1 = L_3$ and $L_2 = L_4$. In this case, the configuration space is a pair of circles with intersections (or, better, *singularities*) – either two or three depending on whether or not all the lengths are the same. These singularities have physical significance: they correspond to configurations which are collinear. Upon building such a linkage, one can *feel* the difference as it passes through a singular point.

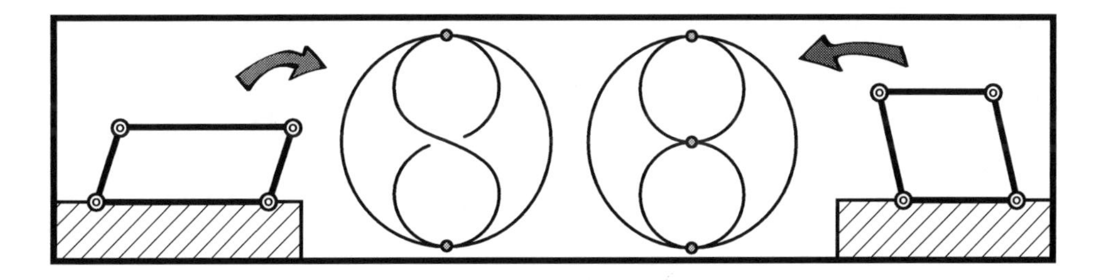

# Notes

1. Homeomorphic manifolds are not always diffeomorphic. In dimensions three and below, they are. Each $\mathbb{R}^n$ has a unique smooth structure except $n = 4$, which has an uncountable number of *exotic* smoothings [162]. Spheres $\mathbb{S}^7$ of dimension seven have exactly 28 distinct smooth structures; see [225] for a survey. It is unknown at the time of publication if $\mathbb{S}^4$ has non-standard smoothings [130].
2. A manifold is **orientable** if it has an atlas such that all transition maps $\phi_\beta \circ \phi_\alpha^{-1}$ between charts preserve orientation (have derivatives with positive determinant). The projective plane $\mathbb{P}^2$ and Klein bottle $K^2$ are well-known non-orientable surfaces. Verifying that all transition maps preserve orientation seems difficult: more efficient algebraic means of orienting manifolds will arise in Chapters 4 and 5.
3. Configuration spaces are for many the entrance to applied topology: it is a subject worthy of its own text. It is instructive and highly recommended to build a complex mechanical linkage and investigate its topology by hand. For planar linkages, flat cardboard, wood, metal or plastic with pin joints works well. For 3-d linkages, the author uses wooden dowels with latex tubing for rotational joints. Sadly, no higher spatial dimensions are available for resolving intersections between edges in spatial linkages. With practice, the user can tell the dimension of the configuration space by feel, without explicit computation (past dimension 10, the author's discernment fails).
4. Other tools of advanced calculus follow in patterns similar to those of derivatives, including differential forms, integration of forms, Stokes' Theorem, partial differential equations, and more: see Chapter 6. The reader interested in calculus on manifolds can find excellent introductions [169], some tuned to applications in mechanics [1, 15].
5. Transversality is a topological approach to genericity. Probability theory offers complementary and, often, incommensurate approaches.
6. Lemma 1.10 is the hidden jewel of this chapter, as it enables so much of the machinery in future chapters. This is the first appearance of an Euler characteristic in this text. It is not the last – there are at least half-a-dozen manifestations of this index in this text.
7. One must be careful in proving genericity results, as the specification of a topology on function spaces is required. Tweaking the topology of the function spaces allows for relative versions, which allow perturbations to one domain while holding the function fixed elsewhere.
8. Higher derivative data associated to a map $f\colon M \to N$ is encoded in the $r$-jet, $j^r f$, taking values in a **jet bundle** $J^r(M, N)$. This $j^r f$ records, for each $p \in M$, the Taylor polynomial of $f$ at $p$ up to order $r$. As always, the computations are done at the chart level and shown to be independent of coordinates. The Jet Transversality Theorem says that, for any submanifold $W \subset J^r(M, N)$ of the jet bundle, the set of $f\colon M \to N$ whose $r$-jets are transverse to $W$ is a residual subset of $C^\infty(M, N)$ [179].
9. The study (and even the definition) of stratified spaces is *much* more involved than here indicated [163]. Various forms of the Transversality Theorem apply to stratified spaces [161, 216]; they are sufficient to derive a form of Whitney's Theorem on embeddings and allow for unique channel response in a transmitter-receiver system outfitted with corners, walls, and reflections [257].
10. The configuration space of **hard spheres** in a domain gives a wonderful class of stratified spaces whose topology can be quite intricate [23]. The topology of these configuration spaces undergoes large-scale qualitative changes as the number of spheres is increased – changes that mimic phase transitions in matter [58].

## Chapter 2

# Spaces: Complexes

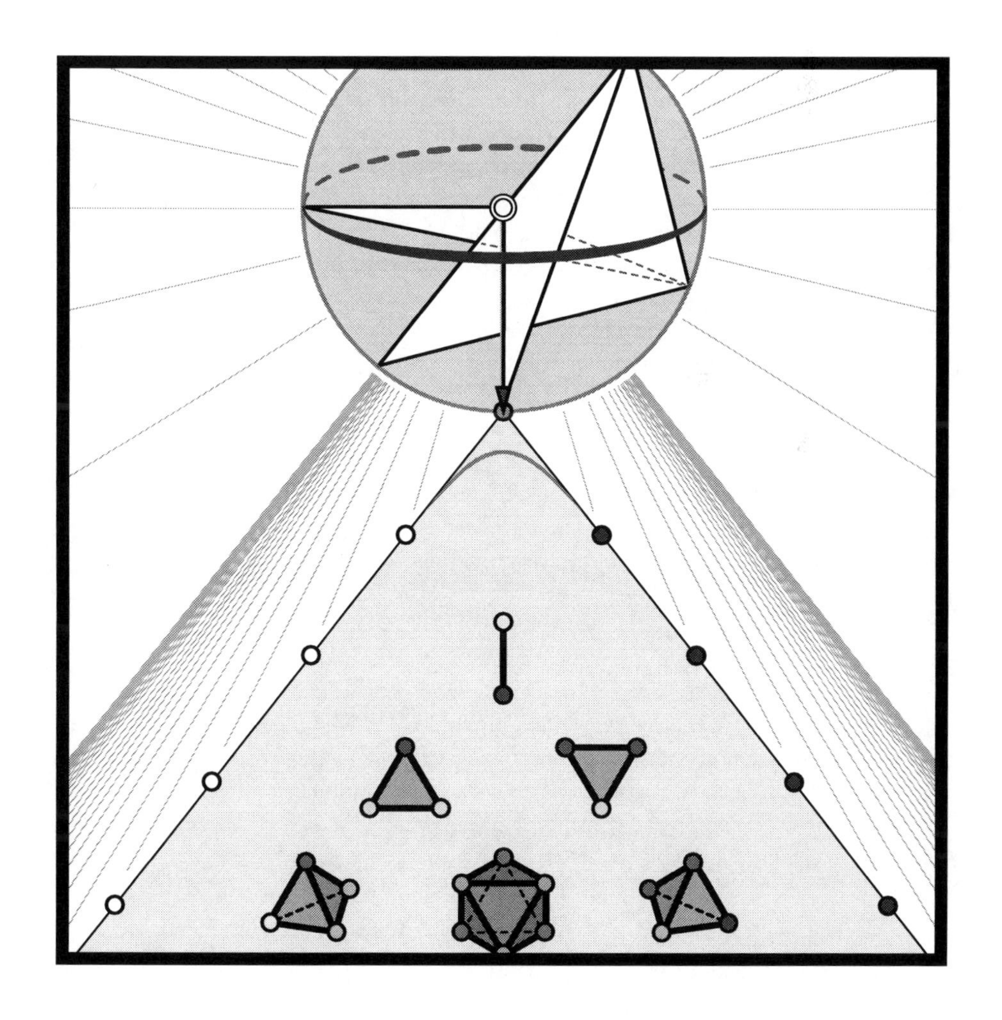

Parts assemble into wholes. A change in focus from smooth spaces patched together with diffeomorphisms to complexes assembled from a bin of simple pieces motivates the eventual transition to algebraic assemblages of spaces and invariants.

## 2.1 Simplicial and cell complexes

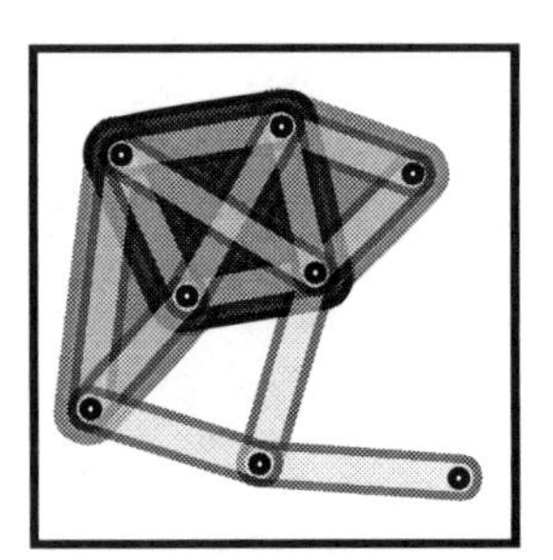

Let $S$ be a discrete set. An **abstract simplicial complex** is a collection $X$ of finite subsets of $S$, closed under restriction: for each $\sigma \in X$, all subsets of $\sigma$ are also in $X$. Each element $\sigma \in X$ is called a **simplex**, or rather a $k$-simplex, where $|\sigma| = k+1$. Given a $k$-simplex $\sigma$, its **faces** are the simplices corresponding to all subsets of $\sigma \subset S$. For example, if $S = \{s_i\}_i$, the subcollection $\{s_2, s_5, s_7\}$ is a 2-simplex with three 1-simplex faces, three 0-simplex faces, and, of course, the empty-set face.

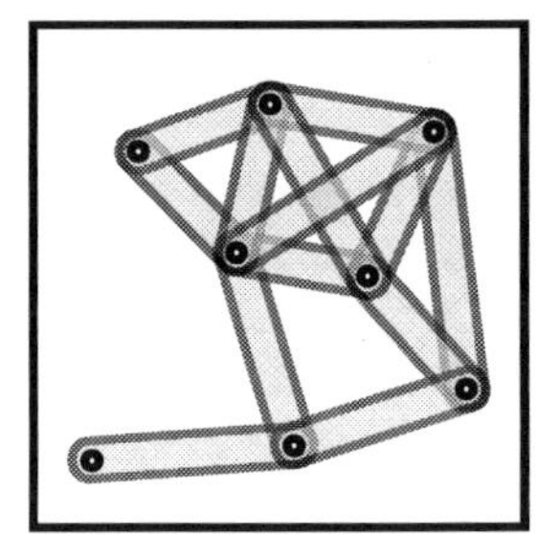

The most familiar simplicial complexes are abstract **graphs**, simplicial complexes without simplices of dimension higher than one. An abstract graph is often presented as a pair $(V, E)$, where $V$ is the set of **vertices** (or 0-simplices) and $E$ is a collection of distinct unordered pairs in $V$. These **edges** are the 1-simplices of the graph. Abstract graphs are ideal for expressing pairwise relations between objects. Relations of a higher order point to simplicial models. One class of elementary but illuminating examples occurs naturally in the form of independence of objects. Given a finite collection of objects $\mathcal{O} = \{x_\alpha\}$, an **independence system** is a collection of unordered subsets of $\mathcal{O}$ declared *independent*. Any such system must be closed with respect to restriction (any subset of an independent set is also independent) and thus defines a simplicial complex: the **independence complex** $\mathcal{I}_{\mathcal{O}}$ is the abstract simplicial complex on vertex set $\mathcal{O}$ whose $k$-simplices are collections of $k+1$ independent objects.

**Example 2.1 (Statistical independence)** ⊚

Independence occurs in multiple contexts, including linear independence of a collection of vectors or linear independence of solutions to linear differential equations. More subtle examples include statistical independence of random variables: recall that the random variables $\mathcal{X} = \{X_i\}_1^n$ are statistically independent if their probability densities $f_{X_i}$ are jointly multiplicative, *i.e.*, the probability density $f_{\mathcal{X}}$ of the combined random variable $(X_1, \ldots, X_n)$ satisfies $f_{\mathcal{X}} = \prod_i f_{X_i}$. The independence complex of a collection of random variables compactly encodes statistical dependencies. ⊚

As with the setting of manifolds, one should rapidly metabolize the formal definition and progress to drawing pictures. One topologizes a simplicial complex $X$ as a quotient space built from topological simplices. The **standard $k$-simplex** is:

$$\Delta^k := \left\{ x \in [0, \infty)^{k+1} : \sum_{i=0}^{k} x_i = 1 \right\}.$$

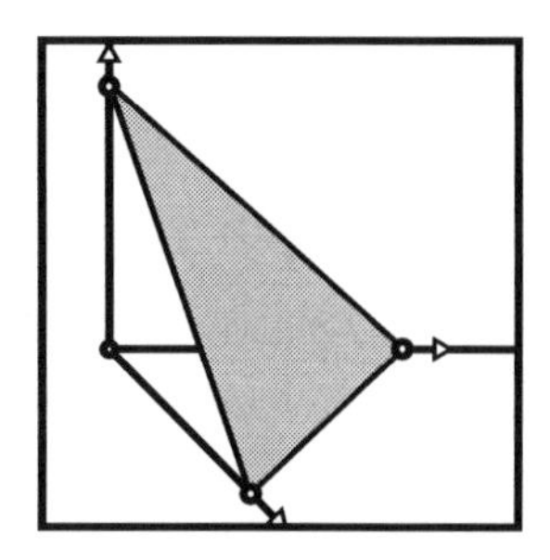

The faces of $\Delta^k$ are copies of $\Delta^j$ for $j < k$ via restriction of $\mathbb{R}^{k+1}$ to coordinate subspaces; via this restriction, one sees how these faces are *attached* along the boundary of $\Delta^k$. One imagines building an abstract simplicial complex $X$ into a space by producing one copy of $\Delta^k$ for each $k$-simplex of $X$, then attaching these together via faces. This ill-named **geometric realization** of $X$ is performed inductively as follows. Define the **$k$-skeleton** of $X$, $k \in \mathbb{N}$, to be the quotient space:

$$X^{(k)} := \left( X^{(k-1)} \bigcup \coprod_{\sigma:\dim \sigma = k} \Delta^k \right) \Big/ \sim, \tag{2.1}$$

where $\sim$ is the equivalence relation that identifies faces of $\Delta^k$ with the corresponding combinatorial faces of $\sigma$ in $X^{(j)}$ for $j < k$.

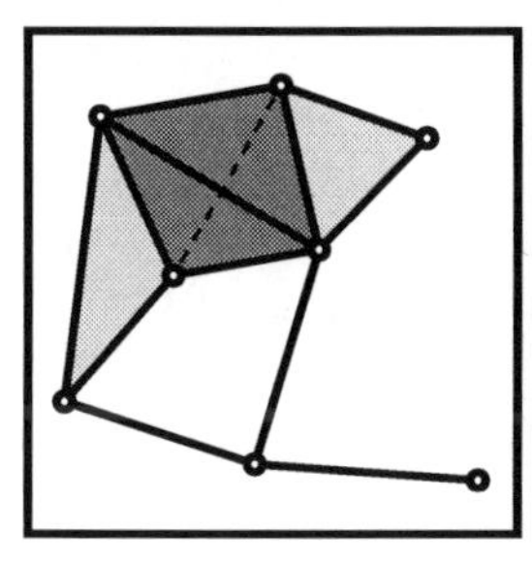

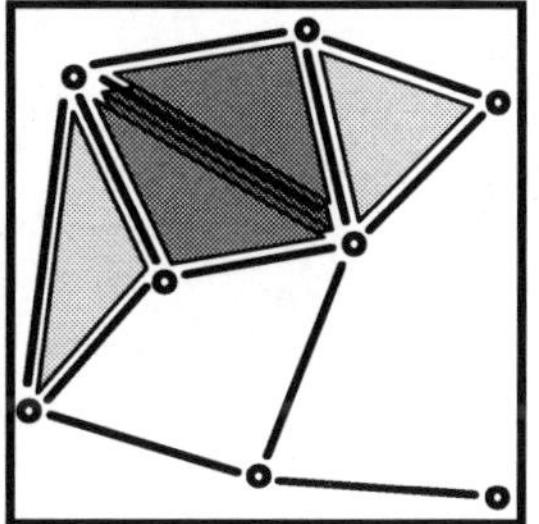

Thus, $X^{(0)} = S$ is a discrete set. The 1-skeleton $X^{(1)}$ is a space homeomorphic to a collection of vertices ($S$) and edges connecting vertices: a topological graph. By construction, $X^{(k)} \supseteq X^{(k-1)}$ and one therefore identifies $X$ with

$$X = \bigcup_{k=0}^{\infty} X^{(k)}.$$

The abuse of notation is intentional; $X$ (the space) will be conflated with $X$ (the abstract simplicial complex). As a space, $X$ is given the **weak topology**: a subset $U \subset X$ is open if and only if $U \cap X^{(k)}$ is open for all $k$.

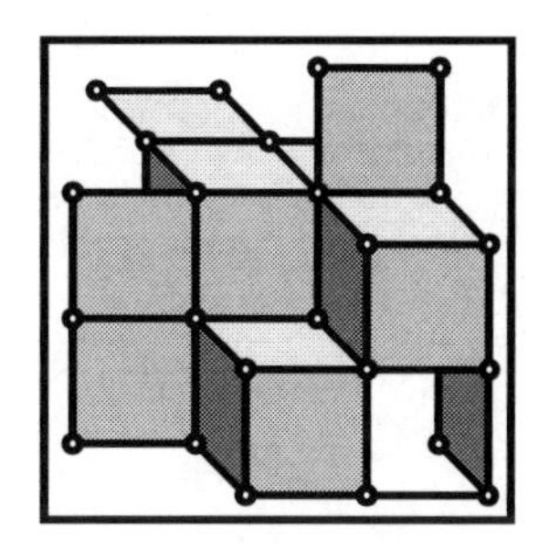

A simplicial complex is one of many possible structures for building a space. For example, one may take the basic $k$-dimensional building block to be a $k$-cube $I^k$, where $I = [0, 1]$. The faces of the cube are obviously smaller-dimensional cubes; attaching maps between faces identify cubes and are likewise clear. The resulting class of **cubical complexes** are natural in many applications. Digital cameras store data in pixels on a 2-dimensional cubical complex. Numerical analysis and finite element methods work well on 3-dimensional cubes (or *voxels*) arranged in a lattice. Cubes and simplices are examples of compact convex polytopes: cell complexes built from arbitrary compact convex polytopes are likewise easy to work with.

More general still is the class of spaces referred to as **CW complexes**. These are built inductively, as per Equation (2.1) with basis cells closed balls $B^k$, and for attaching maps *arbitrary* continuous functions from the boundary $\partial B^k$ to $X^{(k-1)}$. Though this tends to be among the best class of complexes for doing topology, the reader may find it simpler to work at first with **regular cell complexes**, whose cells are balls and whose boundary maps are *homeomorphisms* of $\partial B^k$ to their images in $X^{(k-1)}$.

**Example 2.2 (Unstable manifolds)** ⊙

The flexibility of a CW complex is sometimes needed. Consider the case of a smooth vector field $V$ on a compact 2-manifold $S \subset \mathbb{R}^3$. Assume $V = -\nabla h$ is the gradient field for some height function $h\colon S \to \mathbb{R}$ (so that flowlines of $V$ *flow downhill*), and that there are a finite number of nondegenerate critical points of $h$ (*i.e.*, $V$ is transverse to the zero-section of $T_*S$), so that each is either a source, sink, or saddle of $V$. Recall from §1.4 that each fixed point $p$ has a stable ($W^s(p)$) and unstable ($W^u(p)$) manifold, whose intersection is $p$. By nondegeneracy (and the Stable Manifold Theorem [258]), these are in fact submanifolds of $S$.

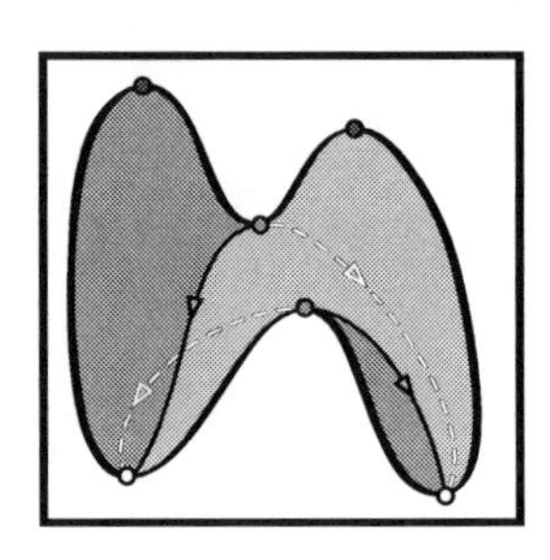

Consider the collection of unstable manifolds $\{W^u(p) : p \in \text{Fix}(V)\}$. These partition the surface $S$ into points (unstable manifolds of sinks), open line segments (unstable manifolds of saddles), and open discs (unstable manifolds of sources). These together define a CW structure on $S$, where: the 0-skeleton $S^{(0)}$ is the union of (unstable manifolds of) sinks; $S^{(1)}$ is the union of unstable manifolds over sinks and saddles obtained by *gluing* intervals to the sinks via the dynamics; and $S^{(2)} = S$ glues the discs defined by unstable manifolds of sources along the 1-skeleton. This simple example can be greatly expanded and forms the basis of **Morse theory**, to come in Chapter 7. ⊙

The distinction between various types of cell complexes is intricate, and the beginner should beware getting lost at this juncture.This text uses simplicial and cubical complexes frequently; regular cell complexes often; and general CW complexes only occasionally. The phrase *for a cell complex* implies that it holds in the fully general CW setting. The remainder of this chapter is a menagerie of interesting simplicial and cellular complexes that have been found useful in applications to date.

## 2.2 Vietoris-Rips complexes and point clouds

Consider a discrete subset $\mathcal{Q} \subset \mathbb{R}^n$ of Euclidean space. If one receives $\mathcal{Q}$ as a collection of data points – a **point cloud** – sampled from a submanifold, it may be desirable to reconstruct the underlying submanifold from the point cloud. A graph based on a point cloud $\mathcal{Q}$ can be efficacious in discerning shape (as any child doing a dot-to-dot can attest), but a simplicial complex may improve matters.

Choose a constant $\epsilon > 0$. The **Vietoris-Rips complex** of scale $\epsilon$ on $\mathcal{Q}$, $\text{VR}_\epsilon(\mathcal{Q})$, is the simplicial complex whose simplices are all those finite collections of points in

$\mathcal{Q}$ of pairwise-distance $\leq \epsilon$ (which is clearly closed under subsets). The *hope* is that $\mathrm{VR}_\epsilon(\mathcal{Q})$ gives a good approximation to the structure underlying the point cloud. Note the difficulty in specifying what that means, as well as choosing the *correct* $\epsilon$ to best approximate. Although this complex is based on points in $\mathbb{R}^n$, there may be simplices of dimension much larger than $n$ (when sufficiently many points are crowded together).

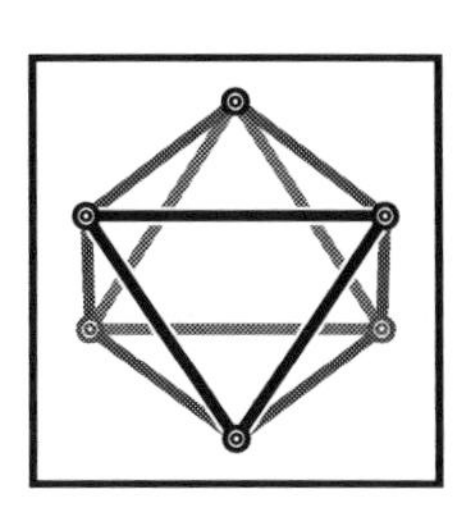

The degree to which a Vietoris-Rips complex VR (of some fixed but unwritten $\epsilon$) accurately captures the topology of the point cloud is not obvious. Consider a point cloud $\mathcal{Q}$ in $\mathbb{R}^n$ and the projection map $\mathcal{S}\colon \mathrm{VR} \to \mathbb{R}^n$ taking the vertices $\mathrm{VR}^{(0)}$ to $\mathcal{Q}$ and taking a $k$-simplex of VR to the convex hull of the associated vertices in $\mathcal{Q}$. The image of $\mathcal{S}$ in $\mathbb{R}^n$ is called the **shadow**, Sh, of the Vietoris-Rips complex. There is no hope of VR and Sh being homeomorphic, as the domain of the projection $\mathcal{S}$ is likely to have higher dimension than the codomain. Even homotopy equivalence is too much to ask, as it is possible to arrange data points in $\mathcal{Q}$ in $\mathbb{R}^2$ so as to have a simplicial sphere $\mathbb{S}^k$ for arbitrary $k > 1$ which projects in Sh to a convex set: such bubbles are artifacts of the Vietoris-Rips complex and not representative the the point cloud topology. Nevertheless the Vietoris-Rips complex *seems* to capture the topology of large-scale holes correctly: see [64].

## 2.3 Witness complexes and landmarks

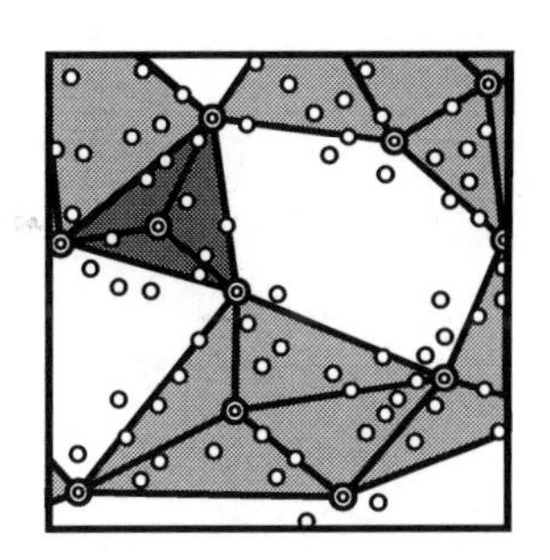

The Vietoris-Rips complex of a large data set may be too unwieldy to store and manipulate. To address this concern, the **witness complex** constructions of Carlsson and de Silva [86] greatly reduce size at the potential expense of topological precision. This has become an intricate subject, with a variety of related definitions built to suit different data sets. The following is a elementary version, suitable for an introduction.

Given a (large) point cloud of nodes $\mathcal{Q}$ in a Euclidean space $\mathbb{R}^n$, choose a (small) set of **landmarks** $\mathcal{L} \subset \mathcal{Q}$. The **weak witness complex**, $W_{\mathcal{L}}$, is an abstract simplicial complex on the vertex set $\mathcal{L}$ defined as follows. A subcollection of points $S \subset \mathcal{L}$ is a simplex in the witness complex if and only if for each subset $T \subset S$ there is a **weak witness** $x_T \in \mathcal{Q}$ with every point in $T$ closer to $x_T$ than to any point in $\mathcal{L} - T$. Simple examples indicate that for a reasonable choice of landmarks, the witness complex provides a decent topological approximation to the underlying structure of the point cloud. More recent generalizations to *strong* and *parameterized* witness complexes span the gap between computability and rigorous recovery of the correct topological type [55].

## 2.4 Flag complexes and networks

The Vietoris-Rips complex is an example of a more general class of simplicial complexes that fill a frame. One of the advantages of this construction is its parsimonious representation with regards to input and storage of the data. To encode a typical simplicial complex into a form that computational software can manage, it is necessary to list all the simplices. In contrast, note that the Vietoris-Rips complex is entirely specified by knowing the vertices $\mathcal{Q}$ and the edges (depending on $\epsilon$). The remaining, higher-dimensional, simplices are all accounted for: the graph determines the complex.

Given any graph, the **flag complex** (or **clique complex**) is the maximal simplicial complex having the graph as its 1-skeleton. Otherwise said, whenever there appears to be the outline of a simplex, one fills it in. The Vietoris-Rips complex is the flag complex associated with the distance-$\epsilon$ proximity graph of a point cloud vertex set $\mathcal{Q}$.

Flag complexes also arise as higher-dimensional models of communications networks. Consider a collection of points $\mathcal{Q} \subset \mathbb{R}^2$ that represent the locations of nodes in a wireless communications network. Ignoring the details of the communications protocols, consider the following simple model of communication: nodes broadcast their unique identities, received by nodes which are within range; these neighbors then establish communication links and in so doing assemble an *ad hoc* network.

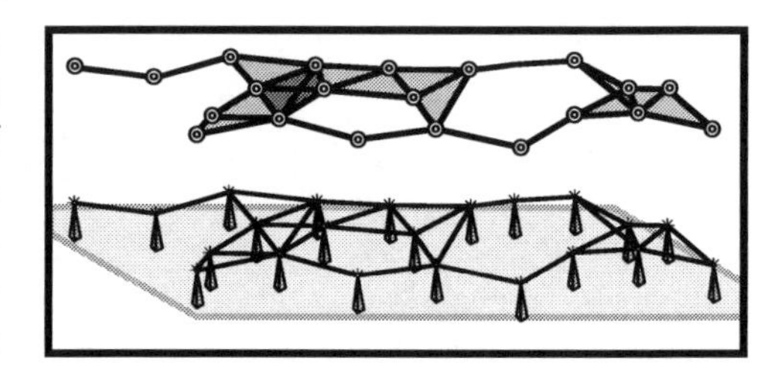

Due to irregularities in transmission characteristics, ambient characteristics of the domain, and signal bounce, the communication links formed may not be solely a function of geometric distance between nodes. Nevertheless, forming the flag complex associated with these edges may recover some of the rough structure implicit in the global network. In §5.6, a method for addressing coverage-type problems is considered using these flag complexes.

## 2.5 Čech complexes and random samplings

The problem of building topologically accurate simplicial models is classical. In the context of the Vietoris-Rips complexes of §2.2, it is frustrating to have higher-dimensional features appearing. One way to circumvent this is to consider the higher-dimensional simplices of the Vietoris-Rips complex more carefully. Given a point cloud $\mathcal{Q}$ in $\mathbb{R}^n$ and a length parameter $\epsilon > 0$, define the **Čech complex** $\check{C}_\epsilon$ to be the simplicial complex built on $\mathcal{Q}$ as follows. A $k$-simplex of $\check{C}_\epsilon$ is a collection of $k+1$ distinct elements $x_i$ of $\mathcal{Q}$ such that the net intersection of *diameter* $\epsilon$ balls at the $x_i$'s is nonempty. The Čech complex is a subcomplex of the Vietoris-Rips complex $\mathrm{VR}_\epsilon$; it is sometimes a proper subcomplex. The disadvantages of a Čech complex are its storage requirements (one cannot simply store the 1-skeleton and fill the rest in) and its construction (one must check many intersections to build the full complex). These are compensated for by a topological accuracy: the Čech complex $\check{C}_\epsilon$ is *always* homotopic to the union of diameter-$\epsilon$ balls about $\mathcal{Q}$ (see Theorem 2.4).

Niyogi, Smale, and Weinberger [241] have used this property of Čech complexes to prove a result about randomly sampled point clouds. Consider a smooth submanifold $M \subset \mathbb{R}^n$. Let $\mathcal{Q}$ be a collection of points sampled at random from $M$. It is intuitively obvious that for a sufficiently dense sampling, the union of properly sized balls about the points is homotopic to $M$: this union of balls can be captured by the Čech complex of $\mathcal{Q}$. The ability of the Čech complex to approximate $M$ is conditional upon the density of sampling and the manner in which $M$ is geometrically embedded in $\mathbb{R}^n$. Let $\tau$ denote the **injectivity radius** of $M$: the largest number such that all rays orthogonal to $M$ of length $\tau$ are mutually nonintersecting.

**Theorem 2.3 ([241]).** *Assume a collection $\mathcal{Q}$ of points on a smooth compact submanifold $M \subset \mathbb{R}^N$, where $M$ has injectivity radius $\tau$. If the density of $\mathcal{Q}$ is such that the minimal distance from any point of $M$ to $\mathcal{Q}$ is less than $\epsilon/2$ for $\epsilon < \tau\sqrt{3/5}$, then the Čech complex $\check{C}_{2\epsilon}(\mathcal{Q})$ deformation retracts to $M$.*

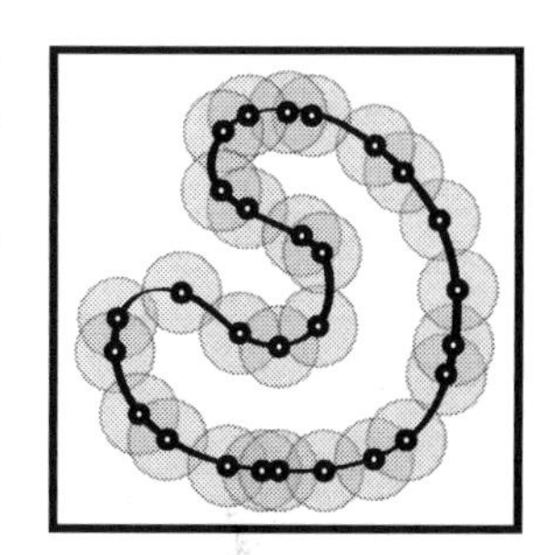

Additional results, such as criteria for these hypotheses for uniformly randomly distributed points on $M$ (with or without various types of noise) are also available and relevant to the statistics of point clouds. This result has been useful in validating data analysis from experiments in, *e.g.*, materials science [82, 83].

## 2.6 Nerves and neurons

As the Vietoris-Rips complex is a metric-ball version of the more general flag complex construction, so also is the Čech complex the metric-ball version of a more general object. Given a collection $\mathcal{U} = \{U_\alpha\}$ of (say, compact) subsets of a topological space $X$, one builds the **nerve** of $\mathcal{U}$, $\mathcal{N}(\mathcal{U})$, as follows. The $k$-simplices of $\mathcal{N}(\mathcal{U})$ correspond to nonempty intersections of $k+1$ distinct elements of $\mathcal{U}$. For example, vertices of the nerve correspond to elements of $\mathcal{U}$; edges correspond to pairs in $\mathcal{U}$ which intersect nontrivially. This definition respects faces: the faces of a $k$-simplex are obtained by removing corresponding elements of $\mathcal{U}$, leaving the resulting intersection still nonempty. Examples of nerves based on, *e.g.*, convex subsets of $\mathbb{R}^n$, suffice to suggest the following classical generalization of the result for Čech complexes.

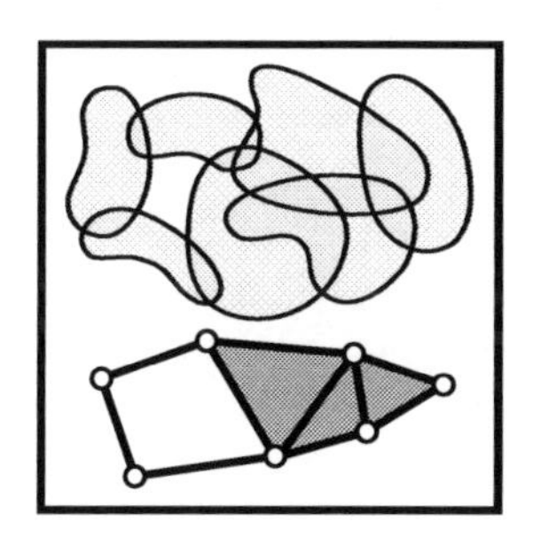

**Theorem 2.4 (Nerve Lemma).** *If $\mathcal{U}$ is a finite collection of open contractible subsets of $X$ with all non-empty intersections of subcollections of $\mathcal{U}$ contractible, then $\mathcal{N}(\mathcal{U})$ is homotopic to the union $\cup_\alpha U_\alpha$.*

One of the more recent and interesting applications of nerves is in neuroscience. The work of Curto and Itskov [80] considers how neural activity can represent external environments. In particular, the authors consider the impacts of external stimuli on certain **place cells** in the dorsal hippocampus of rats. These cells experience dynamic electrochemical activity which is known to strongly correlate to the rat's location in physical space. Each such cell group is assumed to determine a specific location in space, and the collection of such location patches, or **place fields**, forms a collection $\mathcal{U}$ satisfying the hypotheses of the Nerve Lemma, assuming that place fields exist and are stable, omni-directional, and have firing fields that are convex. Curto and Itskov argue that these assumptions are generally satisfied for place fields of dorsal hippocampal place cells recorded from a freely foraging rat in a familiar open field environment.

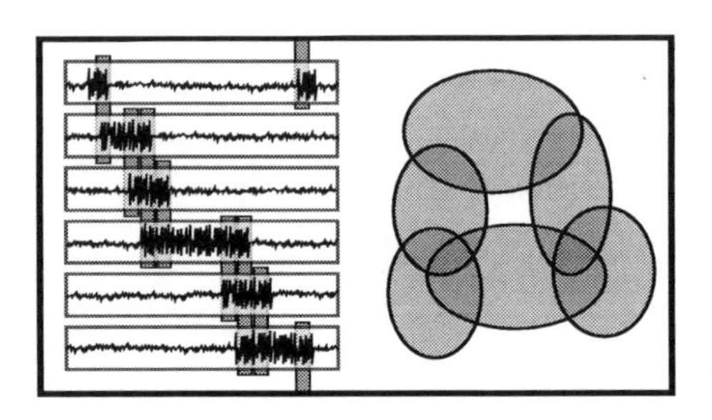

In an experiment, the place cells are monitored, with activity recorded as a time-series. These time series show occasional bursts of activity, such **spike trains** being common data forms in neuroscience. Cell groups are identified by correlating spike train firings that occur at the same time (within a window of time based on some parameter chosen appropriately). The correlation of spiking activity provides the intersection data of the place fields. This, in turn, allows for the computation of the nerve of the place fields based on spike train correlation. These investigations suggest that rats may build a structured representation of their external environment within an abstract space of stimuli. Of note is the lack of metric data – the construction of the physical environment is purely topological and can be effected without reference to coordinates.

## 2.7 Phylogenetic trees and links

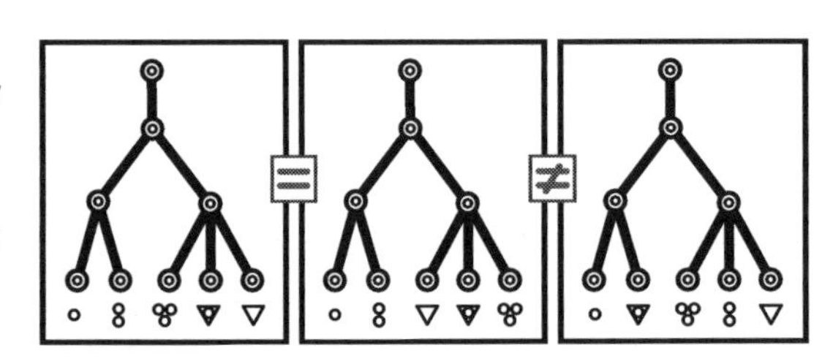

Mathematical biology is ripe for topological tools, given the noisy nature of Nature. In addition to the neuroscience example above is an analogue of configuration spaces associated to genetic data in the form of **phylogenetic trees** – data structures for organizing and comparing taxonomic and genetic data. The relevant objects of study are rooted, end-labeled, metric tress. A **tree** is a graph without cycles. The **leaves** of a tree are vertices of degree 1 (only one edge is attached), and a **root** is a dedicated leaf. A phylogenetic tree is a rooted tree with (1) leaves labeled by data (genes, phenotypes, or other taxonomica), and (2) all interior edges (those not connected to a leaf) labeled with positive real numbers (and represented as the length of the edge). These trees can be readily illustrated as planar graphs, but the embedding of the tree into the plane is *not* part of the data.

Let $\mathcal{T}_n$ denote the following **tree space** of [37]: the space of rooted metric trees with $n$ labeled (non-root) leaves. For simplicity, assume that the label on the root is 0 and labels on the leaves are $1, 2, 3, \ldots, n$. This space has a natural cube complex structure as follows. All non-leaf vertices of the graphs are assumed to be essential in that the number of edges attached is greater than two.

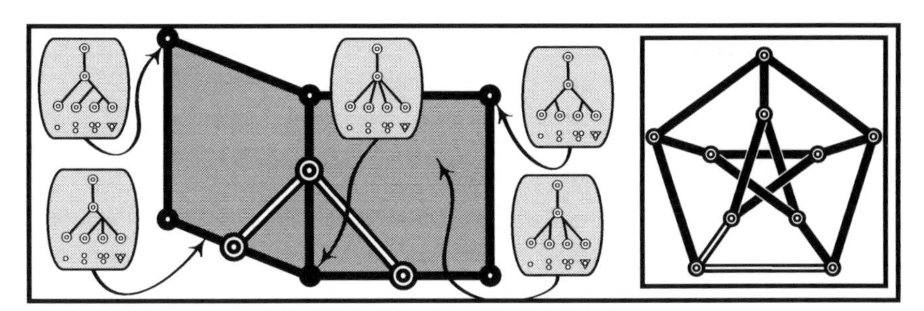

Two such trees are *equivalent* if there is a mapping between them that preserves the graph structure (takes vertices to vertices and edges to edges), the leaf labels, and the interior edge lengths; they are *isomorphic* if there is a mapping between them that preserves the graph structure and the labels. A fixed isomorphism class of phylogenetic trees is parameterized by the interior edge lengths: this yields an open cube of dimension equal to the number of interior edges.

To compactify this to a closed cube, reparameterize all interior edge lengths to the open interval $(0, 1)$ and add the boundary faces. Those faces with one or more 0 factors may be thought of as having the corresponding interior edges collapsed ("zero length edges"). These edge-collapse faces are identifiable as trees of a different isomorphism class. By gluing together all such cubes according to isomorphism class, a finite cube complex $\mathcal{T}_n$ is obtained.

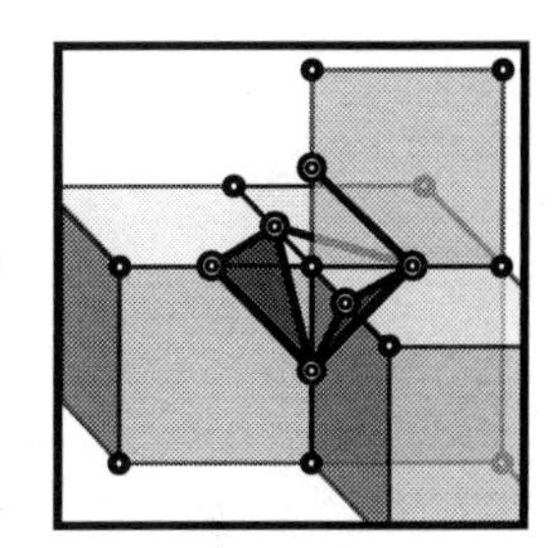

As topological spaces, $\mathcal{T}_n$ may seem too simple: it is contractible to an *origin*, the radial graph with no interior edges, by shrinking each interior edge continuously to length zero. Despite this simplicity, the manner in which the various cubes are assembled is topologically (and geometrically) of interest. The best way to analyze this assembly is through a simplical model called a **link**. Let $X$ denote a cell complex built from simplices, cubes, or other simple polyhedral basis cells, and let $v \in X^{(0)}$ be a 0-cell. The link of $v$ in $X$, $\ell k_X(v)$, is the simplicial complex whose $k$-simplices consists of $(k+1)$-dimensional cells attached to $v$, with faces inherited from faces of cells in $X$.

**Example 2.5 (Tree spaces)** ⊚

The tree space $\mathcal{T}_3$ is a "letter Y." The space $\mathcal{T}_4$ consists of 15 2-dimensional squares (corresponding to the 15 possible binary rooted trees with 4 labeled leaves) glued together so as to have a common corner (corresponding to the radial rooted tree with 4 leaves) and a link equal to the 10-vertex, 15-edged **petersen graph**. As one might guess, links of tree spaces $\mathcal{T}_n$ grow quickly in size and complexity with $n$; their topology, however, is classifiable.

**Theorem 2.6 ([37]).** *The link of the origin in $\mathcal{T}_n$ is a flag complex that has the homotopy type of $(n-1)!$ copies of $\mathbb{S}^{n-3}$ glued together at a single point.* ⊚

There are numerous reasons why scientists and engineers want to work with *spaces* of objects rather than merely the objects themselves. The tree spaces $\mathcal{T}_n$ (though arising in other contexts in topology and algebraic geometry as certain moduli spaces) were generated in response to challenges in statistics associated to genetic data. Experimental data is used to generate phylogenetic trees, and this data does not spring forth fulled formed from the mind of the researcher: the data are often noisy and not perfectly repeatable. A collection of numerical or vector-valued data can be averaged, but what does it mean to average a sequence of phylogenetic trees in a scientifically meaningful manner? This requires a geometry (and thus a topology) on the space of all possible phylogenetic trees.

## 2.8 Strategy complexes and uncertainty

Many of the complexes of this chapter approximate manifolds or provide models of data; however, complexes also have broad applicability in engineering systems as spaces that collate states. Consider a **transition graph**, a directed graph $X$ whose vertices $V$ represent states of an abstract system, and whose edges $E$ represent deterministic actions. For concreteness, consider the example [109] of a transportation network, where vertices are locations and edges represent flights, freeways, subways, trains, or other discrete enter-exit modes of transport. Other examples include motion-planning in robotics with discrete states (robots move from landmark to landmark) or manufacturing processes (states are subassemblies; edges are assembly steps).

Planning on the transition graph is straightforward: find a directed path from initial to goal vertices, then execute the appropriate actions sequentially. However, stepping onto a flight bound for Chicago at noon does not guarantee arrival, either at Chicago or at noon. Worse still are the kinds of uncertainties precipitated by an adversary. Ill weather and air travel does not *exactly* fit that scheme, but other situations do exhibit adversarial forms of uncertainty for which, if something *may* go amiss, it *must* be mitigated.

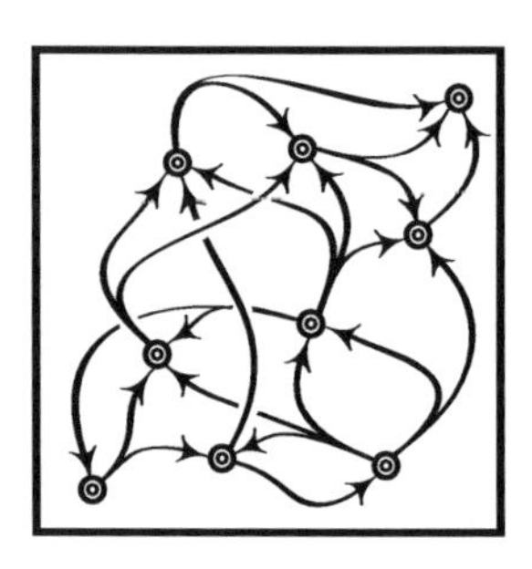

Erdmann [108, 109] models adversarial uncertainty in the transition graph via *branched* edges. Actions are deterministically chosen, but once chosen, the outcome is one of several possibilities, out of the control of the chooser. These possibilities are *not necessarily* chosen probabilistically; an adversary may determine any outcome among the terminal edges. The state chosen by the adversary is known to the game-player, but only *after* the nondeterministic action has been executed. In such a setting, the notion of a guaranteed strategy for reaching a goal state despite adversarial meddling seems difficult if not impossible.

The question of path-planning within a nondeterministic transition graph is one amenable to simplicial data structures. Given a nondeterministic transition graph $X$, define the **strategy complex**, $\Delta_X$, as the abstract simplicial complex defined on the set of actions $E(X)$, in which a $k$-simplex is a collection of $k+1$ disjoint actions whose unions (including all branches of any chosen nondeterministic action) contain *no directed cycles.* This relation is closed under restriction (a smaller collection of such edges clearly also contains no directed cycle), and is sensible, as a directed cycle means that a skilled adversary can cycle the player through states *ad infinitum*. The intuition is that $\Delta_X$ consists of *trap-free* strategies: each simplex of $\Delta_X$ is a strategy for progressing to one or more states.

For $g \in V(X)$ a goal state, say that $X$ has a *complete* strategy for attaining $g$ if there exists a mapping from $V(X)$ to $E(X)$ which associates to $v$ an action based at $v$, so that one attains $g$ with the execution of at most $N$ actions for $N$ sufficiently large. Note: the goal is certain, but the path is not. The obstruction to attainability lies in the strategy **loopback** complex: denote by $\Delta_{X \leftarrow g}$ the strategy complex of $X$ augmented by deterministic arrows from $g$ to each $v \in V(X) - g$.

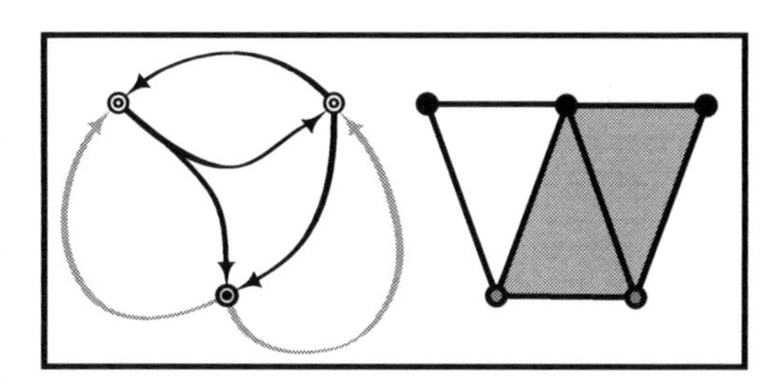

**Theorem 2.7 ([108, 109]).** *A nondeterministic transition graph $X$ possesses a complete strategy for attaining a goal $g \in V(X)$ if and only if*

$$\Delta_{X \leftarrow g} \simeq \mathbb{S}^{\#V-2}.$$

*Otherwise, it is contractible.*

The proofs of this and related theorems use nerves extensively. Additional results include realizability of simplicial complexes as strategy complexes, incorporation of stochastic nondeterminism, and decompositions of strategy complexes by means of deterministic subsystems.

## 2.9 Decision tasks and consensus

A related use of simplicial structures assists in multi-agent decision-making. Consider the problems of consensus or distribution among collaborative agents, a common scenario in biology (animal flocking), sociology (voting, beliefs), engineering systems (synchronized network clocks), and robotics (cooperative action). The problem of collaborative consensus has a large literature: mathematical approaches depend sensitively on the modeling assumptions. Most work on distributed consensus phenomena [227, 287] focuses on graph-theoretical methods. Hidden beneath these 1-dimensional models is a rich higher-dimensional topological structure underlying influence, consensus, and division.

**Example 2.8 (Allocation)** ⊚

Consider the problem of $N$ transmitters attempting to broadcast without interference over a frequency spectrum with $C$ distinct, non-interfering channels. How may these channels be assigned? What if certain transmitters have frequency constraints? What does the space of solutions to this problem look like?

Build a simplicial complex on a vertex set of choices: vertices correspond to transmitter $n$ choosing channel $c$, for each allowable assignment $(n, c)$. A $k$-simplex consists of $k+1$ transmitters having made compatible (distinct channel) choices, and the other transmitters having unassigned frequencies. This is closed under restriction, since letting one or more assigned transmitter release a frequency leaves the remaining transmitters still compatible. The resulting simplicial complex is rightly viewed as a *configuration space* of solutions to the allocation problem. For example, three transmitters choosing from four channels gives rise to a simplicial 2-torus $\mathbb{T}^2$. ⊚

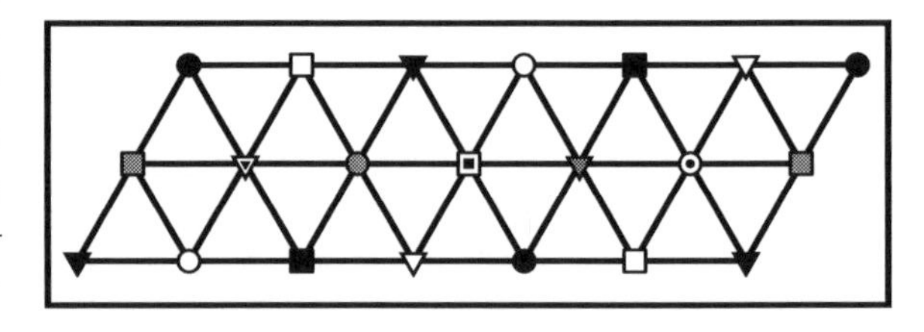

Work of Herlihy and Shavit [178] pioneered a topological approach to consensus/decision problems in the context of distributed, fault-tolerant computation. In this setting, it is presumed that there are $N$ independent processors $P_i$ who share a common read/write memory element for communication purposes. For simplicity and concreteness, assume that there are $M > 1$ possible labels or states and that each processor has an initial preference. The problem is whether there exists an algorithm for the processors, via read/write communication to the common memory, to come to deterministic finite-time consensus. The interesting catch is that all computations are *asynchronous* (each processor works at its own speed) and *faulty* (sometimes, a processor crashes and can engage no further in communication). The problem in reasoning about such systems is that processors which take a very long time to reach decisions are not readily distinguished from processors which have failed and will never respond. A *fault tolerant* distributed algorithm is one which will terminate in finite time even in the event that some (but not all) processors fail irrevocably.

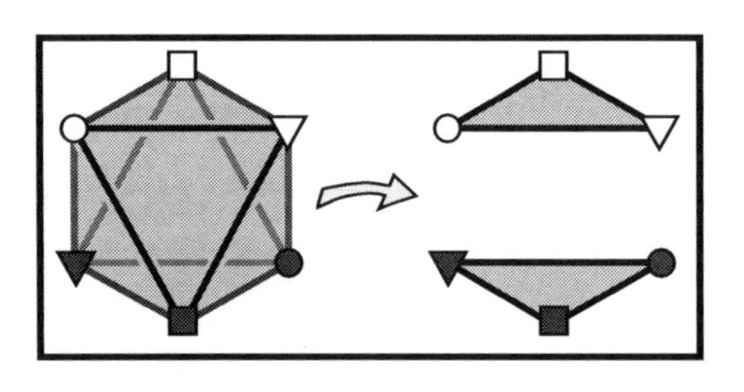

The approached used in [178] begins with a simplicial model of the problem. Given $N$ processors with $M$ possible initial states, define the **input complex** $\mathcal{I}$ to be the abstract simplicial complex whose $k$-simplices are labelings of the states of $k+1$ distinct processors. The **output complex** $\mathcal{O}$ is the abstract simplicial complex whose $k$-simplices are labelings of the states of $k+1$ distinct processors which are consistent with a desired output. The intuition behind a $k$-simplex is that $k+1$ of the processors exhibit a legal system state and the $N-(k+1)$ remaining processors have failed to report. Because of this, the topology of the complexes (encoded by the simplex faces) correlates with failure of processors. A **decision task** consists of the input and output complexes, along with a set of constraints. For example, if all $N$ processors begin in consensus with the same label, then the putative consensus algorithm must of necessity output that precise label as the consensus state. Because of the relationship between faces and failures, it is possible to reduce questions of existence of distributed fault-tolerant consensus algorithms to the existence of certain maps from the input complex to the output complex.

**Example 2.9 (Consensus)** ⊚

In the case of $N$ processors trying to come to binary consensus (*i.e.*, two labels, zero and one), the input complex $\mathcal{I} \simeq \mathbb{S}^{N-1}$ is a simplicial sphere and the output complex $\mathcal{O} = \Delta^{N-1} \sqcup \Delta^{N-1}$ is a disjoint pair of simplices: all zeros and all ones. The decision task is to come to consensus, with the proviso that initial consensus terminates the program immediately. Thus, any wait-free fault-tolerant protocol would have to induce a surjective continuous map from $\mathcal{I}$ to $\mathcal{O}$. This is impossible, since $\mathcal{I}$ is connected and $\mathcal{O}$ is not. Thus, there is no distributed asynchronous fault-tolerant solution to the problem of binary consensus. That in itself is no surprise; yet this approach has resolved more complex consensus problems for which other solutions were unknown. These deeper examples rely not on common-sense notions of connectivity but rather on holes of higher order – the homology of Chapter 4. ⊚

One of the key theorems [178] gives *necessary and sufficient* conditions for consensus stated in terms of existence of maps (with details concerning simplicial subdivisions and colorings). The salient feature of the result is that the topology of the decision task can determine whether an algorithm *exists*. This type of implicit inference is a hallmark of topological methods.

## 2.10 Discretized graph configuration spaces

The strategy complexes of §2.8 and the allocation, input, and output complexes of §2.9 are all viewable as a simplicial sort of configuration space. The types of configuration spaces used in robotics (from §1.2 and §1.5) are neither simplicial nor cellular complexes. One can, however, approximate such spaces with cell complexes. This is most easily done on the already-discrete structure of a graph.

Consider a finite graph $X$, now thought of not as an abstract system of states, but rather as a physical, geometric highway of paths. Keeping track of vehicles or robots (or *tokens*) on $X$ leads, as in §1.5, to configuration spaces. The configuration space of $n$ labeled points on $X$, $\mathcal{C}^n(X)$, is relevant to motion planning in robotics when the automated vehicles are constrained by tracks, guidewires, or optical paths [149]; however, $\mathcal{C}^n(X)$ is in general neither a manifold nor a simplicial/cell complex. The following constructions [2] are approximations to these spaces by cubical complexes.

Define the **discretized configuration space** of $X$ as:

$$\mathcal{D}^n(X) := (X \times \cdots \times X) - \tilde{\Delta}, \tag{2.2}$$

where $\tilde{\Delta}$ denotes the set of all open cells in $X^n$ whose closures intersect the topological diagonal $\Delta$. Equivalently, $\mathcal{D}^n(X)$ is the set of configurations for which, given any two tokens on $X$ and any path in $X$ connecting them, the path contains at least one entire (open) edge. Thus, instead of restricting tokens to be at least some intrinsic distance $\epsilon$ apart, one now restricts tokens on $X$ to be at least one full edge apart. Note that $\mathcal{D}^n(X)$ is a subcomplex of the cubical complex $X^n$ and a subset of $\mathcal{C}^n(X)$ (it does not contain partial cells that arise when cutting along the diagonal), and is, in fact, the largest subcomplex of $X^n$ that does not intersect $\Delta$.

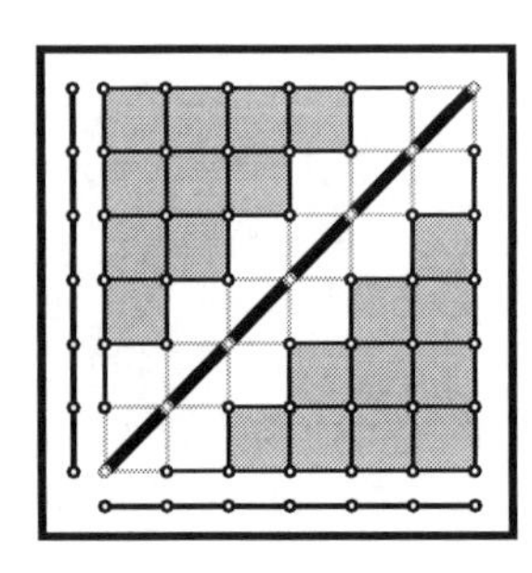

With this natural cell structure, one can think of the 0-cells of $\mathcal{D}^n(X)$ as *discretized* configurations – arrangements of labeled tokens at the vertices of the graph. The 1-cells of $\mathcal{D}^n(X)$ indicate which discrete configurations can be connected by moving one token along an edge of $X$. Each 2-cell in $\mathcal{D}^n(X)$ represents two physically independent edges: one can move a pair of tokens independently along disjoint edges. A $k$-cell in $\mathcal{D}^n(X)$ likewise represents the ability to move $k$ tokens along $k$ closure-disjoint edges in $X$ in an independent manner.

**Example 2.10 (Digital microfluidics)** ⊚

An excellent physical instantiation of the discretized configuration space appears in work on **digital microfluidics** [111], in which small (~1mm diameter) droplets of fluid can be quickly and accurately manipulated in an inert oil suspension between two plates. The plates are embedded with a grid of wires, providing discrete localization of droplets via **electrowetting** – a process that exploits current-induced dynamic surface tension effects to propel and position a droplet. Applying an appropriate current drives the droplet a discrete distance along the wire grid. The goal of this and related microfluidics research is to create an efficient *lab on a chip* in which droplets of various chemicals or liquid suspensions can be positioned, mixed, and then directed to the appropriate outputs. Using the grid, one can manipulate many droplets in parallel. The discretized configuration space of droplets on the graph of the electrical grid captures the topology of the multi-droplet coordination problem. Note, however, that collisions are not always to be avoided – chemical reactions depend on controlled collisions. ⊚

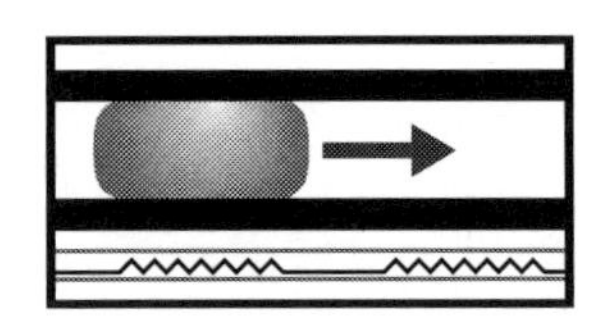

**Example 2.11 (A nonplanar graph)** ⊚

Discretization can untangle a complex-looking configuration space. One interesting example is the configuration space of two labeled points on a complete connected graph of five vertices, $K_5$. The discretized configuration space, $\mathcal{D}^2(K_5)$, is a 2-dimensional cube complex.

Fix a single vertex $v \in \mathcal{D}^2(K_5)$: this corresponds to a pair of distinct labeled tokens on vertices of $K_5$. From this state $v$ emanate six edges in $\mathcal{D}^2(K_5)$: each token can move to any of the three open vertices of $K_5$. For each edge $e$ incident to $v$, the token not implicated by $e$ can move to two of the three available vertices without interfering with $e$; thus, $e$ is incident to exactly two 2-cells of $\mathcal{D}^2(K_5)$. There are a total of six such 2-cells incident to $v$, arranged cyclically. Thanks to the symmetry of $K_5$, the complex looks the same at each vertex, and $\mathcal{D}^2(K_5)$ is thus a topological 2-manifold: a compact surface. With the tools of the next chapter, it will be shown that this surface is of genus six. ⊙

The discretized configuration space is an accurate cubical model of the configuration space.

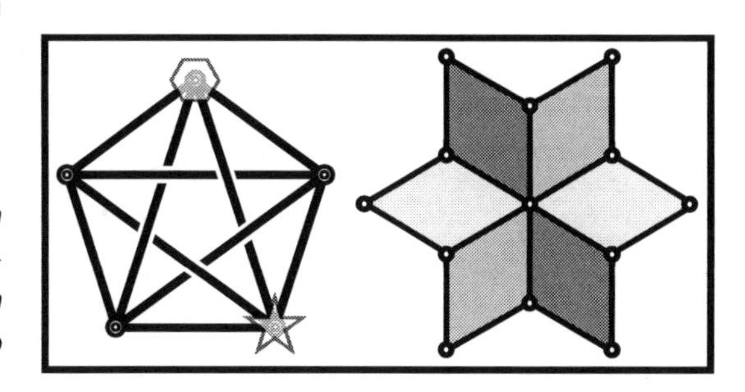

**Theorem 2.12 ([2, 245]).** *Let $X$ be a finite topological graph (with no edges connecting a vertex to itself) and let $\tilde{X}$ be the subdivision of $X$ that inserts vertices so as to split each edge of $X$ into $n-1$ edges. Then $\mathcal{D}^n(\tilde{X})$ is homotopic to $\mathcal{C}^n(X)$.*

## 2.11 State complexes and reconfiguration

Discretized configuration spaces of graphs are generalizable to a very broad class of systems – including metamorphic robots, protein chains, and more – which are reconfigurable based on local rules. The resulting configuration spaces, called **state complexes**, are cube complexes with interesting topological and geometric properties.

Fix a graph $X$ and an alphabet $\mathcal{A}$, used to label the vertices $V$, each such labeling $u \colon V \to \mathcal{A}$ comprising a state $u$. A **reconfigurable system** is a collection of states $\{u_\alpha\}$, closed under the actions of a fixed set of local reconfigurations, or **generators**. Each generator $\phi$ consists of a support subgraph $\text{supp}_\phi \subset X$ and an *unordered* pair of **local states**, labelings of the vertex set of $\text{supp}_\phi$ by elements of $\mathcal{A}$. A generator is **admissible** at a given state $u$ if one of the generator's local states matches the restriction of $u$ to $\text{supp}_\phi$. The **trace** of a generator is the (nonempty) subset of the support $\text{trace}_\phi \subset \text{supp}_\phi$ on which the local states differ.

A simple example comes from discretized configuration spaces of a graph from §2.10. An alphabet $\mathcal{A} = \{0, 1, \ldots, N\}$ labels the vertices of $X$ as empty (0) or occupied with one of $N$ (labeled) tokens. Local reconfigurations are supported on the closure of an edge in $X$; each of the $N$ generators $\phi$ has $\text{supp}_\phi = \text{trace}_\phi$ equal to a single closed edge with local state a labeling of one vertex with 0 and the other $0 < i \leq N$ for some $i$, the other local state reversing these labelings. Applying a sequence of admissible generators is identical to moving labeled agents discretely along $X$. This system has a well-defined notion of a configuration space, $\mathcal{D}^N(X)$, in which $k$-dimensional cubes correspond to $k$ local moves that can be executed *in parallel*: this generalizes to any reconfigurable system as follows.

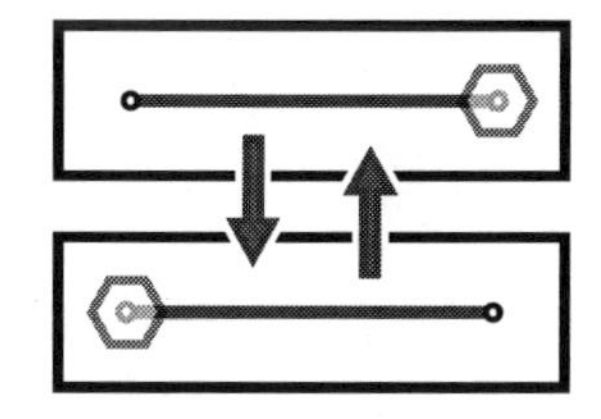

A collection of generators $\{\phi_\alpha\}$ admissible at a given state $u$ is said to **commute** if their supports and traces are pairwise non-intersecting:

$$\text{trace}_{\phi_\alpha} \cap \text{supp}_{\phi_\beta} = \varnothing \qquad \forall \alpha \neq \beta.$$

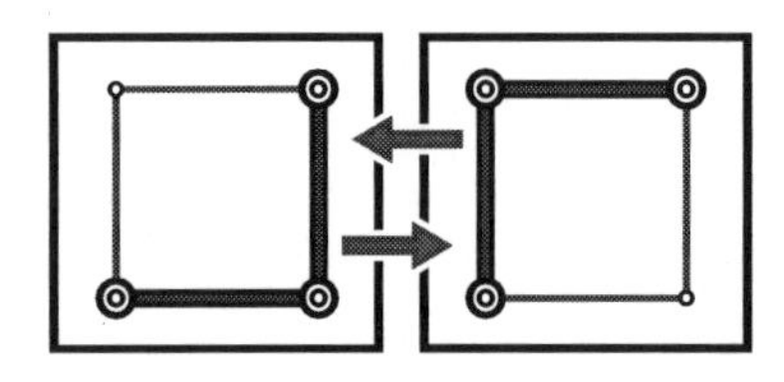

This means that the changes implicated by a generator (its trace) do not impact the applicability of the other generators (their supports): *commutativity encodes physical independence.* Define the **state complex** of a reconfigurable system to be the cube complex $\mathcal{S}$ with an abstract $k$-cube for each collection of $k$ admissible commuting generators. That is, the 0-skeleton consists of the states in the reconfigurable system; the 1-skeleton is the graph whose edges pair two states which differ by the local states of a single generator; 2-cells correspond to commutative pairs of generators acting on a state, *etc.* For reconfigurable systems that admit only finitely many admissible generators at any given state, the state complex $\mathcal{S}$ is a locally compact cubical complex.

**Example 2.13 (Articulated planar robot arm)** ⊚

A reconfigurable system models the position of an articulated robotic arm with one end fixed at the origin and which can (1) rotate at the end and (2) "flip" a 2-segment corner from up-right to right-up and vice versa. This arm is *positive* in the sense that it may extend up and to the right only. A grammatical model is simplest: states are length $n$ words in two letters. Specifically, let $X$ be a linear graph on $n$ vertices with an alphabet $\mathcal{A} = \{U, R\}$ encoding *up* and *right* respectively. Generators are of two types: (1) exchange the subwords $UR$ and $RU$ along an edge; (2) change the terminal letter in the word. Supports (and traces) are (1) a closed edge, and (2) the terminal vertex, respectively. Generators commute when the associated subwords are disjoint. The state complex in the case $n = 5$ has cubes of dimension at most three. In this case also, although the transition graph (the 1-skeleton of $\mathcal{S}$) for this system is complicated, the state complex itself is contractible. ⊚

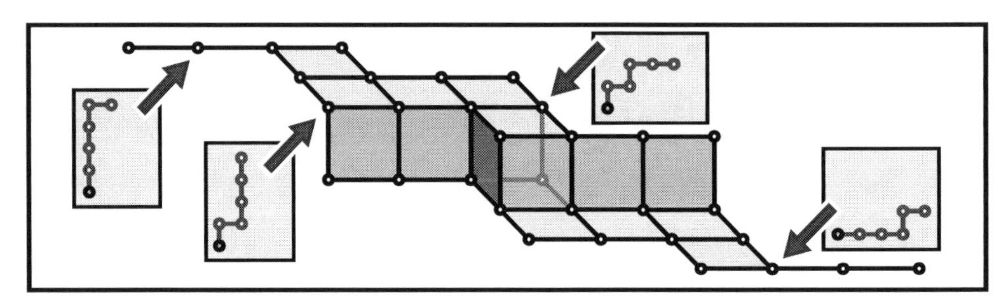

**Example 2.14 (Robot path coordination)** ⊚

The following example is inspired by robot coordination problems. Consider a collection of $n$ graphs $(X_i)_1^n$, each embedded in the plane of a common workspace (with intersections permitted). On each $X_i$, a robot $R_i$ with some particular fixed size/shape is free to translate along edges of $X_i$. The **coordination space** of this system is defined to be the space of all configurations in $\prod_i X_i$ for which there are no collisions – the geometric robots $R_i$ have no overlaps in the workspace.

To model this with a reconfigurable system, let the underlying graph be the disjoint union $X = \sqcup_i X_i$. The generators for this system are as follows. For each edge $\alpha \in E(X_i)$, there is exactly one generator $\phi_\alpha$. The trace is the (closure of the) edge itself, $\text{trace}(\phi_\alpha) = \alpha$, and its generator corresponds to sliding the robot $R_i$ from one end of the edge to the other. The support, $\text{supp}(\phi_\alpha)$ consists of this trace union any other edges $\beta$ in $X_j$ ($j \neq i$) for which the robot $R_j$ sliding along the edge $\beta$ can collide with $R_i$ as it slides along $\alpha$. The alphabet is $\mathcal{A} = \{0, 1\}$ and the local states for $\phi_\alpha$ have zeros at all vertices of all edges in the support, except for a single 1 at the boundary vertices of $\alpha$. Any state of this reconfigurable system has all vertices labeled with zeros except for one vertex per $X_i$ with a label 1. The resulting state complex is a cubical complex that approximates the coordination space. In the case where $X_i = X$ is the same for all $i$ and the robots $R_i$ are sufficiently small, this reconfigurable system has $\mathcal{S} = \mathcal{D}^n(X)$. ⊚

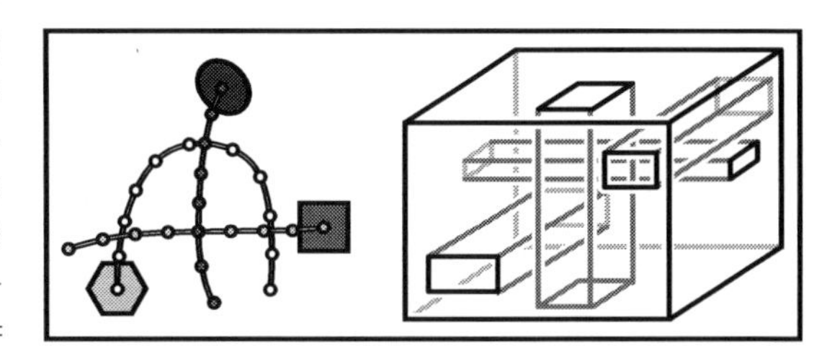

State complexes encompass a wide variety of different systems and can appear to be full of holes, as seen in Examples 2.11 and 2.14 Nevertheless, the types of holes that arise are unusually restricted:

**Theorem 2.15 ([153]).** *Every state complex $\mathcal{S}$ is* **aspherical**: *any map $f\colon \mathbb{S}^n \to \mathcal{S}$ for $n > 1$ is homotopic to a constant map $\mathbb{S}^n \to \star$.*

Note the recurring theme: state complexes, strategy complexes, tree spaces, and more all seem to be characterized or clarified by the absence of presence of *spheres*, leading one to wonder if these are not the primary building-blocks of topology.

# Notes

1. A cell complex structure is a type of finite approximation to a space and, as such, is central to the ability to do computation. However, the type of cell structure matters. An $n$-dimensional sphere $\mathbb{S}^n$ has a CW structure with exactly two cells (dimensions 0 and $n$). As a regular cell complex, $2n$ cells suffice. As a cubical or simplicial complex, the minimal total number of cells is exponential in $n$.
2. One virtue of topological methods is their coordinate-free nature: it is not necessary to know exact locations. If, however, one does possess strong geometric data in the form of coordinates, there are numerous ways to improve on the Vietoris-Rips construction.

One popular approach is the **alpha complex** of Edelsbrunner [104] which restricts a Čech complex by a Delaunay triangulation to produce a topologically accurate simplicial model with computationally small footprint.

3. Theorem 2.3 is of interest primarily for its explicit computation of density bounds – it was previously known [204] that a sufficiently dense sampling recovers the topology of the submanifold.
4. The Nerve Lemma is ubiquitous in topology and its applications. It is often attributed to Leray [206], whose version, concerning sheaves, is much deeper than that stated. It echoes antecedents in the (independent) work of Čech [63] and Alexandrov [8]. The version stated is for open covers, but covers by compact contractible sets also suffice. It is often stated that $\mathcal{U}$ be comprised of convex sets, as these and all nonempty intersections thereof are contractible. It will sometimes be mentioned that the Nerve Lemma breaks down if the sets involved are not contractible. This statement is accurate but should propel the reader to more sophisticated methods (such as the **Leray spectral sequence**) rather than to despair.
5. The survey article [242] gives a broad treatment of applications of geometric and topological combinatorics to a broad class of problems in phylogenetics. This is a very active branch of applied topology and geometry.
6. Many results about the topology of state complexes are highly dependent on their geometric features [153]. There is a precise sense in which these complexes are devoid of positive curvature, and it is this property that most influences their global topology.
7. State complexes are *undirected* versions of the **high-dimensional automata** of Pratt [244]. They may also be viewed as weaker versions of the very interesting class of combinatorial cell complexes known as **hom complexes**. The monograph of Kozlov [198] has a wealth of information on this latter class of spaces.

## Chapter 3

# Euler Characteristic

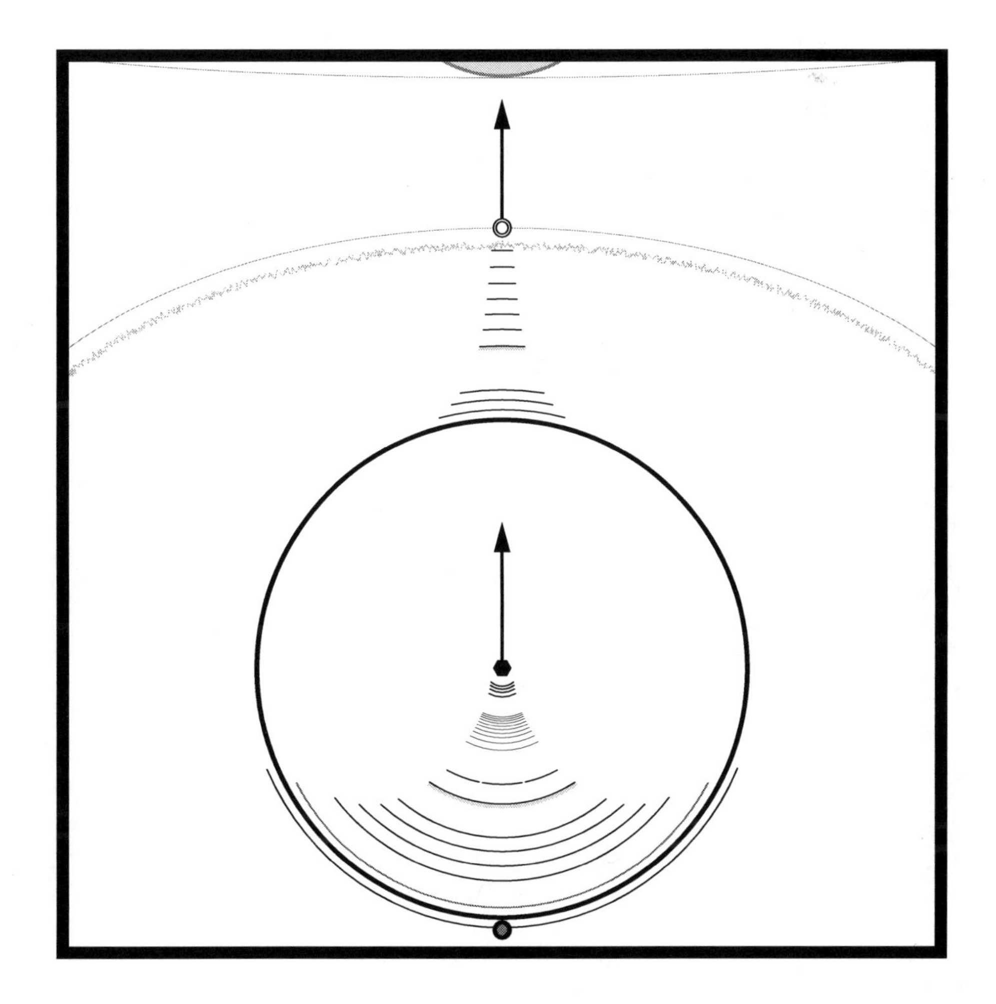

The simplest and most elegant non-obvious topological invariant is the Euler characteristic, an integer-valued invariant of suitably nice spaces. This chapter is comprised of observations and applications of this invariant, meant to serve as motivation for the algebraic tools to follow that justify its remarkable properties.

## 3.1 Counting

Euler characteristic is a generalization of counting. Given a finite set $X$, the Euler characteristic is its cardinality $\chi(X) = |X|$. Connect two points together by means of an edge (in a cellular/simplicial structure); as the resulting space has one fewer component, the Euler characteristic is decreased by one. Continuing inductively, the Euler characteristic counts vertices with weight $+1$ and edges with weight $-1$.

This intuition of counting connected components works at first; however, for certain examples, the addition of an edge does not change the count of connected components. Note that this occurs precisely when a cycle is formed. To fill in such a cycle in the figure with a 2-cell would return to the setting of counting connected components again, suggesting that 2-cells be weighted with $+1$. This intuition of counting with weights inspires the combinatorial definition of Euler characteristic. Given a space $X$ and a partition thereof into a finite number of open cells $X = \sqcup_\alpha \sigma_\alpha$, where each $k$-cell $\sigma_\alpha$ is homeomorphic to $\mathbb{R}^k$, the **Euler characteristic**[1] of $X$ is defined as

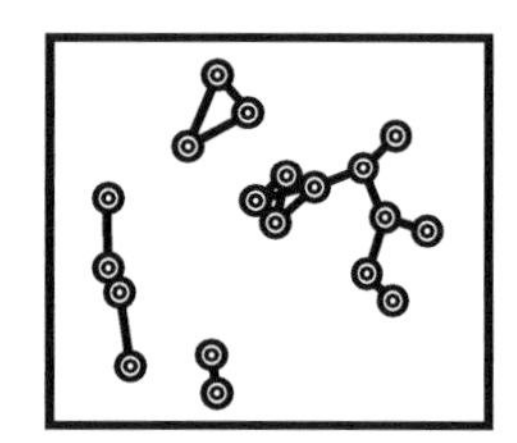

$$\chi(X) := \sum_\alpha (-1)^{\dim \sigma_\alpha}. \tag{3.1}$$

This quantity is well-defined for a reasonably large class of spaces (see §3.5) and is independent of the decomposition of $X$ into cells: $\chi$ is a homeomorphism invariant. It is *not* a homotopy invariant for non-compact cell complexes, as, *e.g.*, it distinguishes $\chi((0,1)) = -1$ from $\chi([0,1]) = 1$. Among compact finite cell complexes, $\chi$ *is* a homotopy invariant, as will be shown in Chapter 5.

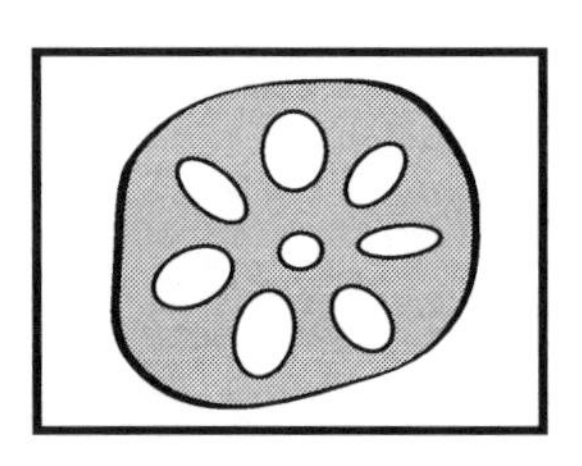

Euler characteristic can determine the homotopy type of a compact connected graph; *e.g.*, such is a tree (a contractible graph) if and only if $\chi = 1$. Euler characteristic is also sharp invariant among connected compact orientable 2-manifolds: $\chi = 2 - 2g$, where $g$ equals the genus. Any compact convex subset of $\mathbb{R}^n$ has $\chi = 1$. Removing $k$ disjoint convex open sets from such a convex superset results in a compact space with Euler characteristic $1 - k(-1)^n$.

**Example 3.1 (Configuration spaces of graphs)** ◎

Recall from Example 2.11 that the discrete configuration space $\mathcal{D}^2(K_5)$ of two points on a complete connected graph of five vertices is a cubical 2-manifold. Being built from finitely many cells, it is clearly compact. It is also orientable and connected, as the reader may check. To determine the genus of this configuration space, one

[1] This is sometimes called the *combinatorial*, *geometric*, or *motivic* Euler characteristic.

computes the Euler characteristic. Vertices of $\mathcal{D}^2(K_5)$ correspond to ordered pairs of distinct vertices of $K_5$, of which there are $(5)(4) = 20$. Edges correspond to pairs of one closed edge and one disjoint vertex: there are $(2)(10)(5-2) = 60$ such. Faces correspond to ordered pairs of closure-disjoint edges in $K_5$, of which there are $(10)(3) = 30$. Thus, $\chi(\mathcal{D}^2(K_5)) = 20 - 60 + 30 = -10$ and this surface has genus $g = 1 - \frac{1}{2}\chi = 6$. ⊙

The remainder of this chapter complements the enumerative interpretation of $\chi$ with geometric, dynamical, analytic, and probabilistic perspectives.

## 3.2 Curvature

A blend of Euler characteristic and integration is prevalent in geometry. The classical result that initiated the subject is the **Gauss-Bonnet Theorem**. Let $M$ be a smooth surface embedded in $\mathbb{R}^3$. The **Gauss map** is the map $\gamma\colon M \to \mathbb{S}^2$ that associates to each point of $M$ the direction of the unit vector normal to $M$ in $\mathbb{R}^3$. The **Gauss curvature** $\kappa = \det(D\gamma)$ is the determinant of the derivative of the Gauss map. Note that the curvature is a geometric quantity: rigid translations and rotations leave it invariant, but stretching and deformation of $M$ change $\kappa$. This change is local, but not global, thanks to the classical:

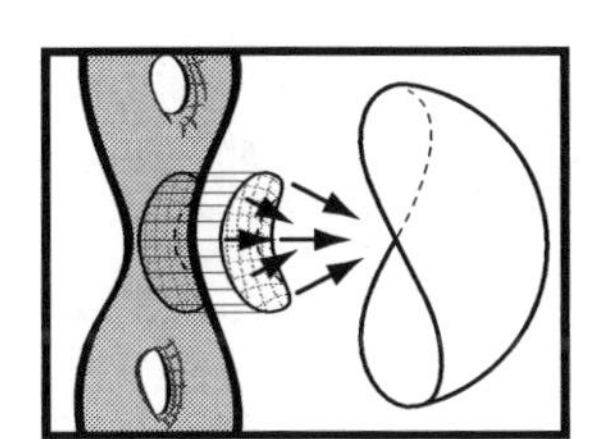

**Theorem 3.2 (Gauss-Bonnet Theorem).** *For $M$ a compact smooth oriented surface in $\mathbb{R}^3$, the integral of Gauss curvature with respect to area on $M$ equals*

$$\int_M \kappa \, dA = 2\pi\chi(M). \tag{3.2}$$

That this is the beginning of a much larger story is evidenced by the following mild generalization. Let $M$ be a compact oriented surface in $\mathbb{R}^3$, piecewise smooth over a cell structure, perhaps with piecewise-smooth boundary curve(s). Again, $2\pi\chi(M)$ equals the integral of a certain curvature over $M$, but this curvature measure, like $M$, is stratified.

1. On 2-cells of $M$, $d\kappa$ means $\kappa \, dA$, Gauss curvature times the area element;
2. On 1-cells of $M$, $d\kappa$ means $k_g \, ds$, geodesic curvature times the length element;
3. On 0-cells of $M$, $d\kappa$ means the angle defect.

With this interpretation, the Gauss-Bonnet formula can be written as

$$\int_M d\kappa = \int_{M^{(0)}} d\kappa + \int_{M^{(1)}} d\kappa + \int_{M^{(2)}} d\kappa = 2\pi\chi(M), \tag{3.3}$$

*i.e.*, the integral over the 2-cells of Gauss curvature plus the integral over 1-cells of geodesic curvature, plus the sum over 0-cells of angle defect equals $2\pi\chi(M)$. There are several corollaries of this result relevant to discrete and differential geometry:

1. A smooth, closed surface has total (integrated) Gauss curvature constant, no matter how the surface is deformed.
2. For a geodesic triangle, $d\kappa$ vanishes along the geodesic edges and the sum of the angles of the triangle equals $\pi$ plus the integral of Gauss curvature over the triangle face. This recovers classical notions of angle-sums for triangles on spheres and other curved surfaces.
3. For $M$ the boundary of a compact convex polyhedron in $\mathbb{R}^3$, all faces and edges are flat, $\chi(M) = \chi(\mathbb{S}^2) = 2$, and the sum of the vertex angle defects (in this case, $2\pi$ minus the sum of the face angles) equals $4\pi$.

The reader may rightly suspect whether an embedding in $\mathbb{R}^3$ is required. Recall, again, from Example 2.11, the cubical 2-manifold built from arranging six squares around each of 20 vertices. Placing a flat Euclidean metric on each square yields a space with an intrinsic geometry, locally flat except at the vertices. Since each vertex has angle defect $2\pi - 6\frac{\pi}{2} = -\pi$, Gauss-Bonnet implies that the surface has Euler characteristic $\chi = 20(-\pi)/2\pi = -10$: *cf.* Example 3.1.

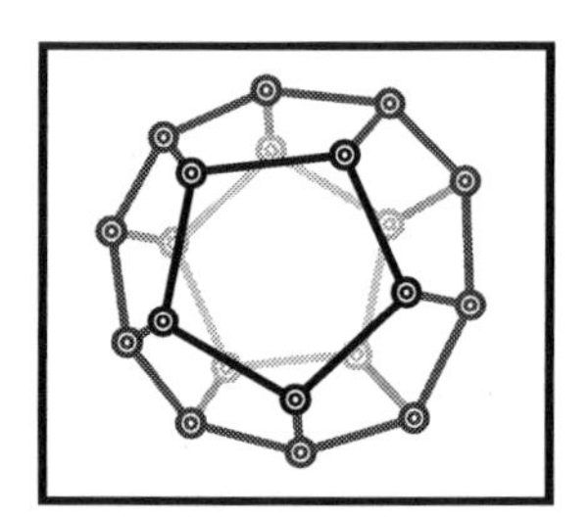

## 3.3 Nonvanishing vector fields

Euler characteristic has many uses, one of which is as an **obstruction** to the existence of certain vector fields. Recall from §1.4 that a vector field is a continuous assignment of a tangent vector to each point in a manifold, and that a vector field vanishes at its zeros.

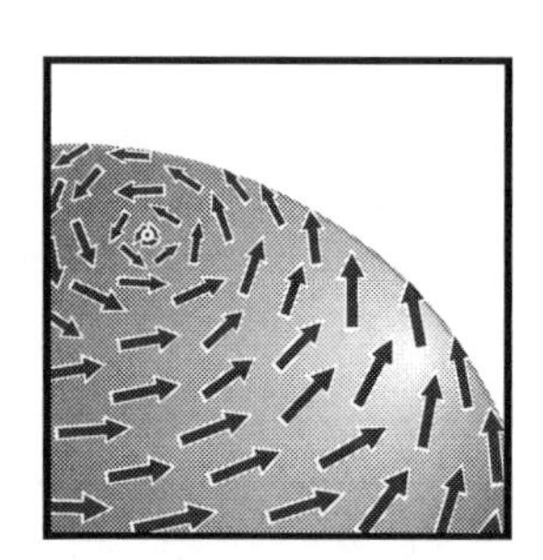

**Theorem 3.3 (Hairy Ball Theorem).** *A connected compact manifold $M$ without boundary possesses a nonvanishing vector field if and only if $\chi(M) = 0$.*

This is not terribly useful in dimensions 1 and 3, since *all* compact manifolds in these dimensions have Euler characteristic zero. However, since $\chi(\mathbb{S}^2) = 2$, there are no fixed-point-free vector fields on a 2-sphere.

**Proof.** *(sketch)* The interesting direction (only if) reveals the obstructive nature of $\chi$. The easiest proof without invoking advanced tools involves imposing a dense cellular mesh and working cell-by-cell, assuming smoothness when needed. Assume that $M$ is a compact closed $n$-manifold that admits a smooth nonvanishing vector field $V$. Place a smooth cell structure on $M$ with all cells represented in charts as convex polytopes of sufficiently small diameter. Invoking transversality and refining as needed, perturb the cell structure so that the vector field $V$ is transverse to all cells. As the cells are sufficiently small and the vector field $V$ is nonvanishing, $V$ is approximately linear in a neighborhood of each cell. On each $n$-dimensional cell $\sigma$, consider the collection, $I(\sigma)$, of open faces of $\sigma$ on which $V$ points to the interior of $\sigma$.

By transversality, $I(\sigma)$ is nonempty for each cell $\sigma$ and contains both the interior cell of $\sigma$ (homeomorphic to an open disc $D^n$) along with some boundary faces, the union of which (by linearity of $V$ and convexity of the polyhedral cell structure) is homeomorphic to the open disc $D^{n-1}$. Thus, $\chi(I(\sigma)) = (-1)^n + (-1)^{n-1} = 0$ and, since the sets $I(\sigma)$ partition all cells of $M$, $\chi(M) = \sum_\sigma \chi(I(\sigma)) = 0$. ⊙

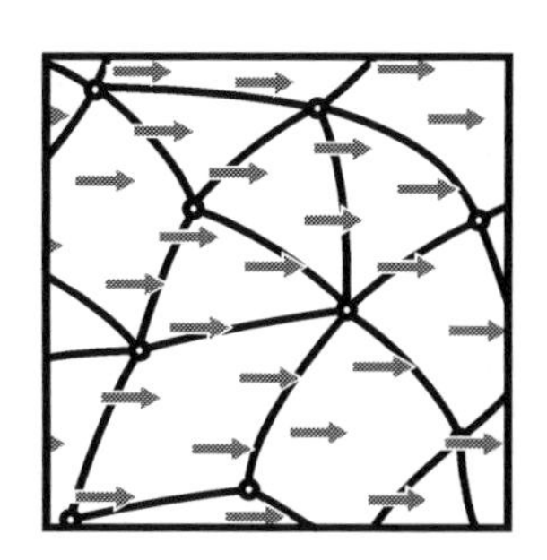

These invocations of transversality and convexity are admittedly glib. With better tools (from Chapters 4-5), it will be possible to drop the assumptions about smoothness, manifold, and cell structures: see Theorem 5.19. Note, however, how the proof constructed a type of local *index*, $I(\sigma)$, and built up a global inference by means of a sort of integration. These themes are recurrent in topology.

## 3.4 Fixed point index

Euler characteristic is the basis for numerous topological indices, the simplest of which applies to vector fields. Consider a vector field $V$ on an oriented 2-manifold $\Sigma$. Let $p \in \mathrm{Fix}(V)$ be an isolated fixed point of $V$. Let $B_p$ denote a sufficiently small ball about $p$ with boundary $\gamma = \partial B_p$. The **index** of $V$ at $p$, $\mathfrak{I}_V(p)$, is defined to be the following line integral:

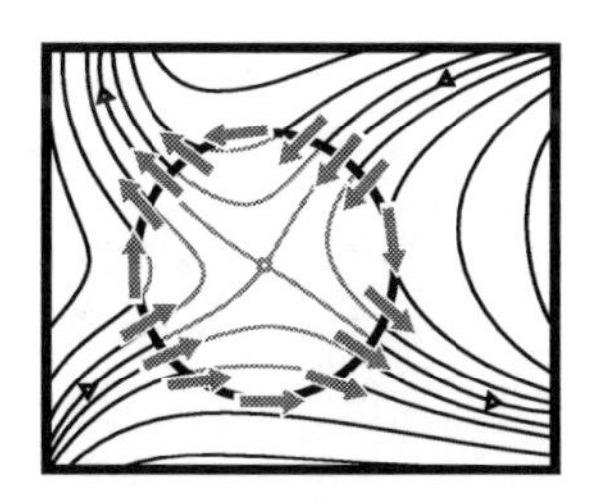

$$\mathfrak{I}_V(p) := \frac{1}{2\pi} \oint_\gamma d\theta_V, \tag{3.4}$$

where $\theta_V$ is the angle made by $V$ in local coordinates. More specifically, if in $(x, y)$ coordinates based at $p$, the vector field is of the form $V = (v_x, v_y)$, then the integrand $d\theta_V$ is

$$d\theta_V = \frac{v_x\,dy - v_y\,dx}{v_x^2 + v_y^2},$$

so that $\mathfrak{I}_V(p)$ represents the (signed integer) number of turns the vector makes in a small curve about $p$. This index is well-defined and independent of (a sufficiently small) $B_p$, thanks to Green's Theorem. Among the nondegenerate fixed points, sources and sinks have index $+1$; saddle points have index $-1$. It is clear that Equation (3.4) extends to define an index $\mathfrak{I}_V(\gamma)$ to any closed curve $\gamma$ which avoids $\mathrm{Fix}(V)$. One argues (in a manner not unlike that used in Green's Theorem or in contour integration) that index is additive. Let $D$ be a disc whose boundary $\gamma = \partial D$ avoids $\mathrm{Fix}(V)$. Then,

$$\mathfrak{I}_V(\gamma) = \sum_{p \in D \cap \mathrm{Fix}(V)} \mathfrak{I}_V(p). \tag{3.5}$$

This important result will be revisited and reinterpreted in §5.10 and §7.7.

**Example 3.4 (Population dynamics)** ⊚

Consider the following differential equation model of competing species with (normalized) population sizes $x(t)$, $y(t)$ as functions of time:

$$\frac{dx}{dt} = 3x - x^2 - 2xy \quad ; \quad \frac{dy}{dt} = 2y - xy - y^2. \tag{3.6}$$

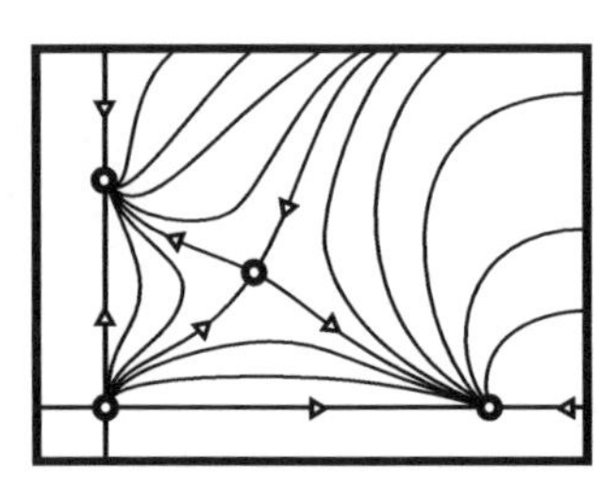

The equilibrium solutions consist of $(0,0)$ [mutual death], $(3,0)$ [species $x$ survives], $(0,2)$ [species $y$ survives], and $(1,1)$ [coexistence]. Linearization reveals that the coexistence solution is a saddle point; thus, $\mathcal{I}(1,1) = -1$. It is clear (from the equations and from the interpretation) that the $x$ and $y$ axes are invariant sets. Can there be a periodic orbit? *No.* The index of a periodic orbit $\gamma$ must by Equation (3.4) equal $\mathcal{I}_V(\gamma) = +1$, since $\gamma$ is everywhere tangent to the vector field. However, by Equation (3.5), $\gamma$ must surround a collection of fixed points whose indices sum to $+1$. If $\gamma(t)$ is a nontrivial periodic solution, then it cannot intersect the fixed point set or the (invariant) $x$ or $y$ axes, and the only remaining enclosable fixed point has negative index. ⊚

The additivity present in Equation (3.5) is reminiscent of the Euler characteristic. The rationale for this equation (and the proper definition of the index of a vector field in all dimensions) will become clearer in Chapters 4-7: see Example 4.23, §5.10, and §7.7. The following classical theorem is a hint of these deeper connections:

**Theorem 3.5 (Poincaré-Hopf Theorem).** *For a continuous vector field $V$ with isolated fixed points on a compact manifold $M$,*

$$\sum_{\mathrm{Fix}(V)} \mathcal{I}_V(p) = \chi(M). \tag{3.7}$$

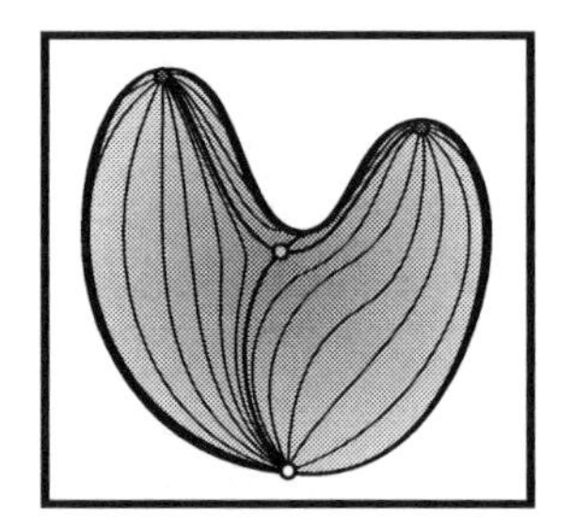

The proofs of Theorems 3.3 and 3.5 compute $\chi$ locally (on cells and fixed points respectively), then add up this local data to return the global $\chi$. This *integrative* technique motivates an extension of Euler characteristic from sets to certain functions over sets: an integration theory. Its definition requires an excursion into exactly which sets have a well-defined Euler characteristic.

## 3.5 Tame topology

Euler characteristic is well-defined for spaces with a decomposition into a finite number of cells. Though an explicit cell structure is often present in, say, the simplicial setting, not all *"organic"* spaces come with a natural cell decomposition: *e.g.*, configuration space of points in a domain, level sets of a smooth functions $f\colon \mathbb{R}^n \to \mathbb{R}$, *etc.* Besides the lack of an explicit cell structure, worse things can occur, as with the graph of $\sin(1/x)$ for $0 < x < 1$. Though this is homeomorphic to an interval, it is a wild type of equivalence, as the closure of this set in $\mathbb{R}^2$ is *not* homeomorphic to a closed interval. In applications, one wants to avoid such oddities and focus on spaces (and mappings) that are for all intents and purposes *tame*.

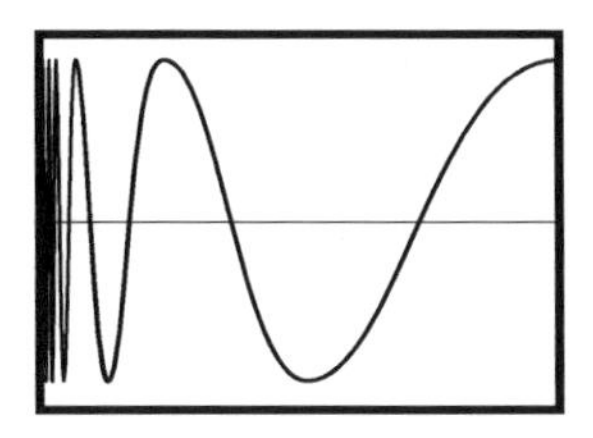

Different mathematical communities focus on, *e.g.*: piecewise linear (PL) spaces, describable in terms of affine sets and matrix inequalities; **semialgebraic** sets, expressible in terms of a finite set of polynomial inequalities; or **subanalytic** sets, defined in terms of images of analytic mappings [277]. Logicians have created an axiomatic reduction of such classes of sets in the form of an **o-minimal structure**.[2]

An o-minimal structure $\mathcal{O} = \{\mathcal{O}_n\}$ (over $\mathbb{R}$) is a sequence of Boolean algebras $\mathcal{O}_n$ of subsets of $\mathbb{R}^n$ (families of sets closed under the operations of intersection and complement) which satisfies certain axioms:

1. $\mathcal{O}$ is closed under cartesian products;
2. $\mathcal{O}$ is closed under axis-aligned projections $\mathbb{R}^n \to \mathbb{R}^{n-1}$;
3. $\mathcal{O}_n$ contains diagonals $\{(x_k)_1^n \colon x_i = x_j\}$ for each $i \neq j$;
4. $\mathcal{O}_2$ contains the subdiagonal $\{x_1 < x_2\}$; and
5. $\mathcal{O}_1$ consists of all finite unions of points and open intervals.

Elements of $\mathcal{O}$ are called **tame** or, more properly, **definable** sets. Canonical examples of o-minimal structures are semialgebraic sets and subanalytic sets. The finiteness of the final axiom is the crucial piece that drives the theory.

Given a fixed o-minimal structure, one can work with tame sets with relative ease. Tame mappings are likewise easily defined: a (not necessarily continuous) function between tame spaces is *tame* (or *definable*) if its graph (in the product of domain and range) is a tame set. A **definable homeomorphism** is a tame bijection between tame sets. To repeat: *definable homeomorphisms are not necessarily continuous.* Such a convention makes the following theorem concise:

**Theorem 3.6 (Triangulation Theorem).** *Any definable set is definably homeomorphic to a finite disjoint union of open standard simplices. The intersection of the closures of any two of the simplices in this definable triangulation is either empty, or the closure of another open simplex in the triangulation.*

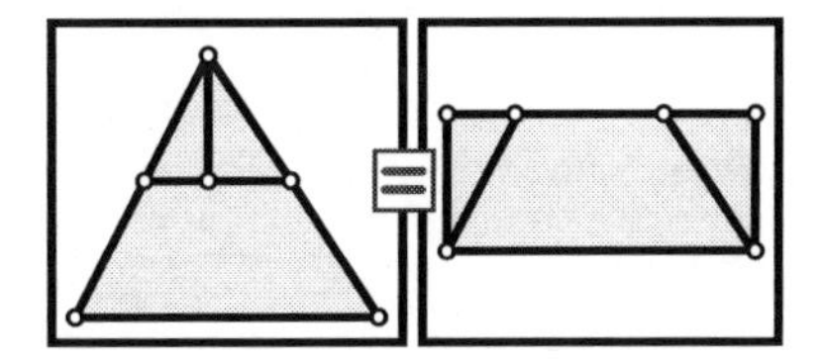

This result implies that tame sets always have a well-defined Euler characteristic and a well-defined dimension (the max of the dimensions of the simplices in a triangulation). The surprise is that these two quantities are not only topological invariants with respect to definable homeomorphism, they are *complete* invariants.

[2]The term derives from *order minimal*, in turn coming from model theory. The text of Van den Dries [293] is a beautifully clear reference.

**Theorem 3.7 ([293]).** *Two definable sets in an o-minimal structure are definably homeomorphic if and only if they have the same dimension and Euler characteristic.*

This result reinforces the idea of a definable homeomorphism as a **scissors equivalence**. One is permitted to cut and rearrange a space with abandon. Recalling the utility of such scissors-work in computing areas of planar sets, the reader will not be surprised to learn of a deep relationship between tame sets, the Euler characteristic, and integration.

## 3.6 Euler calculus

It is possible to build a topological calculus based on Euler characteristic. The integral in this calculus depends on the following **additivity**:

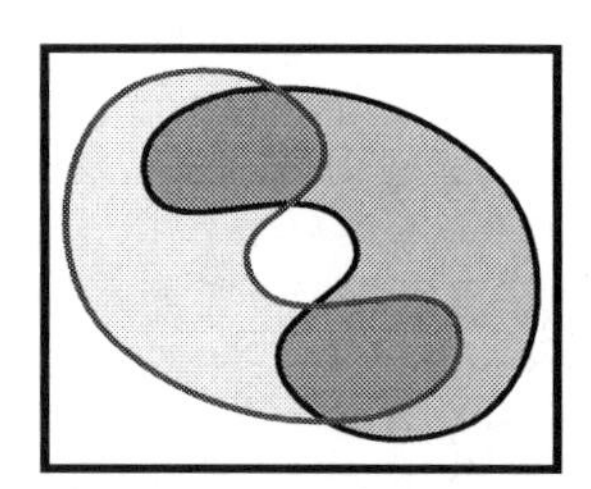

**Lemma 3.8.** *For $A$ and $B$ definable sets,*

$$\chi(A \cup B) = \chi(A) + \chi(B) - \chi(A \cap B). \tag{3.8}$$

This additivity has been foreshadowed in the imagery of $\chi$ as counting in §3.1, in the Gauss-Bonnet Theorem of §3.2, in the proof of Theorem 3.3, and in the additivity of the fixed point index of §3.4. The reader may prove (3.8) via triangulation, induction, and toil, until the more automatic homological tools of Chapters 4-5 are available. The similarity between Equation (3.8) and the definition of a measure is no coincidence. Following ideas that date back to Blaschke (at least), one constructs a measure[3] $d\chi$ over definable sets $A \subset X$ via:

$$\int_X \mathbb{1}_A \, d\chi := \chi(A). \tag{3.9}$$

Measurable functions in this integration theory are integer-valued and **constructible**, meaning that for $h\colon X \to \mathbb{Z}$, all level sets $h^{-1}(n) \subset X$ are tame. Denote by $\mathrm{CF}(X)$ the set of bounded compactly supported constructible functions on $X$. The **Euler integral** is defined to be the homomorphism $\int_X\colon \mathrm{CF}(X) \to \mathbb{Z}$ given by:

$$\int_X h \, d\chi = \sum_{s=-\infty}^{\infty} s\chi(\{h = s\}) = \sum_{s=0}^{\infty} \chi(\{h > s\}) - \chi(\{h < -s\}), \tag{3.10}$$

where the last equality is a manifestation of a discrete fundamental theorem of integral calculus.[4] Alternately, using the definition of tame sets, one may write $h \in \mathrm{CF}(X)$ as $h = \sum_\alpha c_\alpha \mathbb{1}_{\sigma_\alpha}$, where $c_\alpha \in \mathbb{Z}$ and $\{\sigma_\alpha\}$ is a decomposition of $X$ into a disjoint union of open cells, then

$$\int_X h \, d\chi = \sum_\alpha c_\alpha \chi(\sigma_\alpha) = \sum_\alpha c_\alpha (-1)^{\dim \sigma_\alpha} \tag{3.11}$$

[3]The proper term is a **valuation**, not a measure; the abuse of terminology is to prompt the reader to think explicitly in terms of integration theory.

[4]That is, a telescoping sum.

That this sum is invariant under the decomposition into definable cells is a consequence of Lemma 3.8 and Theorem 3.7.

## 3.7 Target enumeration

A simple application of Euler integration to data aggregation demonstrates the utility of this calculus. Consider a finite collection of **targets**, represented as discrete points in a space $W$. Assume a field of sensors, each of which observes some subset of $W$ and counts the number of targets therein. The sensors will be assumed to be distributed over a region so densely as to be approximated by a topological space $X$.

There are many modes of sensing: infrared, acoustic, optical, magnetometric, and more are common. To best abstract the idea of sensing away from the engineering details, the following topological approach is used. In a particular system of sensors in $X$ and targets in $W$, let the **sensing relation** be the relation $\mathcal{S} \subset W \times X$ where $(w, x) \in \mathcal{S}$ iff a sensor at $x \in X$ detects a target at $w \in W$. The horizontal and vertical **fibers** (inverse images of the projections of $\mathcal{S}$ to $X$ and $W$ respectively) have simple interpretations. The vertical fibers – **target supports** – are those sets of sensors which detect a given target in $W$. The horizontal fibers – **sensor supports** – are those target locations observable by a given sensor in $X$.

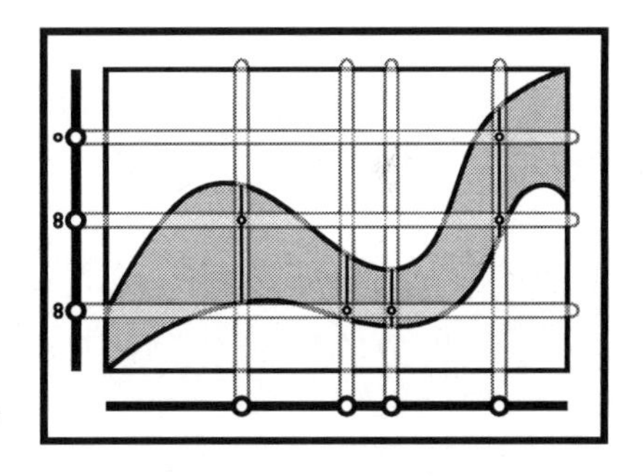

Assume that the sensors are additive but anonymizing: each sensor at $x \in X$ counts the number of targets in $W$ detectable and returns a local count $h(x)$, but the identities of the sensed targets are unknown. This counting function $h\colon X \to \mathbb{Z}$ is, under the usual tameness assumptions, constructible. A natural problem in this context is to aggregate the redundant anonymous target counts: given $h$ and some minimal information about the sensing relation $\mathcal{S}$, determine the total number of targets in $W$. This is not as easy as it sounds, and it is even less easy when $X$ is not a continuum space, but rather a discretization thereof. It is therefore remarkable that a purely topological solution exists, independent of knowing a decomposition of the counting function $h$.

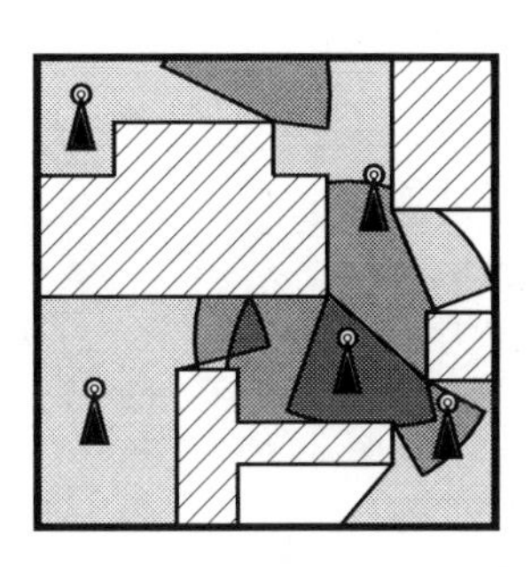

**Proposition 3.9 ([24]).** *If $h\colon X \to \mathbb{N}$ is a counting function for target supports $U_\alpha$ of uniform Euler characteristic $\chi(U_\alpha) = N \neq 0$ for all $\alpha$, then*

$$\#\alpha = \frac{1}{N} \int_X h \, d\chi.$$

**Proof.**

$$\int_X h\,d\chi = \int_X \left(\sum_\alpha \mathbb{1}_{U_\alpha}\right) d\chi = \sum_\alpha \int_X \mathbb{1}_{U_\alpha}\,d\chi = \sum_\alpha \chi(U_\alpha) = N\,\#\alpha.$$

⊙

For contractible supports (such as in the setting of beacons visible on star-convex domains), the target count is, simply, the Euler integral of the function. This solves a problem in the aggregation of redundant data, since many nearby sensors with the same reading are detecting the same targets; in the absence of target identification (an expensive signal processing task), it is nontrivial to aggregate the redundancy. Notice that the restriction $N \neq 0$ is nontrivial. If $h \in \mathrm{CF}(\mathbb{R}^2)$ is a finite sum of characteristic functions over annuli, it is not merely inconvenient that $\int_{\mathbb{R}^2} h\,d\chi = 0$, it is a fundamental obstruction to disambiguating sets. Given this, it is all the more remarkable that sets with $\chi \neq 0$ can be enumerated easily. Since the solution is in terms of an integral, local and distributed computations may be used in practice.

## 3.8 A Fubini Theorem

Euler characteristic is like a volume in another aspect: it is multiplicative under cartesian products.

**Lemma 3.10.** *For $X$ and $Y$ definable, $\chi(X \times Y) = \chi(X)\chi(Y)$.*

**Proof.** The product $X \times Y$ has a definable cell structure using products of cells from $X$ and $Y$. For cells $\sigma \subset X$ and $\tau \subset Y$, the lemma holds via the exponent rule since $\dim(\sigma \times \tau) = \dim \sigma + \dim \tau$. Additivity of the integral completes the proof. ⊙

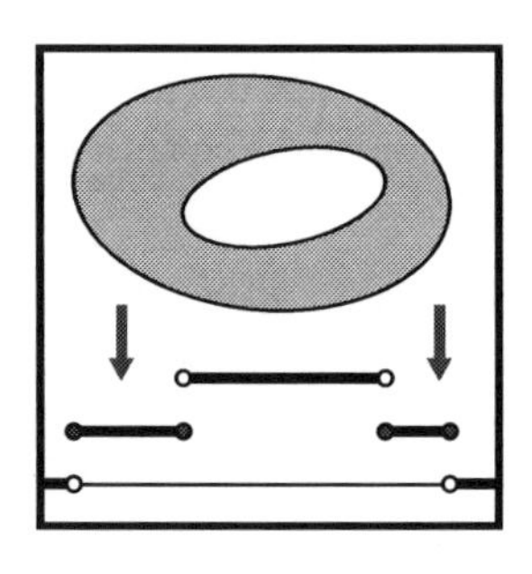

The assertion that $d\chi$ should be regarded as an honest topological measure is supported by this fact and its corollary: the Euler integration theory admits a Fubini Theorem. Calculus students know that:

$$\iint f(x,y)dx\,dy = \iint f(x,y)dy\,dx.$$

Real-analysis students learn to pay attention to finer assumptions on measurability (*cf.* tameness assumptions). This familiar result is the image of a deeper truth about integrations and projections. Given $F\colon X \to Y$, one can integrate over the **fibers**, or level sets, of $F$ first, then integrate the resulting function over the projected **base** $Y$.

**Theorem 3.11 (Fubini Theorem).** *Let $F\colon X \to Y$ be a definable mapping. Then for all $h \in CF(X)$,*

$$\int_X h\, d\chi = \int_Y \left( \int_{F^{-1}(y)} h(x)\, d\chi(x) \right) d\chi(y). \tag{3.12}$$

**Proof.** If $X = U \times Y$ and $F$ is projection to the second factor, the result follows from Lemma 3.10. The o-minimal Hardt Theorem [293] says that $Y$ has a definable partition into tame sets $Y_\alpha$ such that $F^{-1}(Y_\alpha)$ is definably homeomorphic to $U_\alpha \times Y_\alpha$ for $U_\alpha$ definable, and that $F\colon U_\alpha \times Y_\alpha \to Y_\alpha$ acts via projection. Additivity of the integral completes the proof. ⊙

**Example 3.12 (Enumerating vehicles)** ◎

The following example demonstrates an application of the Fubini Theorem to time-dependent targets. Consider a collection of vehicles, each of which moves along a smooth curve $\gamma_i\colon [0, T] \to \mathbb{R}^2$ in a plane filled with sensors that count the passage of a vehicle and increment an internal counter. Specifically, assume that each vehicle possesses a *footprint* – a support $U_i(t) \subset \mathbb{R}^2$ which is a compact contractible neighborhood of $\gamma_i(t)$ for each $i$, varying tamely in $t$. At the moment when a sensor $x \in \mathbb{R}^2$ detects when a vehicle comes within proximity range – when $x$ crosses into $U_i(t)$ – that sensor increments its internal counter. Over the time interval $[0, T]$, the sensor field records a counting function $h \in \mathrm{CF}(\mathbb{R}^2)$, where $h(x)$ is the number of times $x$ has entered a support. As before, the sensors do not identify vehicles; nor, in this case, do they record times, directions of approach, or any ancillary data.

**Proposition 3.13 ([24]).** *The number of vehicles is equal to $\int_{\mathbb{R}^2} h\, d\chi$.*

**Proof.** Each target traces out a compact tube in $\mathbb{R}^2 \times [0, T]$ given by the union of slices $(U_i(t), t)$ for $t \in [0, T]$. Each such tube has $\chi = 1$. The integral over $\mathbb{R}^2 \times [0, T]$ of the sum of the characteristic functions over all $N$ tubes is, by Proposition 3.9, $N$, the number of targets. Consider the projection map $p\colon \mathbb{R}^2 \times [0, T] \to \mathbb{R}^2$. Since $p^{-1}(x)$ is $\{x\} \times [0, T]$, the integral over $p^{-1}(x)$ records the number of (necessarily compact) connected intervals in the intersection of $p^{-1}(x)$ with the tubes in $\mathbb{R}^2 \times [0, T]$. This number is precisely the sensor count $h(x)$ (the number of times a sensor detects a vehicle coming into range). By the Fubini Theorem, $\int_{\mathbb{R}^2} h\, d\chi$ must equal the integral over the full $\mathbb{R}^2 \times [0, T]$, which is $N$. ⊙ ◎

## 3.9 Euler integral transforms

Euler integration admits a variety of operations which mimic the analytic tools so useful in signal processing, imaging, and inverse problems.

**Example 3.14 (Convolution and Minkowski sum)** ◎

On a real vector space $V$, a **convolution** operation with respect to Euler characteristic is straightforward. Given $f, g \in \mathrm{CF}(V)$, one defines

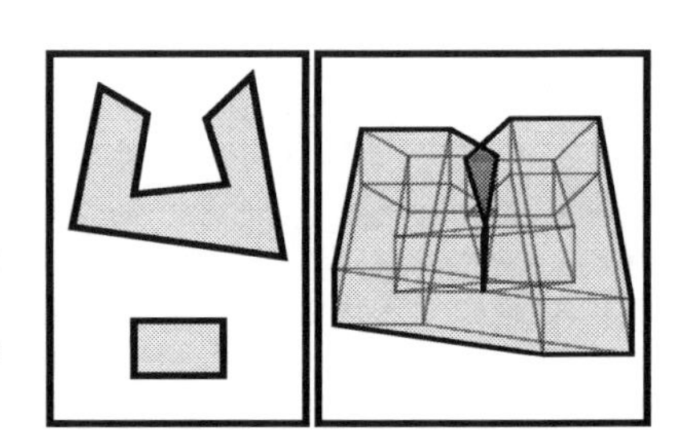

$$(f * g)(x) := \int_V f(t)g(x-t)\, d\chi(t). \quad (3.13)$$

There is a close relationship between convolution and the **Minkowski sum**: for $A$ and $B$ convex, $\mathbb{1}_A * \mathbb{1}_B = \mathbb{1}_{A+B}$, where $A + B$ is the set of all vectors expressible as a sum of a vector in $A$ and a vector in $B$ [297, 269]. This, in turn, is useful in applications ranging from computer graphics to motion-planning for robots around obstacles [168]. For non-convex shapes, convolution of indicator functions can take on values larger than 1 in regions where the intersections of translations of $A$ with $B$ are disconnected. Unlike non-convex Minkowski sum, Euler-convolution is always invertible [27, 269]. ◎

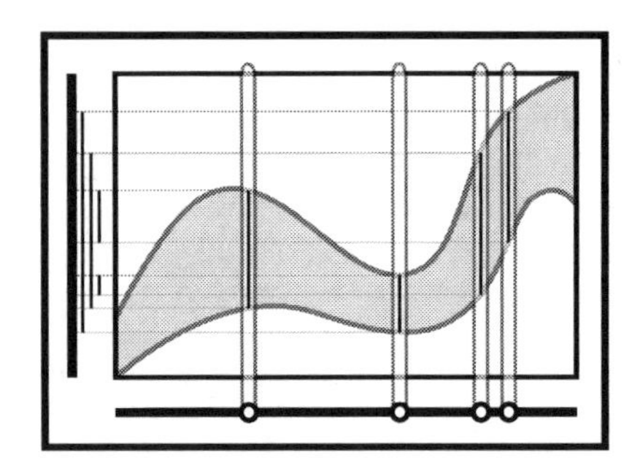

One of the most general integral transforms is the **Radon transform** of Schapira [48, 67, 270]. Consider a locally closed definable relation $\mathcal{S} \subset W \times X$ (that may or may not come from sensing), and let $\pi_W$ and $\pi_X$ denote the projection maps of $W \times X$ to their factors. The Radon transform with kernel $\mathcal{S}$ is the map $\mathcal{R}_\mathcal{S} \colon \mathrm{CF}(W) \to \mathrm{CF}(X)$ given by lifting $h \in \mathrm{CF}(W)$ from $W$ to $W \times X$, filtering with the kernel $\mathbb{1}_\mathcal{S}$, then integrating along the projection to $X$ as follows:

$$(\mathcal{R}_\mathcal{S} h)(x) := \int_W h(w)\mathbb{1}_\mathcal{S}(x, w)\, d\chi(w). \quad (3.14)$$

**Example 3.15 (Target enumeration)** ◎

Consider the sensor relation $\mathcal{S} \subset W \times X$, and a finite set of targets $T \subset W$ as defining an indicator function $\mathbb{1}_T \in \mathrm{CF}(W)$. Observe that the counting function which the sensor field on $X$ returns is precisely the Radon transform $\mathcal{R}_\mathcal{S}\mathbb{1}_T$. In this language, Proposition 3.9 is equivalent to the following: assume that $\mathcal{S} \subset W \times X$ has vertical fibers $\pi_W^{-1}(w) \cap \mathcal{S}$ with constant Euler characteristic $N$. Then, $\mathcal{R}_\mathcal{S} \colon \mathrm{CF}(W) \to \mathrm{CF}(X)$ scales integration by a factor of $N$: $\int_X \circ \mathcal{R}_\mathcal{S} = N \int_W$. ◎

A similar regularity in the Euler characteristics of fibers allows a general inversion formula for the Radon transform [270]. One must choose an 'inverse' relation $\mathcal{S}' \subset X \times W$.

**Proposition 3.16 (Schapira inversion formula).** *Assume that $\mathcal{S} \subset W \times X$ and $\mathcal{S}' \subset X \times W$ have fibers $\mathcal{S}_w$ and $\mathcal{S}'_w$ satisfying (1) $\chi(\mathcal{S}_w \cap \mathcal{S}'_w) = \mu$ for all $w \in W$; and (2) $\chi(\mathcal{S}_w \cap \mathcal{S}'_{w'}) = \lambda$ for all $w' \neq w \in W$. Then for all $h \in \mathrm{CF}(W)$,*

$$(\mathcal{R}_{\mathcal{S}'} \circ \mathcal{R}_{\mathcal{S}})h = (\mu - \lambda)h + \lambda \left( \int_W h \right) \mathbb{1}_W. \tag{3.15}$$

**Proof.** The conditions on the fibers of $\mathcal{S}$ and $\mathcal{S}'$ imply that $\int_X \mathcal{S}(w,x)\mathcal{S}'(x,w')d\chi = (\mu-\lambda)\delta_{w-w'} + \lambda$ for all $w, w' \in W$, where $\delta$ is the Dirac delta function. Thus, for any $w' \in W$,

$$\begin{aligned}
(\mathcal{R}_{\mathcal{S}'} \circ \mathcal{R}_{\mathcal{S}} h)(w') &= \int_X \left[ \int_W h(w) \mathbb{1}_{\mathcal{S}}(w,x)\, d\chi \right] \mathbb{1}_{\mathcal{S}'}(x,w')\, d\chi \\
&= \int_W h(w) \left[ \int_X \mathbb{1}_{\mathcal{S}}(w,x)\mathbb{1}_{\mathcal{S}'}(x,w') d\chi \right] d\chi \\
&= \int_W \left[ (\mu-\lambda)h(w)\delta_{w-w'} + \lambda h(w) \right] d\chi \\
&= (\mu-\lambda)h(w') + \lambda \int_W h\, d\chi,
\end{aligned}$$

where the Fubini Theorem is used in the second equality. ⊙

Recall that the sensor counting field $h\colon X \to \mathbb{Z}$ is equal to $\mathcal{R}_S \mathbb{1}_T$, where $T \subset W$ is the set of targets. If the conditions of Proposition 3.16 are met *and* if $\lambda \neq \mu$, then the inverse Radon transform $\mathcal{R}_{S'} h = \mathcal{R}_{S'} \mathcal{R}_S \mathbb{1}_T$ is equal to a (nonzero) multiple of $\mathbb{1}_T$ plus a multiple of $\mathbb{1}_W$. Thus, one can localize the targets by performing the inverse transform. It is remarkable that enumerative data alone can yield not only target counts but target positions as well. By changing $T$ from a discrete set to a collection of (say, contractible) compact sets, one notes that Radon inversion has the potential to not merely localize but recover shape. This *topological tomography* is the motivation for Schapira's incisive paper [270].

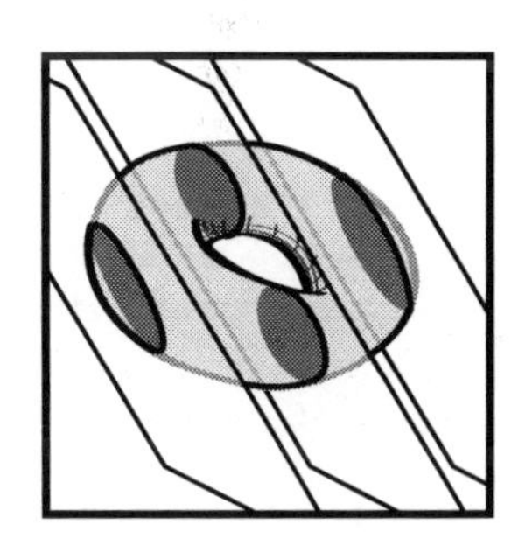

**Example 3.17 (Topological tomography)** ⊚

Assume that $W = \mathbb{R}^3$ and that one scans a compact subset $T \subset W$ by slicing $\mathbb{R}^3$ along all flat hyperplanes, recording simply the Euler characteristics of the slices of $T$. Since a compact subset of a plane has Euler characteristic the number of connected components minus the number of holes (which, in turn, equals the number of bounded connected components of the complement – see §6.5), it is feasible to compute an accurate Euler characteristic, even in the context of noisy readings.

This yields a constructible function on the sensor space $X = \mathbb{AG}_2^3$ (the affine Grassmannian of all planes in $\mathbb{R}^3$) equal to the Radon transform of $\mathbb{1}_T$. Using the same sensor relation to define the inverse transform is effective. Since $\mathcal{S}_w \cong \mathbb{P}^2$ and $\mathcal{S}_w \cap \mathcal{S}_{w'} \cong \mathbb{P}^1$, one has $\mu = \chi(\mathbb{P}^2) = 1$, $\lambda = \chi(\mathbb{P}^1) = 0$, and the inverse Radon transform, by (3.15), yields $\mathbb{1}_T$ exactly: one can recover the shape of $T$ based solely on connectivity data of black and white regions of slices. ◎

**Example 3.18 (Fourier transforms and curvature)** ◎

One relationship between Euler and Lebesgue measures is encoded in the Gauss-Bonnet theorem of §3.2. This, too, can be lifted from manifolds to definable sets and then to $\mathrm{CF}(\mathbb{R}^n)$. The mechanism is the **Fourier-Sato transform**. Given $h \in \mathrm{CF}(\mathbb{R}^n)$, a point $x \in \mathbb{R}^n$, and a unit *frequency* vector $\xi \in \mathbb{S}^{n-1}$, the Fourier-Sato transform is defined via:

$$(\mathcal{F}_{\mathcal{S}}\, h)_x(\xi) := \lim_{\epsilon \to 0^+} \int_{B_\epsilon(x)} \mathbb{1}_{\xi \cdot (y-x) \geq 0}\, h\, d\chi(y), \tag{3.16}$$

where $B_\epsilon$ denotes the *open* ball of radius $\epsilon$. Like the classical Fourier transform, $\mathcal{F}_{\mathcal{S}}$ takes a *frequency* vector $\xi$ and integrates over *isospectral sets* defined by a dot product. For $Y$ a compact tame set, $\mathcal{F}_{\mathcal{S}}\mathbb{1}_Y(\xi)$ is the constructible function on $Y$ that, when averaged over $\xi$, yields a curvature measure $d\kappa_Y$ on $Y$ implicated in the Gauss-Bonnet Theorem (as shown by Bröcker and Kuppe [48]). For any $U \subset \mathbb{R}^n$ open, define the net curvature of $Y$ on $U$ to be:

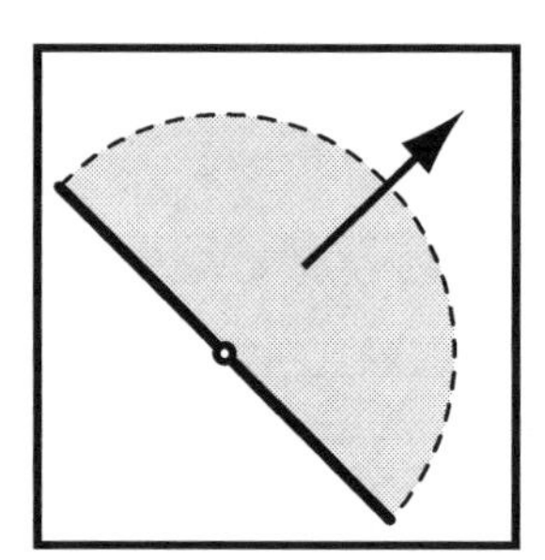

$$\int_U d\kappa_Y := \frac{1}{\mathrm{Vol}\, \mathbb{S}^{n-1}} \int_{\mathbb{S}^{n-1}} \int_U (\mathcal{F}_{\mathcal{S}}\mathbb{1}_Y)_u(\xi)\, d\chi\, d\xi.$$

This curvature measure has support on $Y$ and, up to a constant, agrees with the three notions of curvature mentioned in §3.2. The Euler-calculus interpretation of the Gauss-Bonnet Theorem says that this rescaled net curvature of $Y$ is precisely its Euler characteristic.

$$\int_Y d\kappa_Y = \chi(Y).$$ ◎

## 3.10 Intrinsic volumes

The humble combinatorial definition of $\chi$ has matured in this chapter to play the role of a measure. This has precedent in the subject of **integral geometry**. There is a family of *"measures"* on Euclidean $\mathbb{R}^n$ that mediate Euler and Lebesgue while entwining topological and geometric data. The $k^{\text{th}}$ **intrinsic volume**[5] $\mu_k$ is characterized uniquely by the following: for all $A$ and $B$ tame subsets of $\mathbb{R}^n$,

1. **Additivity:** $\mu_k(A \cup B) = \mu_k(A) + \mu_k(B) - \mu_k(A \cap B)$;

[5] Intrinsic volumes are also known as **Hadwiger measures**, **quermassintegrale**, **Lipschitz-Killing curvatures**, **Minkowski functionals**, and, likely, a few more names unknown to the author.

2. **Euclidean invariance:** $\mu_k$ is invariant under rigid motions of $\mathbb{R}^n$;
3. **Homogeneity under scaling:** $\mu_k(c \cdot A) = c^k(A)$ for all $c \geq 0$; and
4. **Normalization:** $\mu_k$ of a closed unit ball in $\mathbb{R}^n$ equals 1.

These measures (more properly, *valuations*) generalize Euclidean $n$-dimensional volume ($\mu_n = \mathrm{dvol}_n$) and Euler characteristic ($\mu_0 = d\chi$). There are several equivalent definitions, all revolving about the notion of an average Euler characteristic. One way to define the intrinsic volume $\mu_k(A)$ is in terms of the Euler characteristic of all slices of $A$ along affine codimension-$k$ planes:

$$\mu_k(A) = \int_{\mathbb{AG}^n_{n-k}} \chi(A \cap P)\, d\lambda(P) = \int_{\mathbb{AG}^n_{n-k}} \int_P \mathbb{1}_A \, d\chi \, d\lambda(P), \tag{3.17}$$

where $d\lambda$ is an appropriate measure[6] on $\mathbb{AG}^n_{n-k}$, the space of affine $(n-k)$-planes in $\mathbb{R}^n$. The details of this construction are not elementary [234]: the point here stressed is that all the intrinsic volumes are certain Lebesgue-averaged Euler integrals.

These intrinsic volumes are more than isolated examples. The classical theorem of Hadwiger [174] characterizes those Euclidean-invariant valuations which are continuous with respect to the Hausdorff metric on compact convex sets (the *compact-continuous* valuations):

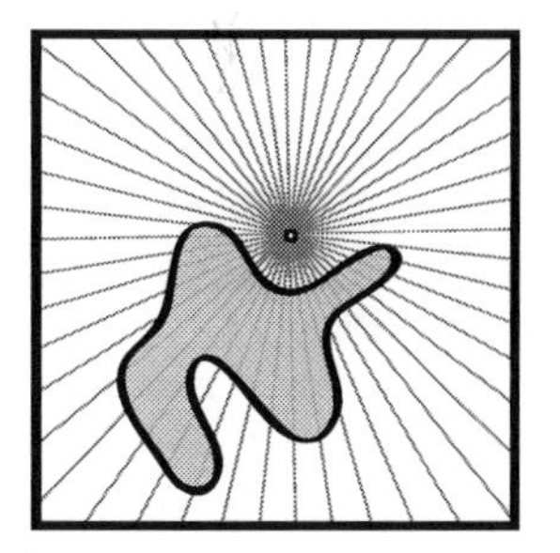

**Theorem 3.19 (Hadwiger Theorem).** *The space of all compact-continuous Euclidean-invariant valuations on $\mathbb{R}^n$ is a vector space of dimension $n+1$ with basis $\{\mu_k\}_{k=0}^n$.*

These amalgamations of Euler and Lebesgue measure are the key to several interesting applications.

**Example 3.20 (Microstructure coarsening)** ⊙

Anyone who has observed foam in a pilsner glass knows that the *cell walls* of foam evolve over time so that some cells grow, while others shrink unto disappearance (neglecting pops). The same processes are ubiquitous in the microstructure coarsening of froth, metals, and some ceramics. It has long been known how idealized cells evolve and coarsen in the 2-dimensional setting. The **von Neumann - Mullins formula** states that the area $A(t)$ of a cell in an ideal dynamic 2-d microstructure evolves as:

$$\frac{dA}{dt} = -2\pi M\gamma \left(1 - \frac{|C_0|}{6}\right), \tag{3.18}$$

where $M$ is a mobility constant, $\gamma$ is a surface tension constant, and $|C_0|$ equals the number of corners (vertices) of the cell. Thus, cells with fewer than six corners shrink: those with more grow. The proper extension of this formula to three-dimensional cell structures was recently discovered by MacPherson and Srolovitz [214]. The key to the

[6] This measure is derived from the Haar measure on the Grassmannian $\mathbb{G}^n_{n-k}$ and Lebesgue measure on the orthogonal $\mathbb{R}^k$.

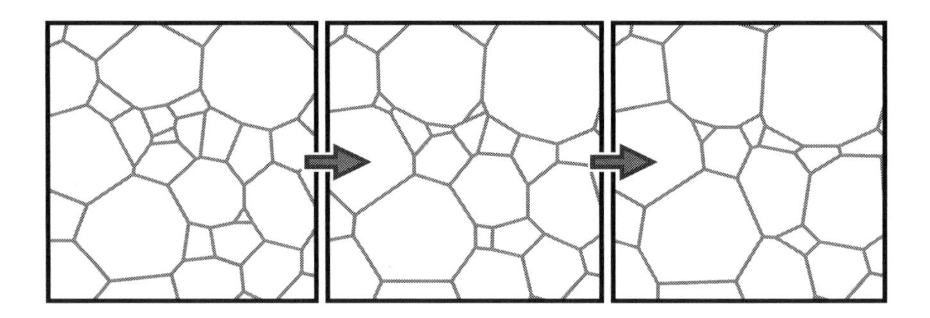

extension was to note that the "1" in Equation (3.18) is in fact the Euler characteristic of the (contractible, compact) cell. To lift from 2-d to 3-d requires lifting the relevant interior volumes from $\mu_0$ ($= \chi$) to $\mu_1$. The MacPherson-Srolovitz formula for the volume $V(t)$ of a cell in a 3-d microstructure is:

$$\frac{dV}{dt} = -2\pi M\gamma \left( \mu_1 - \frac{|C_1|}{6} \right), \tag{3.19}$$

where $\mu_1$ is the intrinsic 1-volume of the cell and $|C_1|$ is the total length of all the edges (1-dimensional boundary curves) of the cell. When contemplated in the light of intrinsic volumes, it is clear that Equation 3.19 is in fact an elegant relationship between $\mu_3$ and $\mu_1$. ⊚

## 3.11 Gaussian random fields

Most readers will be comfortable with the utility (if not the details) of stochastic processes. A real-valued **stochastic process** is, in brief, a collection $\{h(x) \colon x \in X\}$ of random variables $h(x)$ parameterized by points in a space $X$. A **random field** is a stochastic process whose arguments vary continuously over the parameter space $X$. For example, a random field may be used to model the height of water on a noisy sea or the magnitude of noise in a sea of cell-phone towers, both examples having as parameter space a geometric domain of dimension two (or three, with the addition of time). As the simplest random variables are **Gaussian** (having well-defined finite mean and variance with normal distribution), so are the simplest random fields. A **Gaussian random field** with parameter space $X \subset \mathbb{R}^n$ is a random field $h$ over $X$ such that, for each $k \in \mathbb{N}$ and each $k$-tuple $\{x_i\}_0^k \subset X$, the collection $\{h(x_i)\}_0^k$ of random variables has a multivariate Gaussian distribution. For simplicity, attention will be restricted to Gaussian distributions with the following features:

1. **Centered:** all $h(x)$ have zero mean, $\mathbb{E}\{h(x)\} = 0$;
2. **Common variance:** all $h(x)$ have fixed variance $\sigma^2 > 0$;
3. **Stationary-isotropic:** the covariance $\mathbb{E}\{h(x)h(y)\}$ is a function of distance $|x - y|$.

These last two assumptions imply a well-defined constant $\lambda$, the **second spectral moment**, which, roughly, measures the variance of directional derivatives of $h$ with respect to $x$.

Adler [3], Taylor [5], Worsley [302, 303], and others have led in the exploration of Gaussian random fields from a geometric perspective. There is implicit in this work no small amount of topology, with Euler characteristic featuring prominently. Let $h(x)$, $x \in X \subset \mathbb{R}^n$, be a stationary random Gaussian field. Knowing as much as possible about the tomography of the field $h$ is important in medical imaging, astronomy, and a host of other applications. For example, one might wish to know the expected number of peaks in a field, or some other qualitative properties associated to the (upper) **excursion sets** $\{h \geq s\}$. One of the principal results in this area is that, although the fine details of the expected field excursion sets cannot be computed, the expected Euler characteristics are not merely computable, but computed and commensurate with experimental data.

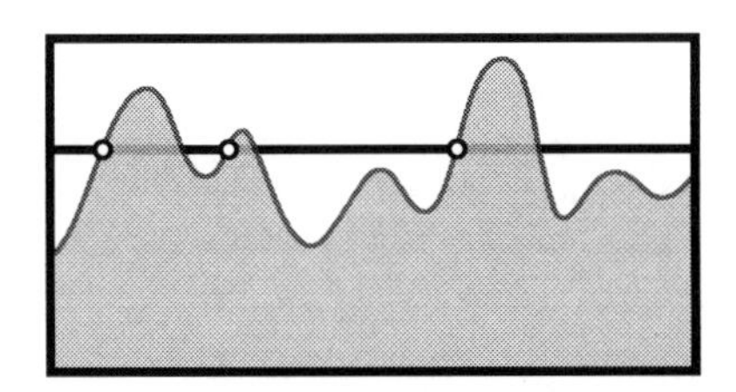

The following classical result of Rice from the 1940s [247] helped initiate the subject: for $h(x)$, $x \in [0, T]$, a stationary, zero mean, $C^1$ Gaussian process with finite variance $\sigma^2$ and second spectral moment $\lambda$, the expected number of up-crossings at $h = s$ (locations where $h$ increases past $s$) is given by

$$\mathbb{E}\#\{h \nearrow s\} = e^{-\frac{1}{2}\left(\frac{s}{\sigma}\right)^2} \frac{\sqrt{\lambda}}{2\pi\sigma} T. \tag{3.20}$$

The observation that (up to a boundary correction term) $\#\{h \nearrow s\}$, the number of up-crossings, is in fact the Euler characteristic $\chi\{h \geq s\}$ of the upper excursion set presages the deeper work of [3] for fields over higher-dimensional domains. It appears very difficult to understand the expected shape or even topological type of these excursion sets; however, the simplification that $\chi$ introduces and its commensurability with integral techniques yields to computational effort. The following result is a highly illustrative example:

**Theorem 3.21 ([3, 5]).** *Assume that $h(x)$, $x \in \mathbb{R}^n$, is a centered, stationary-isotropic Gaussian random field with common variance $\sigma^2$, second spectral moment $\lambda$, and sufficient regularity.*[7] *The expected Euler characteristic of the upper excursion set $\{h \geq s\}$ over a compact definable subset $X \subset \mathbb{R}^n$ is*

$$\mathbb{E}(\chi\{h \geq s\}) = e^{-\frac{1}{2}\left(\frac{s}{\sigma}\right)^2} \sum_{k=0}^{\dim X} (2\pi)^{-\frac{1}{2}(k+1)} \left(\frac{\sqrt{\lambda}}{\sigma}\right)^k \mathrm{He}_{k-1}\left(\frac{s}{\sigma}\right) \mu_k(X), \tag{3.21}$$

*where $\mathrm{He}_k$ is the degree $k$ Hermite polynomial in one variable and $\mu_k(X)$ is the $k^{\text{th}}$ intrinsic volume of $X$.*

Equation (3.21) is a remarkable formula in that it ties together so many ideas from integral geometry, topology, and stochastics in a package that permits honest applications to data. The ability to predict expected Euler characteristic from stochastic

[7] This consists of nondegeneracy assumptions on the joint distributions of the first and second derivatives of $f$ and regularity assumptions on the covariance functions of the second derivatives.

data has had enormous impact in imaging, in everything from medical data to astronomy. Additional work [5] allows one to relax the assumptions of the field being stationary, isotropic, and Gaussian. In addition, expectations for intrinsic volumes $\mu_k$ of excursion sets are likewise derivable.

## Notes

1. The scissors equivalence implicit in Theorem 3.7 is the hint of deeper structures. Euler characteristic and integration over CF is an elementary version of **motivic integration**, of great current interest in algebraic geometry. The papers of Cluckers, Denef, and Loeser [68, 92] detail this somewhat; the exposition of Hales [175] is a good starting point for this theory.
2. Since the Euler measure is the *universal motivic measure* over topological spaces, any other scissors-invariant group-valued measure on definable topological spaces must factor through $\chi$. One way around this depressing result is to extend $\chi$ to a polynomial-valued measure of sequences of spaces. The work of Gal [137] encodes the Euler characteristics of $\mathfrak{UC}^n(X)$ as coefficients of a series in a formal variable $t$. For $X$ tame, the resulting series is the Taylor series of a rational function. Similar algebraic manipulations for Euler characteristics of sequences of spaces appears in work on $\zeta$-functions from algebraic geometry [172]. It is to be suspected that a full theory of integration with respect to $d\chi[t]$ over (say) filtered spaces awaits development.
3. Euler integration is here, as in [262, 268], presented as a combinatorial theory. Higher perspectives are more illuminating. Chapter 5 will unfold the connection to homology; Chapter 7 to Morse theory; and Chapter 9 to sheaf theory. The literature on **normal cycles** [134, 234] allows, thanks primarily to results of Kashiwara [191], an approach in terms of **conormal cycles** and related geometric measure theory: see §6.11.
4. The issue of numerical Euler integration – how to approximate an integral with respect to $d\chi$ based on sampled data – is extremely interesting, technical, and relevant to applications. It appears to have received little treatment from the Mathematics community. See §6.5 for a first step.
5. It is possible and profitable to perform Radon transforms (and inversion thereof) with weighted kernels [27].
6. Intrinsic volumes $d\mu_k$ are well-defined on all of $\mathrm{CF}(\mathbb{R}^n)$ – not merely compact definable sets – thanks to the close relationship with $\int d\chi$. Continuity of these measures is possible, but requires a more delicate function space topology, relying on currents (see §6.11 and [28]).
7. Research in geometric random fields has been, sadly, largely ignored by topologists. It may help matters to employ (1) the language of Euler integration, as it removes much of the mystery as to the natural appearance of $\chi$ in this work, and (2) the o-minimal framework, as it would likely subsume many of the nondegeneracy assumptions. The author has taken the liberty of invoking definable sets in the statement of Theorem 3.21: this is not *quite* what Adler proves. Hopefully it is a true statement.

## Chapter 4

# Homology

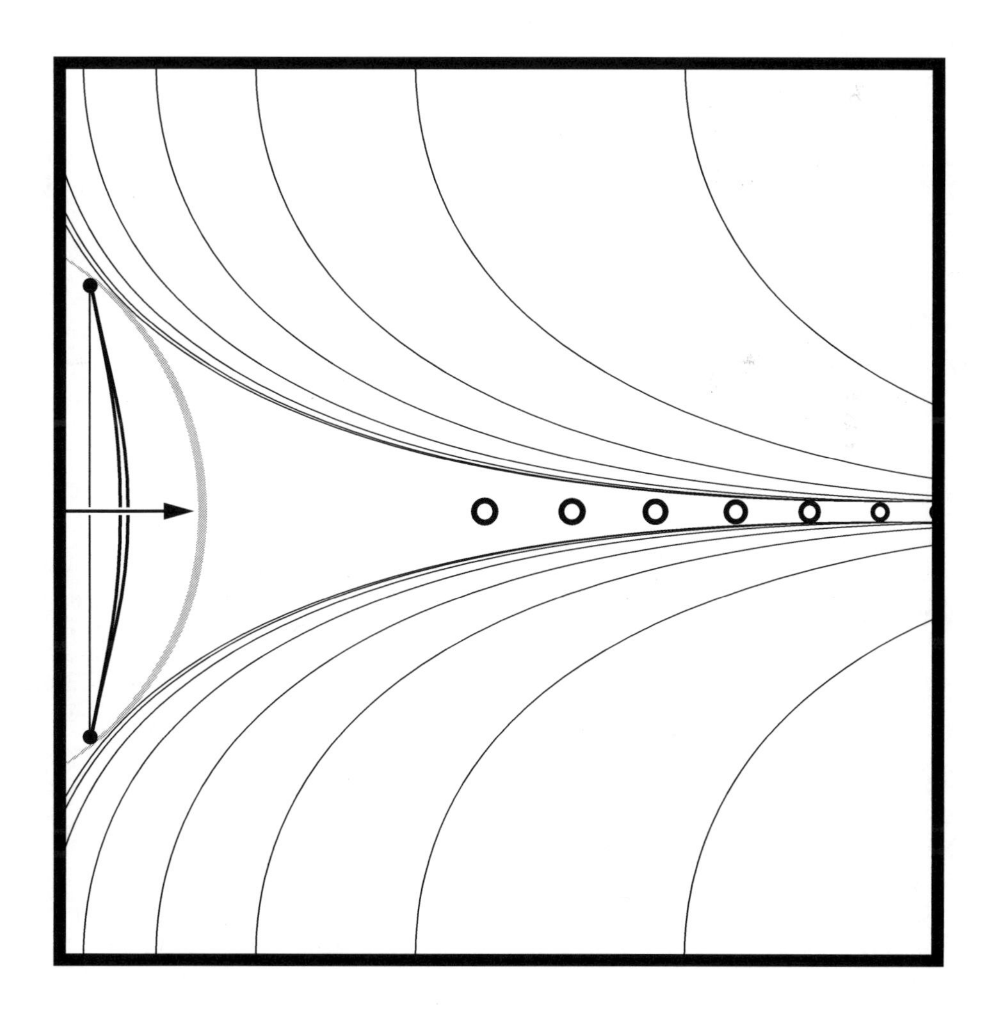

Decomposing a space into cells, counting, and canceling according to parity of dimension, is efficacious for defining one topological invariant: $\chi$. More refined invariants arise upon lifting this enumerative data to the domain of linear algebra. This leads to the notion of **homology**.

## 4.1 Simplicial and cellular homology

The crucial construction of this chapter is that which converts a decomposition of a space in terms of simple pieces into a collection of vector spaces (or modules) and linear transformations (or homomorphisms): an algebraic version of a cell complex. The initial discussion proceeds using the language of linear algebra and passes to more general algebraic constructs. The reader needing review in either linear or abstract algebra is encouraged to consult the Appendix before proceeding.

For simplicity, mod-2 arithmetic is used initially. Recall that $\mathbb{F}_2 = \{0, 1\}$ is the field with two elements. Counting in this field is analogous to flipping a light switch. To build intuition, consider a cell complex $X$, each cell of which is outfitted with a metaphorical light switch. The building blocks of a rudimentary homology for $X$ are as follows.

1. Define ***k*-chains** $C_k$ as the vector space over the field $\mathbb{F}_2$ with basis the $k$-cells of $X$.
2. Consider the **boundary maps**: the linear transformations $\partial_k : C_k \to C_{k-1}$ which send a basis $k$-cell to the abstract sum of basis $(k-1)$-cell faces.

The collection of chains and boundary maps is assembled into a **chain complex**:

$$\cdots \longrightarrow C_k \xrightarrow{\partial_k} C_{k-1} \xrightarrow{\partial_{k-1}} \cdots \xrightarrow{\partial_2} C_1 \xrightarrow{\partial_1} C_0 \xrightarrow{\partial_0} 0\,. \tag{4.1}$$

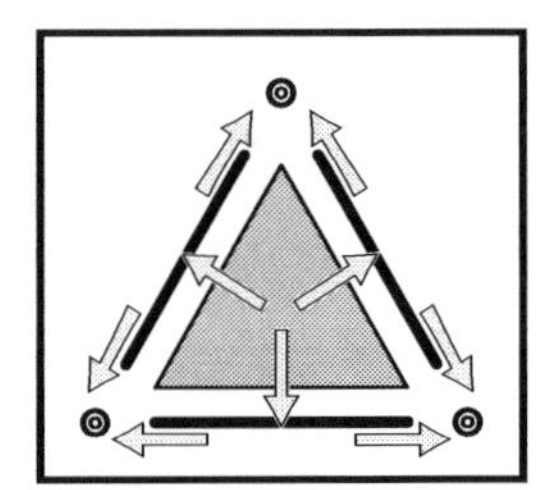

The chain complex is **graded**, in this case by the dimension of the cells. It is beneficial to denote the chain complex as a single object $\mathcal{C} = (C_\bullet, \partial)$ and to write $\partial$ for the boundary operator acting on any chain of unspecified grading. Chain complexes are a representation of a cell complex within linear algebra. It seems at first foolish to algebraicize the problem in this manner – why bother with vector spaces which simply record whether a cell is present (1) or absent (0)? Why express the boundary of a cell in terms of linear transformations, when the geometric meaning of a boundary is clear? By the end of this chapter, probably, and the next, certainly, this objection will have been forgotten.

**Example 4.1 (Simplices and cubes)** ⊚
Consider a single $n$-simplex $\Delta^n$. The resulting chain complex has $C_k$ of dimension $\binom{n+1}{k+1}$. The top-dimensional boundary map $\partial_n$ is a 1-by-$n$ matrix with all entries 1. In contrast, for $I^n = [0,1]^n$, the $n$-dimensional cube with cell structure inherited from the interval $I = [0,1]$ with two endpoints, the chain complex has $C_k$ of dimension $2^{n-k}\binom{n}{k}$. ⊚

The following lemma is deeper than it appears.

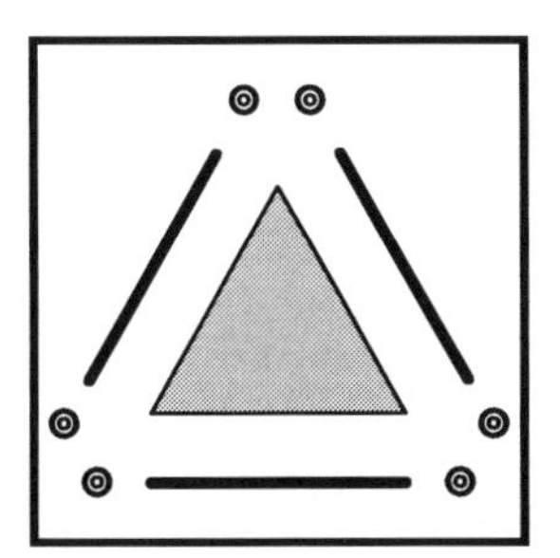

**Lemma 4.2.** *The boundary of a boundary is null:*

$$\partial^2 = \partial_{k-1} \circ \partial_k = 0, \tag{4.2}$$

*for all $k$.*

**Proof.** For simplicity, consider the case of an abstract simplicial complex on a vertex set $V = \{v_i\}$. The **face map** $D_i$ acts on a simplex by removing the $i^{\text{th}}$ vertex $v_i$ from the simplex's list, if present; else, do nothing. The graded boundary operator $\partial\colon C_\bullet \to C_\bullet$ is thus a formal sum of face maps $\partial = \bigoplus_i D_i$. It suffices to show that $\partial^2 = 0$ on each basis simplex $\sigma$. Computing the composition in terms of face maps, one obtains:

$$\partial^2\sigma = \sum_{i \neq j} D_j D_i \sigma. \tag{4.3}$$

Each $(k-2)$-face of the $k$-simplex $\sigma$ is represented exactly twice in the image of $D_j D_i$ over all $i \neq j$. Thanks to $\mathbb{F}_2$ coefficients, the sum over this pair is zero.[1] ⊙

The homology of $\mathcal{C}$, $H_\bullet(\mathcal{C})$, is a sequence of $\mathbb{F}_2$-vector spaces built from the following subspaces of chains.

**Corollary 4.3.** *For all $k$,* im $\partial_{k+1}$ *is a subspace of* ker $\partial_k$.

A **cycle** of $\mathcal{C}$ is a chain with empty boundary, *i.e.*, an element of ker $\partial$. Homology is an equivalence relation on cycles of $\mathcal{C}$. Two cycles in $Z_k = \ker \partial_k$ are said to be **homologous** if they differ by an element of $B_k = \operatorname{im} \partial_{k+1}$. The **homology** of $X$ is the sequence of quotient vector spaces $H_k(X)$ over $\mathbb{F}_2$, for $k \in \mathbb{N}$, given by:

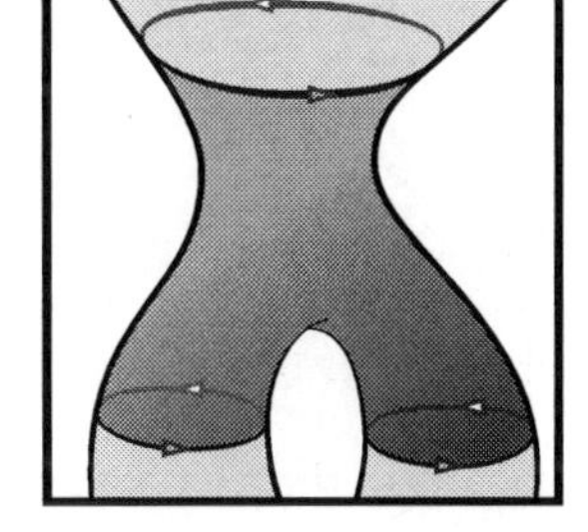

$$\begin{aligned} H_k(X) &:= Z_k/B_k \\ &= \ker \partial_k / \operatorname{im} \partial_{k+1} \\ &= \text{cycles/ boundaries.} \end{aligned} \tag{4.4}$$

To repeat: $Z_k = \ker \partial_k$ is the subspace of cycles in $C_k$, and $B_k = \operatorname{im} \partial_{k+1}$ is the subspace of **boundaries**. Homology inherits the grading of $\mathcal{C}$ and will be therefore denoted $H_\bullet(X)$ when a dimension is

[1] If the reader is reminded by the notation of certain calculus results that rely on mixed partial derivatives commuting and cancelling, then the notation has done its job.

not specified. Elements of $H_\bullet(X)$ are **homology classes** and are denoted $[\alpha]$, where $\alpha \in Z_k$ is a $k$-cycle and $[\cdot]$ denotes the equivalence class modulo elements of $B_k$.

One often distinguishes between **simplicial homology**, $H_\bullet^{\text{simp}}$, and **cellular homology**, $H_\bullet^{\text{cell}}$. Note that only simplicial homology has been given a proper definition above, since Lemma 4.2 was proved only in the simplicial setting. The details of cellular homology will appear in §4.3 and §5.5. Before this chapter is done, several more homology theories will be introduced, including some that do not depend on any cellular structure. It is asserted (with justification in §5.4) that these various homology theories agree (when well-defined) except under unusual circumstances not often seen in applications. The notation $H_\bullet$ will therefore be used to denote whatever homology theory is most convenient or appropriate. In the examples to follow, explicit cellular structures may not be given: let the reader take on faith now what will be explicitly detailed in Chapter 5 – homology is a homotopy invariant of spaces.

## 4.2 Homology examples

### Example 4.4 (Graphs) ⊙

Consider $X$ a topological graph – a compact 1-d cell complex. As $\partial_0 = 0$, all 0-chains are 0-cycles: $Z_0 = C_0$. The boundary subspace $B_0 \subset Z_0$ consists of finite unions of vertices, with an even number in each connected component of the graph. If one chooses the vertex set of $X$ as a basis for $Z_0$, it follows that any two basis elements are homologous if and only if they are the endpoints of an end-to-end sequence of edges in $X$ – that is, if and only if they lie in the same connected path-component of $X$. The dimension of $H_0(X)$ is therefore the number of such path components.

Any element of $Z_1(X)$ consists of a finite union of cyclic end-to-end sequences of edges. As a graph is a 1-dimensional complex, there are no higher dimensional chains with which to form homology classes: for graphs, 1-cycles are homologous if and only if they are identical. Correspondingly, $H_1(X)$ collates the linearly independent cyclic chains of edges, and $H_k(X) = 0$ for $k > 1$. Note that homology gives the collection of 1-cycles in $X$ the structure of a vector space, complete with the notions of sums, spans, and bases. ⊙

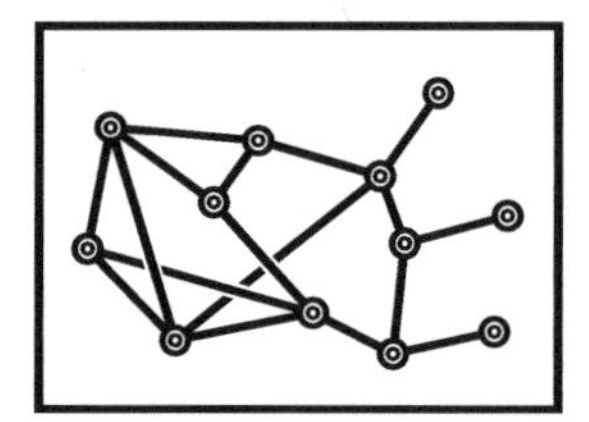

### Example 4.5 (Surfaces) ⊙

For $S_g$ a cell complex homeomorphic to a connected compact orientable surface of genus $g$, the homology is nonzero only in dimensions less than three, since all cells are of dimension at most two. The relation to genus is as follows:

$$\dim H_k(S_g) = \begin{cases} 1 & : \quad k = 0 \\ 2g & : \quad k = 1 \\ 1 & : \quad k = 2 \\ 0 & : \quad k > 2 \end{cases} .$$

It is worth stressing that $H_1(X)$ does not measure whether any cycle in $X$ is *contractible* – whether it can be shrunk to a point in $X$ continuously. For example, on an orientable surface, any simple closed curve which divides the surface into two connected components is **nullhomologous** as a 1-cycle, even though it may not be contractible within the surface. A few facts may help the beginning reader build up an intuition for what it is that homology measures. Justifications will be filled in over subsequent chapters.

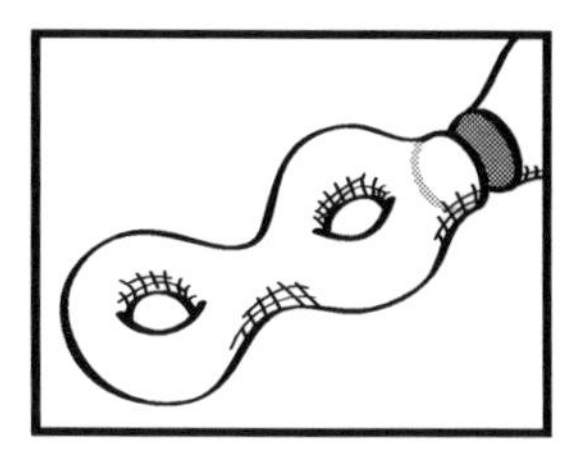

1. For $X$ contractible, $H_k(X) = 0$ for all $k > 0$.
2. The number of path components of $X$ equals dim $H_0(X)$.
3. For $D^2_n$ a disc in the plane with $n$ disjoint discs removed from the interior, dim $H_0(D^2_n) = 1$, dim $H_1(D^2_n) = n$ , and dim $H_k(D^2_n) = 0$ for $k > 1$.
4. The sphere $\mathbb{S}^n$ has $H_k = 0$ for all $k$ except $H_0$ and $H_n$, both of dimension one.
5. For a disjoint union of spaces $X = A \sqcup B$, $H_k(X) = H_k(A) \oplus H_k(B)$.
6. For a **wedge sum** of path-connected spaces $X = A \vee B$ (obtained by identifying a single point of $A$ with a single point of $B$), $H_k(X) = H_k(A) \oplus H_k(B)$ for all $k > 0$ and dim $H_0(X) = 1$.

**Example 4.6 (Products and Künneth)** ⊚

A torus $\mathbb{T}^n = (\mathbb{S}^1)^n$ has homology satisfying

$$\dim H_k(\mathbb{T}^n) = \binom{n}{k}.$$

The reader will rightly suspect a relationship with polynomial algebra, via the the coefficients of $(1+t)^n$: indeed, this foreshadows an algebraic result for products of general spaces. For $X$ a space with finite-dimensional homology, consider the **Poincaré polynomial** of $X$, $P_t(X) := \sum_k \beta_k t^k$, with coefficients $\beta_k :=$ dim $H_k(X)$ the **Betti numbers** of $X$. For example, the Poincaré polynomial of $\mathbb{S}^1$ is $1+t$. The following is a simplistic reduction of the classical **Künneth Theorem**:

**Theorem 4.7 (Künneth Formula).** $P_t(X \times Y) = P_t(X)P_t(Y)$.

The computation for dim $H_k(\mathbb{T}^n)$ above follows. The simplicity of this statement comes from the linear-algebraic approach used. Complications arise when using coefficients not in a field. ⊚

## 4.3 Coefficients

The homology defined in §4.1 is more properly called the simplicial or cellular homology of $X$ *with* $\mathbb{F}_2$ *coefficients*, denoted $H_\bullet(X; \mathbb{F}_2)$. This additional parameter hints at other coefficients. For example, instead of the $C_k$ being vector spaces over a field $\mathbb{F}_2$, other fields may be used. For any field $\mathbb{F}$, one may construct chains $C_k$ for a cell complex $X$

as $\mathbb{F}$-vector spaces. Such coefficients permit expressiveness: $\mathbb{R}$ is effective in describing simplices' *intensities*, in contrast to $\mathbb{F}_2$'s involutive switch. This entails a little more care with directions, but is a well-motivated generalization.

**Example 4.8 (Kirchhoff's current rule)** ⊚

Consider an electric circuit as a 1-d cell complex (or graph), with circuit elements (resistors, capacitors, etc.) located at certain vertices. **Kirchhoff's current rule** states that the current flowing through the edges of the circuit satisfies a conservation principle: at each node, the sum of the incoming currents equals the sum of the outgoing currents, with directionality encoded in the $\pm$ sign. In the language of this chapter: *current is a 1-cycle*. Note the unambiguous need for $\mathbb{R}$ coefficients, since current is measured as a real quantity. ⊚

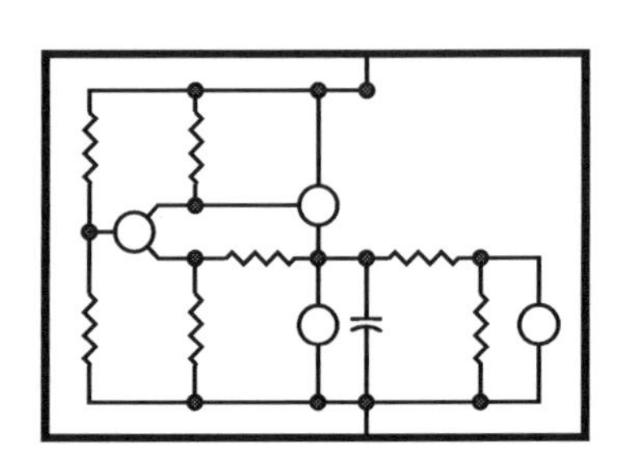

A crucial feature of using coefficients other than $\mathbb{F}_2$ is this: $-1 \neq 1$. This algebraic unfolding necessitates an orientation associated to cells. For electric circuits, graph cycles, and 1-chains in general, this is naturally encoded as a direction. In calculus class, one uses a twist of the right hand for orientations on curves and surfaces: a more principled approach is needed for the present setting. Simplicial complexes yield the most explicit case: recall that an abstract simplicial complex on a vertex set $V$ has each $k$-simplex specified as $\sigma = \{v_0, v_1, \ldots, v_k\} \subset V$. To define an *oriented* $k$-simplex, one begins with an (arbitrary) ordering of vertices, denoted $[v_0, v_1, \ldots, v_k]$.

An **orientation** on $\sigma$ is a choice of equivalence class of orderings up to *even* permutations; thus, $[v_0, v_1, v_2] = -[v_0, v_2, v_1] = [v_2, v_0, v_1]$. This yields a multiplicative action of $\{-1, +1\}$ on oriented chains: multiplication by $-1$ is the chain analogue of reversing orientation. In the algebra of chains, as with contour integrals, adding two simplices of opposite orientation cancel. For homology with coefficients in a field $\mathbb{F}$, one builds $C_n$ on a basis of oriented $n$-simplices. The boundary operator $\partial\colon C_n \to C_{n-1}$ tracks orientation as follows, using the simplicial face map notation from the proof of Lemma 4.2:

$$\partial\sigma = \sum_{i=0}^{k} (-1)^i D_i \sigma. \tag{4.5}$$

The reader may check that the use of $(-1)^i$'s suffices to keep Lemma 4.2 in effect. The resulting simplicial homology with $\mathbb{F}$ coefficients is denoted $H_\bullet^{\mathrm{simp}}(X;\mathbb{F})$. That it is independent of the initial choice of orientations on the simplices is by no means obvious. Chapter 5 will provide a reason for this invariance.

The definition of orientation for abstract simplices is a mechanical index: cellular homology requires more machinery. In the simple setting of a regular cell complex, choose an *"orientation"* for each cell. For the moment, this requires a little imagination – a direction for an edge, a sense of rotation for a 2-cell, a handedness for a 3-cell, *etc.* The boundary of a cell inherits an orientation from the interior, in a manner that is used (if not understood) in calculus courses. Just as in calculus, one compares orientations on a domain and its boundary. For $\sigma \lhd \tau$, with $\sigma$ a codimension-1 face of $\tau$, define the **incidence number** $[\sigma\colon \tau]$, to be $+1$ if the attaching map of $\tau$ preserves the induced orientation from $\partial\tau \to \sigma$; for orientation reversal, $[\sigma\colon \tau] = -1$. The boundary map is then defined as

$$\partial\tau = \sum_{\sigma \lhd \tau} [\sigma\colon \tau]\sigma. \tag{4.6}$$

Defining $[\rho\colon \tau] = 0$ for all faces $\rho$ of codimension greater than one allows the above sum to range over all faces. This bookkeeping yields a chain complex $\mathcal{C}^{\text{cell}} = (C_\bullet, \partial)$ with homology $H_\bullet^{\text{cell}}(X; \mathbb{F})$ once it is shown that $\partial \circ \partial = 0$; namely, that the signs cancel:

$$\sum_{\rho \lhd \sigma \lhd \tau} [\rho\colon \sigma][\sigma\colon \tau] = 0. \tag{4.7}$$

The serendipitous notation of an abstract simplicial complex leads to a simplification in that $[\sigma\colon \tau] = (-1)^i$, where $\sigma$ is the $i^{\text{th}}$ face of $\tau$: these accidents and subsequent formulae such as (4.5) obscure the meaning of the boundary operator $\partial$ – signs are really about orientation, as will be explained in §5.5.

Field coefficients and linear algebra are, still, not the full story. The generalization to $\mathbb{Z}$ coefficients – in which each $C_k$ is a $\mathbb{Z}$-**module** – is particularly relevant. One visualizes chains $C_\bullet(X; \mathbb{Z})$ over a cell complex $X$ with integer coefficients as recording a finite collection of simplices with orientation and multiplicity. Ultimately, one works with a chain complex over an **R**-module (for **R** a ring) with the boundary maps module homomorphisms.

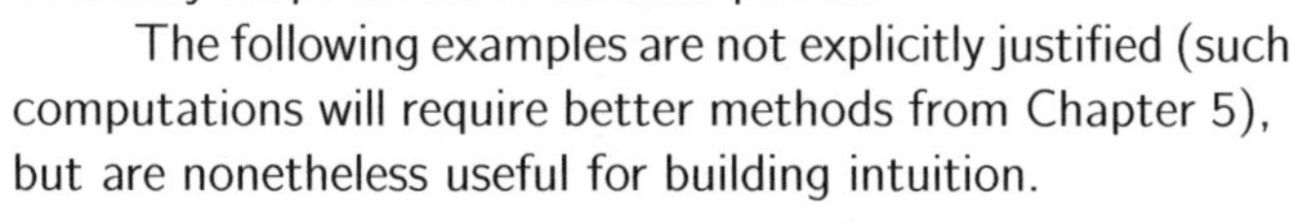

The following examples are not explicitly justified (such computations will require better methods from Chapter 5), but are nonetheless useful for building intuition.

**Example 4.9 (Coefficients)** ⊚

Using different coefficients can lead to genuinely different homology groups. For example, the non-orientable Klein bottle $K^2$ satisfies:

$$H_k(K^2; \mathbb{F}_2) = \begin{cases} \mathbb{F}_2 & : \quad k = 0 \\ \mathbb{F}_2 \oplus \mathbb{F}_2 & : \quad k = 1 \\ \mathbb{F}_2 & : \quad k = 2 \\ 0 & : \quad k > 2 \end{cases} \quad ; \quad H_k(K^2; \mathbb{Z}) = \begin{cases} \mathbb{Z} & : \quad k = 0 \\ \mathbb{Z} \oplus \mathbb{Z}_2 & : \quad k = 1 \\ 0 & : \quad k > 1 \end{cases}$$

In this example, $\mathbb{Z}$ coefficients yields a different rank of homology groups at gradings one and two. The Klein bottle has no 2-cycle (a boundaryless nonzero 2-chain) in $\mathbb{Z}$ coefficients, but it does in $\mathbb{F}_2$: thus, $K^2$ is a non-orientable surface. Note also the presence of **torsion** in $H_1(K^2;\mathbb{Z}) \cong \mathbb{Z} \oplus \mathbb{Z}_2$, where $\mathbb{Z}_2 = \mathbb{Z}/2\mathbb{Z}$ is the quotient of $\mathbb{Z}$ by the subgroup of evens. This indicates that the meridian and longitudinal curves on $K^2$, unlike those of the torus $\mathbb{T}^2$, are qualitatively different: sliding a meridional curve along a longitude reverses orientation. In general, the presence of a torsional element in $H_\bullet(X;\mathbb{Z})$ is indicative of some type of twisting in $X$ which, in surface examples, manifests itself in non-orientability. ⊚

**Example 4.10 (Rotations and projections)** ⊚

Rotations in $\mathbb{R}^3$ offer a fascinating, simple, and useful example of torsional phenomena where coefficients matter. Consider $SO_3$, the group of real 3-by-3 orthogonal matrices with determinant 1. These are precisely the orientation-preserving rotations of Euclidean 3-space. As one can demonstrate using a variety of physical devices (plates, belts, fermions, etc.), there is a torsional core writhing within: $H_1(SO_3;\mathbb{Z}) \cong \mathbb{Z}_2$. Since $SO_3 \cong \mathbb{P}^3$, this is an example of the homology of projective spaces. In general, $\dim H_k(\mathbb{P}^n;\mathbb{F}_2) = 1$ for all $0 \leq k \leq n$. The story for integer coefficients is quite different:

$$H_k(\mathbb{P}^n;\mathbb{Z}) = \begin{cases} \mathbb{Z} & : \quad k=0 \text{ or } k=n, \text{ odd} \\ \mathbb{Z}_2 & : \quad 0<k \text{ odd } <n \\ 0 & : \quad \text{else} \end{cases} .$$

⊚

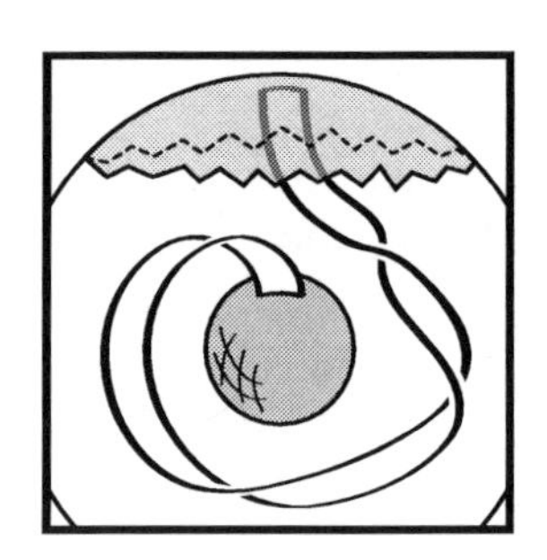

## 4.4 Singular homology

A definition with sufficient strength has emerged: a chain complex $\mathcal{C} = (C_\bullet, \partial)$ is *any* sequence of **R**-modules $C_k$ with homomorphisms $\partial_k : C_k \to C_{k-1}$ satisfying $\partial_k \circ \partial_{k+1} = 0$. Though the generators of $\mathcal{C}$ may be simplices or cells of a space $X$, there are many different objects worthy of being counted and compared. To be noted are the commonalities (*e.g.*, notions of dimension associated to the objects counted) and the distinctions (*e.g.*, auxiliary structures imposed). The next several sections delineate a few major homology theories. The first example is *singular* homology – the most common and generally useful theory.

The reader will have observed that cellular homology is, like Euler characteristic, independent of how the space is decomposed into cells. The best method of proof comes not from serendipitous cancellative combinatorial refinements of cell structures, but rather by limits to uncomputable abundance.

One of the first to take such an approach was Vietoris, who introduced his homology theory for metric spaces [296]. In this theory, one fixes an $\epsilon > 0$ and considers the chain complex generated by $k$-tuples of distinct points in $X$ of pairwise distance $\leq \epsilon$ (something like the Vietrois-Rips complex of all points in $X$). These simplices do not fit together to form a triangulation of $X$: there are far too many. However, for reasonable spaces (*e.g.*, metric finite cell complexes), the homology of the Vietoris chain complex stabilizes as $\epsilon \to 0^+$, and this agrees with the cellular homology of a cell complex. The reason that Vietoris' theory is mostly forgotten lies in the efficacy of **singular homology**, which requires neither a metric space nor the explicit limiting process.

The **singular chain complex** of a topological space $X$ is the complex $\mathcal{C}^{\text{sing}} = (C_\bullet, \partial)$, where the generators of $C_k$ are continuous maps $\sigma\colon \Delta^k \to X$ from the standard $k$-simplex $\Delta^k$ to $X$. The boundary of a singular simplex is the formal sum of restrictions of the map to faces, using the orientation convention of Equation (4.5):

$$\partial\sigma = \sum_{i=0}^{k}(-1)^i D_i\sigma = \sum_{i=0}^{k}(-1)^i\sigma\big|_{D_i\Delta^k}.$$

Notice that there is a decoupling between the grading and the dimension of the *image* of the singular simplex. Indeed, the image of a singular simplex may be very convoluted, with no sensible notion of dimension other than in its preimage. The resulting chain complex $\mathcal{C}^{\text{sing}}$ is *enormous* – certainly of uncountably infinite dimension except in the most trivial cases. However, in this lies its flexibility. For two spaces that are homeomorphic, there is an equivalence between their singular chain complexes that guarantees equivalent homologies. The yet more useful and general homotopy invariance of singular homology in §5.1 is the true reward for the unwieldy bulk of the singular complex.

## 4.5 Reduced homology

It is sometimes convenient to augment a homology theory beyond grading zero. Given a chain complex $\mathcal{C}$ over an **R**-module, define the **reduced** chain complex $\tilde{\mathcal{C}}$,

$$\cdots \longrightarrow C_k \xrightarrow{\partial_k} C_{k-1} \xrightarrow{\partial_{k-1}} \cdots \xrightarrow{\partial_2} C_1 \xrightarrow{\partial_1} C_0 \xrightarrow{\epsilon} \mathbf{R} \longrightarrow 0\,, \qquad (4.8)$$

where $\epsilon\colon C_0 \to \mathbf{R}$ sends each basis 0-chain to the sum of the coefficients.[2] It is advantageous to write $\tilde{C}_k$ for the chain groups, where $k \in \mathbb{Z}$ and:

$$\tilde{C}_k = \begin{cases} C_k & : k \geq 0 \\ \mathbf{R} & : k = -1 \\ 0 & : k < -1 \end{cases},$$

with all boundary maps as above.

**Proposition 4.11.** *For a nonempty chain complex $\mathcal{C}$ with **R**-module coefficients, the reduced homology $\tilde{H}_\bullet(\mathcal{C})$ satisfies $H_k(\mathcal{C}) \cong \tilde{H}_k(\mathcal{C})$ for $k > 0$ and $H_0(\mathcal{C}) \cong \tilde{H}_0(\mathcal{C}) \oplus \mathbf{R}$.*

[2] For $\mathbb{F}$-vector spaces, $\epsilon\colon C_0 \to \mathbb{F}$. So, in the case of $\mathbb{F}_2$, $\epsilon$ records the parity of vertices in a 0-chain.

The reader may think of the reduced complex as having a single basis element at grading $-1$. Reduced homology simplifies certain results: it is convenient to have a nonempty contractible space have $\tilde{H}_\bullet = 0$, as such an **acyclic** space plays the role of a *zero* in homological algebra.

## 4.6 Čech homology

A homology theory counts objects with grading and cancelation. While the objects and gradings are usually indicative of cells and dimensions, the correspondence may be disguised. Consider the case of the **Čech homology** of a cover. Let $X$ be a topological space and $\mathcal{U} = \{U_\alpha\}$ a locally-finite collection of open sets whose union is $X$. The **Čech complex** of $\mathcal{U}$ is the chain complex $\check{\mathcal{C}}(\mathcal{U}) = (\check{C}_\bullet, \check{\partial})$ generated by nonempty intersections of the cover $\mathcal{U}$. That is, the 0-chains $\check{C}_0$ have as basis the elements $U_\alpha$ of $\mathcal{U}$. The $k$-chains $\check{C}_k$ have as basis non-empty intersections of $k+1$ distinct elements of $\mathcal{U}$, denoted $U_J$, where $J = (\alpha_0, \alpha_1, \ldots, \alpha_k)$ is a multi-index, ordered up to even permutations as per the ordered simplices of §4.3. The resulting chain complex is outfitted with a boundary operator of familiar form:

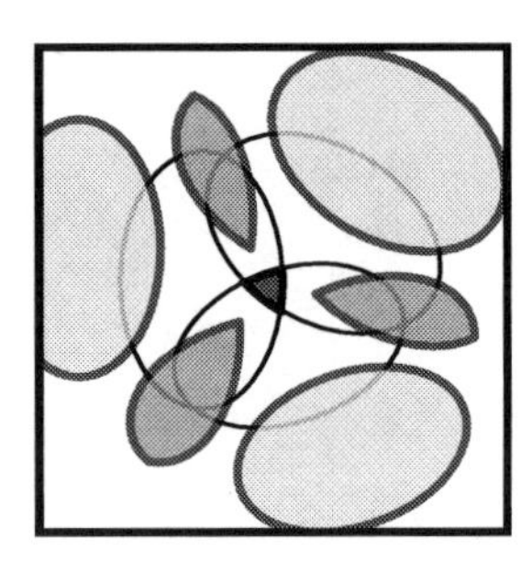

$$\check{\partial}(U_J) = \sum_{i=0}^{k} (-1)^i U_{D_i J}, \tag{4.9}$$

where $D_i$ is the face map of §4.1 that deletes the $i^{\text{th}}$ entry from the multi-index $J$. One checks that $\check{\partial}^2 = 0$ and thus the Čech homology of the cover, $\check{H}_\bullet \mathcal{U} = H_\bullet(\check{\mathcal{C}}(\mathcal{U}))$, is well-defined. The reader will be reminded of the nerve complex of §2.6; the Čech complex $\check{\mathcal{C}}(\mathcal{U})$ is the simplicial chain complex associated to the nerve $\mathcal{N}(\mathcal{U})$ of the cover. A homological version of the Nerve Lemma (Theorem 2.4) holds:

**Theorem 4.12.** *If all nonempty intersections of elements of $\mathcal{U}$ are acyclic ($\tilde{H}_\bullet(U_J) = 0$ for all nonempty $U_J$), then the Čech homology of $\mathcal{U}$ agrees with the singular homology of $X$: $\check{H}_\bullet(\mathcal{U}) \cong H_\bullet(X)$.*

In the same manner that the cellular homology of a (reasonably nice) space is independent of the cell structure used, the Čech homology of a (reasonably nice) space $X$ is well-defined, independent of the (acyclic) cover used to compute it.

## 4.7 Relative homology

Homology makes sense not only for spaces, but for subspaces as well. Let $A \subset X$ be a subset (or subcomplex, as appropriate) and $\mathcal{C}(X)$ a chain complex (singular, cellular, *etc.*) on $X$. There are two natural chain complexes implicating $A$. The first, $\mathcal{C}(A)$, consists of subgroups $C_k(A) < C_k(X)$ with the obvious restriction of $\partial$. The homology of this chain (sub-)complex gives, as expected, the homology of $A$ as a space in its own right.

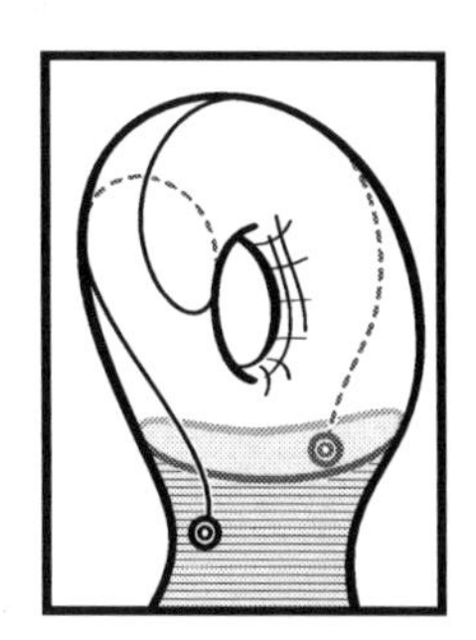

There is, however, a more subtle construct involving the collapse of the subcomplex that yields an important homology complementary to that of $A$. Let $\mathcal{C}(X, A)$ denote the quotient chain complex $(C_k(X, A), \overline{\partial})$, where $C_k(X, A) := C_k(X)/C_k(A)$ consists of *chains on $X$ modulo chains on $A$*, with the induced boundary maps $\overline{\partial}$ on the quotients. These relative chains do not vanish on $A$; rather, they are equivalence classes of chains in $X$ which are identical off of $A$. The resulting **relative homology** $H_\bullet(X, A)$ collates homology classes of relative cycles (chains in $X$ whose boundaries vanish or lie in $A$). A relative 1-cycle may therefore be either a genuine 1-cycle in $X$ or may implicate a chain in $X$ whose boundary points are in $A$.

**Example 4.13 (Reduced homology)** ⊚

The reduced homology of a space $X$ is isomorphic to the homology of $X$ relative to a basepoint $p \in X$: $\tilde{H}_\bullet(X) \cong H_\bullet(X, p)$. The only nontrivial chain in $X$ whose boundary is a nonzero multiple of $p$ is, precisely, a multiple of $p$ and a zero-cycle. Thus, by definition, $H_k(X, p) \cong H_k(X)$ for $k > 0$ and $H_0(X, p)$ has rank one less than $H_0(X)$, since the subspace corresponding to the homology class $[\{p\}]$ has been quotiented out. ⊚

One of the foundational theorems about homology concerns relative homology. The following is stated for the singular theory: for other (cellular, *etc.*) homology theories, the hypotheses must include the relevant structures present.

**Theorem 4.14 (Excision Theorem).** *Let $U \subset A \subset X$ with the closure of $U$ contained in the interior of $A$. Then $H_\bullet(X-U, A-U) \cong H_\bullet(X, A)$.*

Excision implies that for the pair $(X, A)$, what happens inside of $A$ is irrelevant.

**Corollary 4.15.** *For $A \subset X$ a closed subcomplex of a cell complex $X$, $H_\bullet(X, A) \cong \tilde{H}_\bullet(X/A)$.*

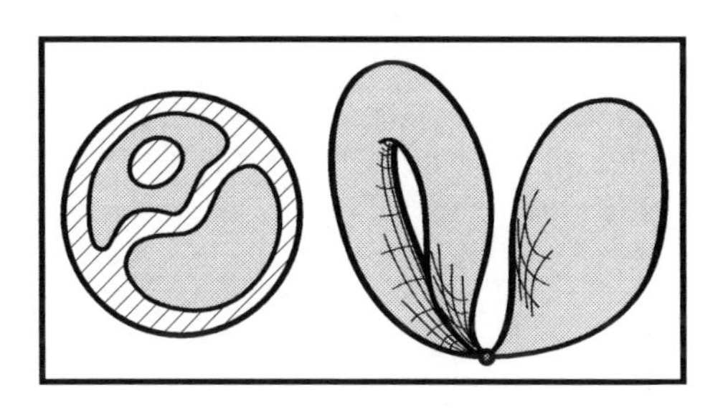

The hypothesis of $A$ a subcomplex is usually too restrictive. In the singular setting, it suffices to have $A$ closed and possessing an open neighborhood which deformation retracts to $A$. This provides a good way to *visualize* relative homology and to compute the homology of quotient spaces.

**Example 4.16 (Relative homology)** ⊚

The homology of a closed disc $\mathbb{D}^n$ relative to its boundary $\partial\mathbb{D}^n$ has as a nontrivial generator $\mathbb{D}^n$ itself, as a relative $n$-cycle. From Corollary 4.15 and the fact that $\mathbb{D}^n/\partial\mathbb{D}^n \cong \mathbb{S}^n$, one sees $H_\bullet(\mathbb{D}^n, \partial\mathbb{D}^n) \cong \tilde{H}_\bullet(\mathbb{S}^n)$. ⊚

## 4.8 Local homology

The **local homology** of $p \in X$ is the (singular) homology $H_\bullet(X, X{-}p)$. This is a little hard to visualize, given that Corollary 4.15 is inapplicable, but a moment's thought yields the intuition that it measures something of the features of $X$ in an arbitrarily small neighborhood of $p$. The correctness of this intuition follows from excision: local homology is local.

**Corollary 4.17.** *Let $p \in X$ be a (closed) point and $V$ an open neighborhood of $p$. Then $H_\bullet(X, X{-}p) \cong H_\bullet(V, V{-}p)$: that is, local homology can be computed from an arbitrarily small neighborhood of $p$.*

**Proof.** Apply Excision, with $U = X{-}V$ and $A = X{-}p$. ⊙

**Example 4.18 (Local orientation)** ⊚

Consider the local homology of an $n$-manifold $M$ at $p$: this is $H_\bullet(M, M{-}p)$. By excision and the definition of a manifold as locally Euclidean, this is isomorphic to $H_\bullet(\mathbb{D}^n, \mathbb{D}^n{-}0)$. As the punctured disc deformation retracts to the boundary, it follows from Example 4.16 that the local homology is isomorphic to $\tilde{H}_\bullet(\mathbb{S}^n)$ and is thus nontrivial of rank one in dimension $n$. This leads to a painless definition of orientation, a concept that usually requires recourse to a dextrous sleight of hand. A **local orientation** at $p \in M$ is a choice of generator for the $n$-dimensional local homology at $p$ (in $\mathbb{Z}$ coefficients). A global **orientation** on a compact $n$-manifold $M$ is a choice of generator for $H_n(M; \mathbb{Z})$. A global orientation, in keeping with one's experience from calculus class, is a consistent choice of local orientations.

This definition of orientation illuminates the cellular homology of §4.3. The orientation of an $n$-dimensional cell $\sigma$ is *defined* to be a choice of generator for $H_n(\sigma, \partial\sigma; \mathbb{Z}) \cong \mathbb{Z}$. The **incidence number** $[\sigma\colon \tau]$ of a face pair $\sigma \trianglelefteq \tau$ is $\pm 1$, depending on whether the orientations of $\sigma$ and $\partial\tau$ agree or disagree. What remains undefined (until §5.5) is how the orientation on $\tau$ induces an orientation on $\partial\tau$. ⊚

## 4.9 Homology of a relation

A **relation** between two sets $X$ and $Y$ is a subset $\mathcal{R} \subset X \times Y$ of their product. A point $x \in X$ is related to a point $y \in Y$ if and only if $(x, y) \in \mathcal{R}$. For example, the sensing modality used for target enumeration in §3.7 is a relation between sensors and targets collating which sensors detect which targets. Other relations are prevalent: in marketing, the *puchase* relation between customers and products; on *Twitter*, the *follows* relation between individuals; or in manufacturing, the *assignment* relation between workers and tasks.

It is possible and profitable to build homologies for relations as follows. Given $\mathcal{R} \subset X \times Y$, define the chain complex $\mathcal{C}(X;\mathcal{R})$ with $\mathbb{F}_2$ coefficients as follows: $C_k(X;\mathcal{R})$ has as basis unordered $(k+1)$-tuples of points in $X$ related to some $y \in Y$. One builds a dual complex $\mathcal{C}(Y;\mathcal{R})$ from columns of $\mathcal{R}$ – unordered tuples of points of $Y$ related to some fixed $x \in X$. The boundary operators for $\mathcal{C}(X;\mathcal{R})$ and $\mathcal{C}(Y;\mathcal{R})$ forget points in the tuples, mimicking the boundary operator of a simplicial complex (the property of being related to a common element is closed under subsets). The homology of the relation, $H_\bullet(\mathcal{R})$, is defined as follows:

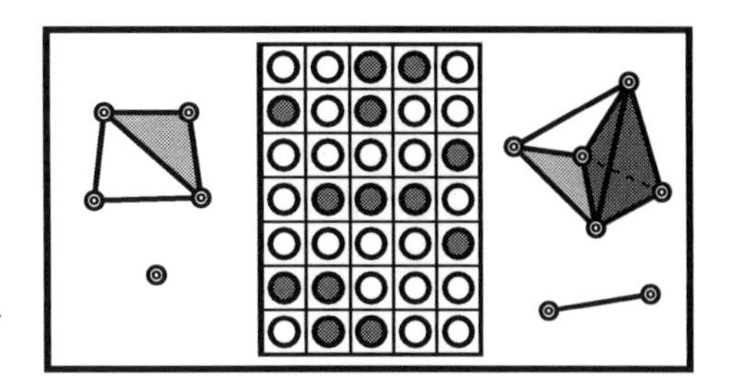

**Theorem 4.19 (Dowker's Theorem [100]).** $H_\bullet(\mathcal{R}) := H_\bullet(X;\mathcal{R}) \cong H_\bullet(Y;\mathcal{R})$.

This result was originally used to prove an equivalence between Čech and Vietoris homology theories, in the case where $X$ is a metric space, $Y$ is a dense net of points in $X$, and $\mathcal{R}$ records which points in $X$ are within $\epsilon > 0$ of points in $Y$. The nerve of the cover of $X$ by metric balls about $Y$ has homology $H_\bullet(Y;\mathcal{R})$, and the Vietoris homology of $X$ is captured by $H_\bullet(X;\mathcal{R})$.

Instead of passing to homology, one can take the intermediate step of assigning to $\mathcal{R}$ a pair of simplicial complexes – the **Dowker complexes**, $\mathcal{R}_X$ and $\mathcal{R}_Y$ – given by the following process. Given $\mathcal{R}$, let $\mathcal{R}_X$ be the nerve complex of the cover of $Y$ by columns of $\mathcal{R}$, and let $\mathcal{R}_Y$ be the nerve complex of the cover of $X$ by rows of $\mathcal{R}$. Otherwise said, $X$ is the vertex set of $\mathcal{R}_X$ and a simplex of $\mathcal{R}_X$ corresponds to a collection of points in $X$ witnessed by some common $y \in Y$ via the relation $\mathcal{R}$. The full version of Dowker's Theorem states that these nerves are homotopic (and, hence, have the same homology). For example, the witness complex of §2.3 is an example of a Dowker complex where $Y$ (the landmarks) is a subset of $X$ (the point cloud) and the relation is via witnessing. There is no reason why, in general, one need constrain landmarks in that manner: they may live in an entirely different space.

**Example 4.20 (Transmitters and receivers)** ◎

Consider a finite set $X$ of transmitters in some domain $\mathcal{D}$ which broadcast their identities (assumed unique) through an unspecified modality (pings, continuous signals, *etc.*). A distinct finite collection of receivers $Y$ can read transmissions and discern identities of transmitters that are within range. The resulting system of transmitters and receivers can be encoded in a relation $\mathcal{R} \subset X \times Y$, where $(x_i, y_j) \in \mathcal{R}$ if and only if signals from device $x_i$ are heard by receiver $y_j$. One may wish or assume or impose that $\mathcal{R}$ has geometric constraints (receivers hear only reasonably nearby transmitters);

such a case permits approximating the common domain $\mathcal{D}$ as a simplicial complex via the Dowker complexes [152]. The duality theorem can assist in computation when there is a disparity in the sizes of $X$ and $Y$. For example, if there are relatively few transmitters as compared to receivers, then the complex $\mathcal{R}_Y$ has $Y$ as its 0-skeleton, and will be a lower-dimensional complex than $\mathcal{R}_X$, whose 0-skeleton is $X$. This leads to an *a priori* bound on the possible dimensions for nonzero homology classes. ⊚

## 4.10 Functoriality

Algebraic topology concerns not only features of spaces but features of *maps* between spaces. This has been hinted at in §1.6 (transversality of *maps* being a crucial tool), §2.9 (where a decision task induces a *map* between input and output complexes), and §3.8 (where *maps* induce actions on Euler integrals). These hints come to fruition in homology, which gives a natural characterization of the qualitative features not only of spaces but also of maps between spaces. This property, at the heart of algebraic topology, is **functoriality**, and it is from this principle the power of the subject emerges.

Consider, for simplicity, a cellular map $f\colon X \to Y$ between cell complexes. In the same manner that $X$ unfolds cell-by-cell into a chain complex $C_\bullet(X)$, one can unfold $f$ to a graded sequence $f_\bullet$ of homomorphisms from $C_k(X) \to C_k(Y)$, generated by basis cells of $X$ being sent to basis cells of $Y$. If a $k$-cell $\sigma$ of $X$ is sent by $f$ to a cell of dimension less than $k$, then the algebraic effect is to send the basis chain in $C_k(X)$ to $0 \in C_k(Y)$. The continuity of the map $f$ induces a **chain map $f_\bullet$** that fits together with the boundary maps of $C_\bullet(X)$ and $C_\bullet(Y)$ to form a **commutative diagram**:

$$\begin{array}{ccccccccc}
\cdots \longrightarrow & C_{n+1}(X) & \xrightarrow{\partial} & C_n(X) & \xrightarrow{\partial} & C_{n-1}(X) & \xrightarrow{\partial} & \cdots \\
& \downarrow{\scriptstyle f_\bullet} & & \downarrow{\scriptstyle f_\bullet} & & \downarrow{\scriptstyle f_\bullet} & & \\
\cdots \longrightarrow & C_{n+1}(Y) & \xrightarrow{\partial'} & C_n(Y) & \xrightarrow{\partial'} & C_{n-1}(Y) & \xrightarrow{\partial'} & \cdots
\end{array} \tag{4.10}$$

Commutativity means the chain map respects the boundary operation, $f_\bullet\partial = \partial' f_\bullet$. Because of this, $f$ acts not only on chains but on cycles and boundaries to yield the **induced homomorphism** $H(f)\colon H_\bullet X \to H_\bullet Y$ on homology. For $\alpha$ a cycle in $X$, define $H(f)[\alpha] := [f_\bullet\alpha] = [f \circ \alpha]$. This is well-defined: if $[\alpha] = [\alpha']$, then, as chains, $\alpha' = \alpha + \partial\beta$ for some $\beta$, and,

$$f_\bullet\alpha' = f \circ \alpha' = f \circ (\alpha + \partial\beta) = f \circ \alpha + f \circ \partial'\beta = f_\bullet\alpha + \partial'(f_\bullet\beta),$$

so that $[f_\bullet\alpha'] = [f_\bullet\alpha]$ in $H_\bullet(Y)$. Homological functoriality means that the induced homomorphisms on homology are an algebraic reflection of the properties of continuous maps between spaces.

This definition was framed with cellular homology and cellular maps for concreteness. Everything carries over naturally to the setting of singular chains and singular homology. The other homology theories of this chapter (relative, reduced, Čech, *etc.*) likewise posses well-defined induced homomorphisms for the appropriate class of maps. The following are simple properties of induced homomorphisms, easily shown from the definitions above:

1. Given $f: X \to Y$, $H(f): H_\bullet X \to H_\bullet Y$ is a (graded) homomorphism.
2. The identity map $\mathrm{Id}: X \to X$ induces the isomorphism $\mathrm{Id}: H_\bullet X \to H_\bullet X$.
3. Given $f: X \to Y$ and $g: Y \to Z$, $H(g \circ f) = H(g) \circ H(f)$.

$$H_\bullet Y_1 \xrightarrow{H(f)} H_\bullet Y_2 \qquad H(f_1): H_\bullet Y_1 \dashrightarrow H_\bullet X \qquad H(f_2): H_\bullet X \dashrightarrow H_\bullet Y_2$$

There is hardly a more important feature of homology than this functoriality. One implication in the sciences is to *inference*. It is sometimes the case that what is desired is knowledge of the homology of an important but unobserved space $X$; the observed data comprises the homology of a pair of spaces $Y_1$, $Y_2$, which are related by a map $f: Y_1 \to Y_2$ that factors through a map to $X$, so that $f = f_2 \circ f_1$ with $f_1: Y_1 \to X$ and $f_2: X \to Y_2$. If the induced homomorphism $H(f)$ is known, then, although $H_\bullet(X)$ is hidden from view, inferences can be made.

**Example 4.21 (Experimental imaging data)** ⊙

The problem of measuring topological features of experimental data by means of imaging is particularly sensitive to threshold effects. Consider, *e.g.*, an open tank of fluid whose surface waves are experimentally measured and imaged. Perhaps the region of interest is the portion of the fluid surface above the ambient height $h = 0$; the topology of the set $A = \{h \geq 0\}$ must be discerned, but can only be approximated by imprecise pixellated images of $\{h \gtrsim 0\}$. Similar situations arise in MRI data, where the structure of a tissue of interest can be imaged as an approximation. Given a reasonably close image, is an observed topological feature (say, a hole, or lack of connectivity) *true*?

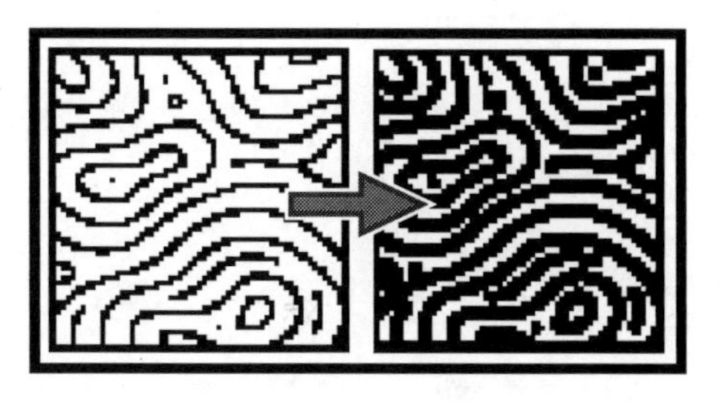

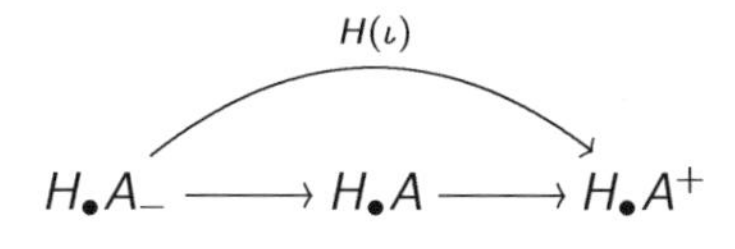


If the goal is to capture the topological features of an exact but unmeasurable set, functoriality may assist. Let, *e.g.*, $A = \{h \geq 0\}$ be the desired but unseen set. If one can measure approximants $A_- \subset A \subset A^+$ from below and above and then match common features of these images, then one has a simple commutative diagram where the map $H(\iota)$ is the induced map on the inclusion $\iota: A_- \hookrightarrow A^+$ that itself factors through inclusion to the invisible desideratum $A$. Any nonzero element in the image of $H(\iota)$ must factor through a nonzero homology class in $A$: one can discern the presence of a true hole with *two* imprecise observations and a map between them. This simple observation is greatly extendable to the concept of *persistence*, as will be seen in §5.13-5.15. ⊙

## 4.11 Inverse kinematics

Other applications of homology are *obstructive*. Consider the idealized **robot arm** of Example 1.5 consisting of rotational joints and rigid rods, grasping a part at the end of the arm. In many applications in manufacturing, one must perform *part placement* – manoeuver the arm so as to locate the part in the correct location and/or orientation. If the arm has, say, $N$ rotational joints, then its configuration space is $\mathbb{T}^N$ (as is common when mathematicians study robot arms, the mechanisms are considered insubstantial, and no thought is wasted on the problem of self-intersection). The critical issue is that of understanding the effect of the arm on the part grasped at the end. Consider, *e.g.*, the orientation problem: given a desired part orientation, is there a sequence of rotations to realize it?

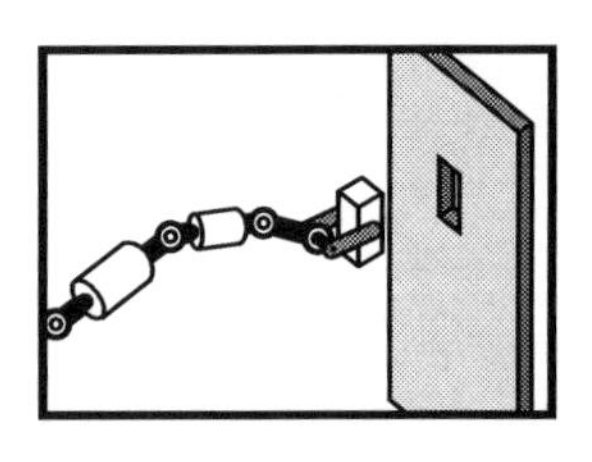

One topologizes the problem [165]: consider the **kinematic map** $\kappa\colon \mathbb{T}^N \to SO_3$ taking the ordered sequence of rotations to the net orientation of the part at the end of the arm. The critical issue is inverting the kinematic map. Given a part orientation in $SO_3$, is there a sequence of rotations to realize it? Certainly, $\kappa$ is onto for $N$ not-too-small and a choice of joint axes which span $\mathbb{R}^3$. It is the inverse kinematic map which is problematic. Given a fixed part orientation, are all nearby part orientations realizable via small changes in the rotations required? This local problem seems to be solvable; the global version is not.

**Proposition 4.22.** *There is no continuous section to $\kappa$. That is, there is no map $s\colon SO_3 \to \mathbb{T}^N$ satisfying $\kappa \circ s = \mathrm{Id}$.*

**Proof.** Consider a putative section $s\colon SO_3 \to \mathbb{T}^N$ with $\kappa \circ s = \mathrm{Id}$. Then on $H_1$ in $\mathbb{Z}$ coefficients one has, from Examples 4.6 and 4.10, $H_1(\mathbb{T}^N) \cong \mathbb{Z}^N$ and $H_1(SO_3) \cong \mathbb{Z}_2$. These fit together in a commutative diagram with induced homomorphisms $H(\kappa)$ and $H(s)$ such that $H(\kappa) \circ H(s) = H(\kappa \circ s) = \mathrm{Id}$, an identity map, thanks to functoriality. This yields a contradiction: since $\mathbb{Z}^N$ has no nonzero elements of finite order, $H(s) = 0$ and $H(\kappa) \circ H(s) = 0$.

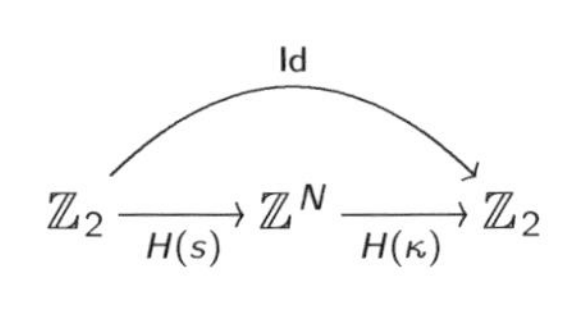


⊙

This result means that a *continuous* assignment of robot arm rotation angles as a function of part orientation is impossible. This decidedly non-intuitive result is a direct consequence of the algebraic topology of configuration spaces, as revealed by $H_1$. Note the role of luck in the appearance of $\mathbb{Z}_2$ in the homology with $\mathbb{Z}$-coefficients: torsion is not always available. Consider the case where the part grasped by the robot arm is rotationally symmetric about the last axis, such as a bolt or pin to be inserted. The desideratum is no longer an orientation in $SO_3$ but rather a point in $\mathbb{S}^2$, the direction in which the last axis of the part points.

Repeating the argument of Proposition 4.22 with the modified kinematic map $\kappa'\colon \mathbb{T}^N \to \mathbb{S}^2$ leads to frustration, as $H_1(\mathbb{S}^2) = 0$. There is no analogous contradiction in assuming that the map $H(\kappa') \circ H(s')$ is the identity, since it is the identity homomorphism on $H_1(\mathbb{S}^2) = 0$. Of course, one does not conclude that the inverse kinematic map $s' \colon \mathbb{S}^2 \to \mathbb{T}^N$ necessarily exists. In fact, $s'$ does not exist.

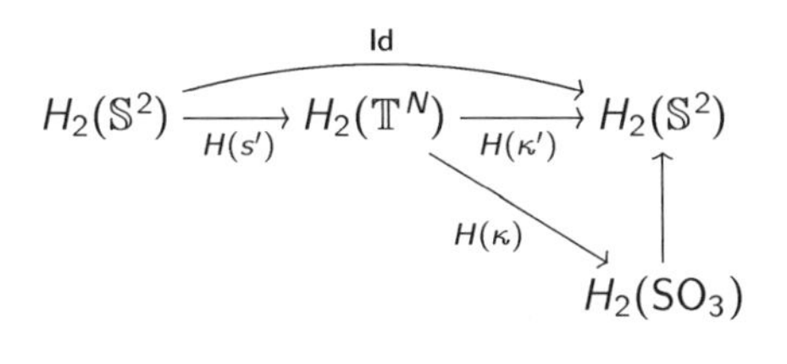


To see this, build the commutative diagram on $H_2$, using $H(s')$ and $H(\kappa')$. Here, one notes that $\kappa'$ factors through the original kinematic map $\mathbb{T}^n \xrightarrow{\kappa} \mathrm{SO}_3$, since the joint angles give the part a true orientation in $\mathrm{SO}_3$. Ignoring all but the last axis in the frame gives a projection map $\mathrm{SO}_3 \to \mathbb{S}^2$. The top-row composition $H(\kappa') \circ H(s')$ cannot be the identity map on $H_2(\mathbb{S}^2)$, since $H_2(\mathrm{SO}_3) \cong 0$ while $H_2(\mathbb{S}^2) \neq 0$. Again, functoriality reveals what individual homology groups do not. There is no continuous assignment of rotation angles to a directional axis for a robot arm manipulating a part.

## 4.12 Winding number and degree

It is helpful to think of an induced map on homology as akin to a winding or linking: to what extent are the holes of $X$ wound about the holes of $Y$ via $f \colon X \to Y$? The full complexity of homomorphisms between abelian groups gives an algebraic picture of the wrapping and winding that maps can execute. Indeed, induced homomorphisms are the right way to express the classical notion of winding numbers.

A continuous simple closed curve $\gamma \colon \mathbb{S}^1 \to \mathbb{R}^2$ in the plane separates the plane into two connected regions: $\dim H_0(\mathbb{R}^2 - \gamma(\mathbb{S}^1)) = 2$. The **mod-2 winding number** of $\gamma$ about a point $p \in \mathbb{R}^2$ not in the image of $\gamma$ is, intuitively, the number which represents whether $p$ is inside (1) or outside (0) the image of $\gamma$. Of the many definitions the reader may have seen (either involving miraculous integrals or the clever counting of intersections), the best is via homology. For $\gamma \colon \mathbb{S}^1 \to \mathbb{R}^2 - p$, consider the induced homomorphism:

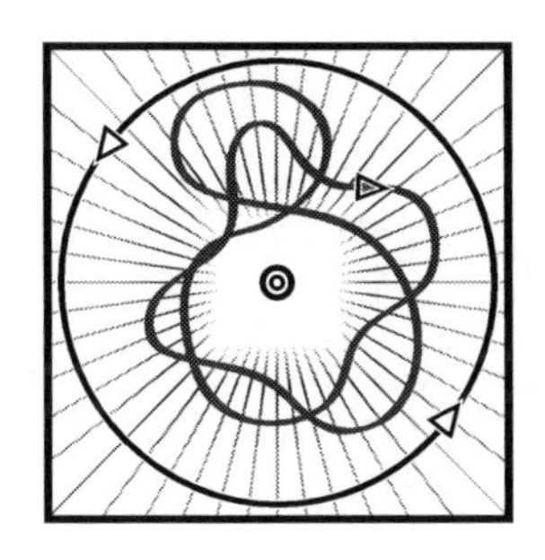

$$H(\gamma) \colon H_1(\mathbb{S}^1) \to H_1(\mathbb{R}^2 - p).$$

Both the domain and codomain of $H(\gamma)$ are of rank one, and the map $H(\gamma)$ is therefore multiplication by a constant: that constant $\deg(\gamma)$ is the **winding number** of $\gamma$ about $p$ (in $\mathbb{F}_2$ or $\mathbb{Z}$ depending on the coefficients used). There is no need to restrict to smooth or non-self-intersecting curves: any map $\gamma \colon \mathbb{S}^1 \to \mathbb{R}^2 - p$ determines a homology class. The winding number measures the algebraic number of times the image of $\gamma$ wraps about $p$, with a choice of orientations (both of $\gamma$ and $\mathbb{R}^2 - p$) determining the sign.

From the fact that $H_n\mathbb{S}^n \cong \mathbb{Z}$, any self-map of $\mathbb{S}^n$ induces a homomorphism on $H_n$ which is multiplication by an integer: this is the **degree** of the map. The resulting degree theory is an important classical topic, some of the important points of which are as follows: for maps $\mathbb{S}^n \to \mathbb{S}^n$,

1. deg is a homotopy invariant;
2. $\deg(f \circ g) = (\deg f)(\deg g)$;
3. $\deg \mathrm{Id} = 1$;
4. $\deg a = (-1)^n$ for the antipodal map $a$ on $\mathbb{S}^n$; and
5. $\deg f = \deg g$ if and only if $f \simeq g$ [the **Hopf Theorem**].

**Example 4.23 (Index theory and vector fields)** ⊚

The index theory for vector fields introduced in §3.3 used line integrals and Green's Theorem to obtain a definition for $\mathbb{R}^2$. This recourse to vector calculus is appropriate only for beginners. Let $V$ be a vector field on an $n$-manifold $M$ with isolated fixed point $p$, and let $B_p$ be a sufficiently small ball about $p$. The **index** of $V$ at $p$, $\mathfrak{I}_V(p)$, is defined to be the degree of the map $V|_{\partial B_p}\colon \partial B_p \cong \mathbb{S}^{n-1} \to \mathbb{R}^n{-}0 \simeq \mathbb{S}^{n-1}$, where $V$ is represented in local coordinates on $B_p$. This index is well-defined, since $p$ is an isolated fixed point. The reader may easily show by the properties of degree that the index is independent of the neighborhood $B_p$ chosen, so long as $B_p \cap \mathrm{Fix}(V) = p$. ⊚

**Example 4.24 (Linking number)** ⊚

One can define a degree for any map $f\colon M \to N$ between oriented compact $n$-manifolds via $H(f)\colon H_n(M) \to H_n(N)$, since, for $M$ and $N$, $H_n \cong \mathbb{Z}$. One classical application of this type of degree is in *knot theory*. A **knot** is an embedding (a smooth injective map) $\mathbb{S}^1 \hookrightarrow \mathbb{S}^3$. It is clear that there are many inequivalent ways to tie a simple closed curve in $\mathbb{R}^3$: §8.3 will show how to distinguish some of them. Degree can be used to characterize how two disjoint oriented knots entwine or link one another. To define the **linking number** of two disjoint oriented knots, $K_1$ and $K_2$, one can parameterize each knot in Euclidean coordinates as $\gamma_1, \gamma_2\colon \mathbb{S}^1 \to \mathbb{R}^3 \subset \mathbb{S}^3$ and set $\ell k(K_1, K_2)$ to be the degree of the map $\phi\colon \mathbb{T}^2 \to \mathbb{S}^2$ given by:

$$\phi\colon \theta_1, \theta_2 \mapsto \frac{\gamma_2(\theta_2) - \gamma_1(\theta_1)}{\|\gamma_2(\theta_2) - \gamma_1(\theta_1)\|}. \tag{4.11}$$

The reader may observe that this quantity is invariant of the parametrization chosen, so long as orientation is respected. Linking number has been found useful in a variety of contexts, from flowlines in fluid- and magnetohydro-dynamics [16] to chemistry [121] and DNA strands [285]: see §8.3. ⊚

**Example 4.25 (Nematic liquid crystals)** ⊚

Liquid crystals are a mesophase of matter which interpolate between ordered and disordered structure. Besides being of great commercial interest, liquid crystals offer fascinating observable defects whose classification is inherently topological. For concreteness, consider the **nematic** liquid crystals, composed of axisymmetric rod-like molecules whose alignment (or **director field**), under a continuum assumption based on average behavior, is a continuous function of position apart from certain defects or **disclinations** [6]. A topological classification of such defects is intricate, but an initial step uses degree as follows. Motivated by Example 4.23, one sets up a map whose induced action on homology yields an appropriate degree.

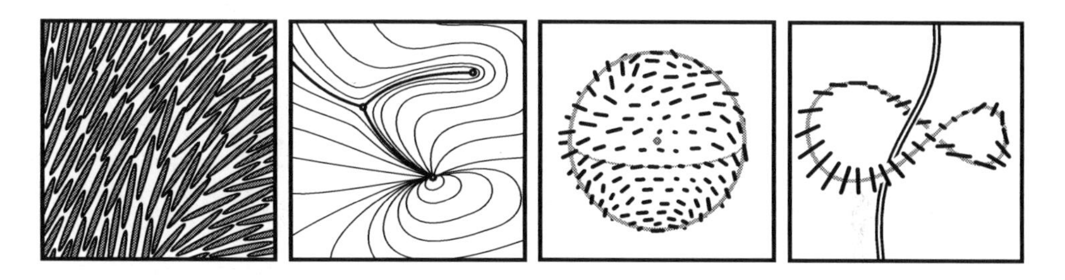

In a nematic liquid crystal that is taken so thin as to be approximated by a plane region to which the molecules align (a so-called **Schlieren texture**), the director field $\xi$ assigns to points in $\mathbb{R}^2$ a planar direction in $\mathbb{P}^1$, well-defined and continuous off the set of singular point-defects. Note that, although $\mathbb{P}^1$ is homeomorphic to $\mathbb{S}^1$, there is no "arrow" and a half-turn suffices to match directors. To assign an index to a defect at $x \in \mathbb{R}^2$, choose an oriented small loop $\gamma$ encircling $x$ and consider the restriction $\xi\colon \gamma \to \mathbb{P}^1 \cong \mathbb{S}^1$. The resulting index $\mathcal{I}_\xi(x) \in \mathbb{Z}$ is an undirected version of that used for vector fields. In the literature, one often divides this integer quantity by 2 so as to obtain agreement with the vector field index when the director field is orientable.

In the case of a nematic in $\mathbb{R}^3$, the director field $\xi$ takes values in $\mathbb{P}^2$. Following the planar case, choose a sufficiently small ball $B$ around an isolated point-defect $x$ and consider the restriction of $\xi\colon \partial B \cong \mathbb{S}^2 \to \mathbb{P}^2$ and its induced map $H(\xi)\colon H_2(\mathbb{S}^2) \to H_2(\mathbb{P}^2)$. From Example 4.10, $H_2(\mathbb{P}^2; \mathbb{Z}) = 0$, and the (integer-valued) degree is trivial; however, passing to $\mathbb{F}_2$ coefficients for $H_2$ yields a well-defined index $\mathcal{I}_\xi(x) \in \mathbb{F}_2$, since $H_2(\mathbb{P}^2; \mathbb{F}_2) \cong \mathbb{F}_2$.

For singularities occurring along a disclination curve, index is measured by a loop $\gamma \cong \mathbb{S}^1$ locally linking the disclination. The resulting map on homology $H(\xi)\colon H_1(\mathbb{S}^1) \to H_1(\mathbb{P}^2)$ is again going to return a degree $\mathcal{I}_\xi \in \mathbb{F}_2$, which, like that of a point-singularity, is sensitive to orientability of the director field about the disclination. ⊚

## 4.13 Fixed points and prices

The following fixed point theorem was one of the earliest triumphs of algebraic topology.

**Theorem 4.26 (Brouwer Fixed Point Theorem).** *Every self-map of the closed disc $\mathbb{D}^n$ has a fixed point.*

As might be expected for so primal a result, there are numerous proofs. The following is of the classical variety.

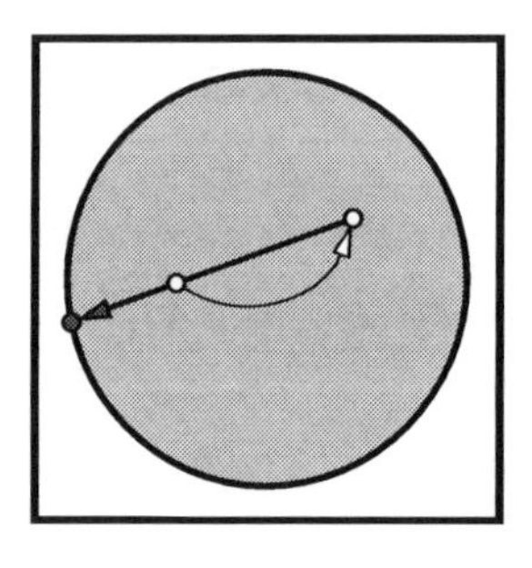

**Proof.** Assume that $f\colon \mathbb{D}^n \to \mathbb{D}^n$ is continuous and fixed-point-free. Then, since $f(x) \neq x$, for any $x$, there is a well-defined point $r(x) \in \partial\mathbb{D}^n$ given by following the ray starting at $f(x)$ and passing through $x$ unto the boundary $\partial\mathbb{D}^n$. This implicitly defines $r\colon \mathbb{D}^n \to \partial\mathbb{D}^n$. Note that $r$ is continuous (since $f$ is) and that it is a retraction: $r(x) = x$ for all $x \in \partial\mathbb{D}^n$. Thus, denoting by $\iota : \partial\mathbb{D}^n \hookrightarrow \mathbb{D}^n$ the inclusion, the diagram on the left commutes:

$$\partial\mathbb{D}^n \xrightarrow[\iota]{} \mathbb{D}^n \xrightarrow[r]{} \partial\mathbb{D}^n \quad (\text{composite } \mathrm{Id}) \qquad \mathbb{Z} \xrightarrow[\iota]{} 0 \xrightarrow[r]{} \mathbb{Z} \quad (\text{composite } \mathrm{Id}).$$

Passing to homology in degree $n-1$ (on the right) reveals the contradiction, since $H_{n-1}(\partial\mathbb{D}^n) \cong \mathbb{Z}$ but $H_{n-1}(\mathbb{D}^n) = 0$: contradiction. ⊙

The reader will likely have seen applications of fixed-point theorems (the efficacy of "YOU ARE HERE" signs, perhaps). Infinite-dimensional versions (*e.g.*, the *Schauder Fixed Point Theorem* for Banach spaces) are important in proving the existence of solutions to differential equations or the termination of certain algorithms. Fixed point theorems for multi-valued mappings (*e.g.*, the *Kakutani Fixed Point Theorem*) are important in proving existence of Nash equilibria in game theory (see §5.11). Perhaps the most well-known applications are in Economics.

**Example 4.27 (Prices)** ⊚

Equilibrium theory – a signature achievement of 20th century economics – asserts that market prices exist. For a simple, explicit example, consider the following result on the existence of equilibrium price distributions in an economy, following Arrow and Debreu [18]. Consider an economy of $N$ items for sale. The space of possible prices is $[0,\infty)^N$. As a way of working with price ratios (and imposing compactness), assume that not all prices are zero (!) and normalize the price space to the closed unit $(N-1)$-simplex $\Delta \subset [0,\infty)^N$ by dividing each vector by the sum of its components. Thus, any point of of $\Delta$ represents a complete set of price ratios in the economy.

Assume a finite set of customers, each with a fixed demand function $D_\alpha\colon \Delta \to \mathbb{R}^N$ and a finite set of suppliers, each with a fixed supply function $S_\beta\colon \Delta \to \mathbb{R}^N$. Both supply and demand have ranges which can be positive or negative. For each choice of price ratio $p \in \Delta$, consider the *excess demand* function $Z\colon \Delta \to \mathbb{R}^N$ given by:

$$Z(p) := \sum_\alpha D_\alpha(p) - \sum_\beta S_\beta(p).$$

Positive components of $Z$ connote an excess of demand, while negative values connote an excess of supply. Assume (for the sake of Mathematics) that $Z$ is continuous and (for the sake of Economics) that $p \cdot Z(p) = 0$: this is known as *Walras' Law* and it represents a balance between net income and net expenditure in the system. Then, it can be shown that the system possesses a **price equilibrium**, meaning a point $p \in \Delta$ such that $Z(p) \leq 0$, meaning that all items with nonzero price have demand equal to supply (since $p \cdot Z(p) = 0$ and $p \geq 0$).

This follows from Theorem 4.26. Consider the map $f\colon \Delta \to \Delta$ given by sending $p$ to $\max\{0, p + Z(p)\}/C$, where the denominator $C := \sum_i \max\{0, p+Z(p)\}$ is chosen to project to $\Delta$. Since $p \cdot Z(p) = 0$, one has $p \cdot (p + Z(p)) = ||p||^2 > 0$, making $C > 0$ and $f$ well-defined. There is a fixed point, $p^*$, of $f$, and this satisfies $p^* = \max\{0, p^* + Z(p^*)\}/C$. Assume the case where $p^* > 0$ (all components are positive); then, since $p^* = (p^* + Z(p^*))/C$ and $Z(p^*)$ is orthogonal to $p^*$, it follows that $C = 1$ and, hence, $Z(p^*) = 0$. On the other hand, if some items have equilibrium price $p_i^* = 0$, then, since $p_i^* + Z_i(p^*) \leq 0$, it follows that $Z_i(p^*) \leq 0$, and one projects to the subspace of items with nonzero price. ⊚

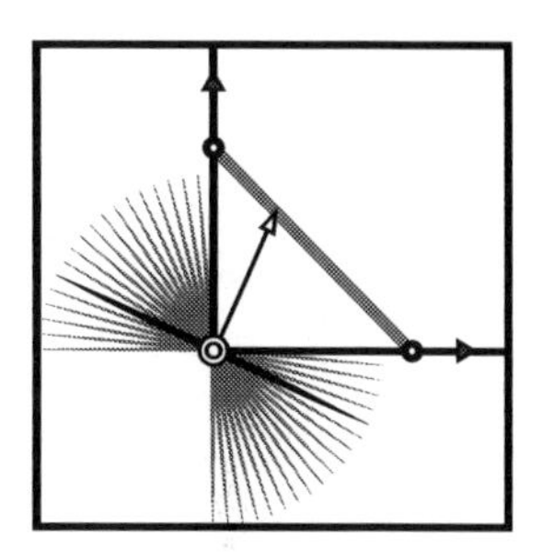

This argument is inelegant (the dot product condition is suspiciously rigid). A more general and beautiful result on equilibria in multi-agent systems – the Nash equilibrium theorem – will be attainable in §5.11 after learning better tools.

## Notes

1. Linear algebra concerns itself very little with vectors and vector spaces: it is with linear transformations that the subject comes to life. The homological algebra hinted at in this chapter is the natural evolution in linear transformations from the single to the sequential.
2. The reader will note, perhaps with displeasure, that computational issued have not been discussed: it is too large and fluid a topic. See [104, 186, 305] for an introduction to the many techniques available for fast computation of homology, some of which are distributable (a near-necessity for realistic scientific applications). The short version seems to be that cellular homology is computable in time near-linear in the number of cells. The curse of dimensionality (high-dimensional complexes have many, many cells) remains a challenge in some contexts, as does the problem of storage (space complexity).
3. Cellular homology has been treated lightly, with an implicit assumption of a regular cell complex, such as with simplicial or cubical complexes. For a general cell complex, the definition of incidence number $[\sigma : \tau]$ is a bit more involved, requiring the tools of §5.5.
4. The Künneth formula of Example 4.6 holds for singular or cellular homology in field coefficients, for spaces with finite dimensional homology groups. There is a more general version which accommodates general coefficients and infinite spaces. In the case of $\mathbb{Z}$ coefficients and for spaces with torsion, it is exact but algebraically delicate: see, *e.g.*, [176].

5. §4.5 is the first hint that taking advantage of *negative*-dimensional chains and homology is permissible. There are homology theories in which $H_\bullet$ takes on significant meaning for negative gradings: see Example 7.18.
6. The Excision Theorem 4.14 is more foundational than the brief treatment of this chapter might indicate. Eilenberg and Steenrod, in their axiomatization of homology, have the excision property as one of the five axioms of a homology theory.
7. This text's treatment of winding numbers begins with an invocation of the classic Jordan Curve Theorem: *Let $A$ be a subset of $\mathbb{S}^n$ homeomorphic to $\mathbb{S}^k$. Then $\tilde{H}_p(\mathbb{S}^n - A) \cong \mathbb{Z}$ for $p = n - k - 1$ and $0$ else.* Note that there is no assumption of tameness in this statement.
8. The linking numbers of Example 4.24 are just the beginning of the algebraic topology of knots and links – a vast and beautiful subject. Knot theory has found several applications, notably in physics, biology, fluid dynamics, and differential equations: see Examples 6.25, 8.10, and 8.11.
9. The treatment of singularities in nematic liquid crystals in Example 4.25 is incomplete. Very interesting phenomena occur as the field is continuously changed and disclination lines are enticed to collide or entangle, or as point-defects move so as to encircle a disclination line [6]. This will be revisited in §8.5.
10. Baryshnikov has announced applications of homology and linking numbers to problems of **caging** in robotics, where, given a fixed geometric object in $\mathbb{R}^2$, one wishes to choose the smallest discrete set $D \subset \mathbb{R}^2$ that prevents the object from being able to be moved 'to infinity' by means of Euclidean motions (translations and rotations) in $\mathbb{R}^2 - D$ [259]. The related problem of performing motion-planning in robotics about obstacles by means of specifying homology classes leads to computable optimization methods [35].
11. The applications of fixed point theorems to economics is worthy of a text in itself. Example 4.27 is of limited value as many important considerations (time-dependence, absolute-versus-relative price, zero-price demand, etc.) have been ignored. Nor is it to be denied that equilibria can be proved (and in some cases computed) via combinatorial means. The power of the fixed point method is the ability to analyze large collections of agents with diverse goals.

Chapter 5

# Sequences

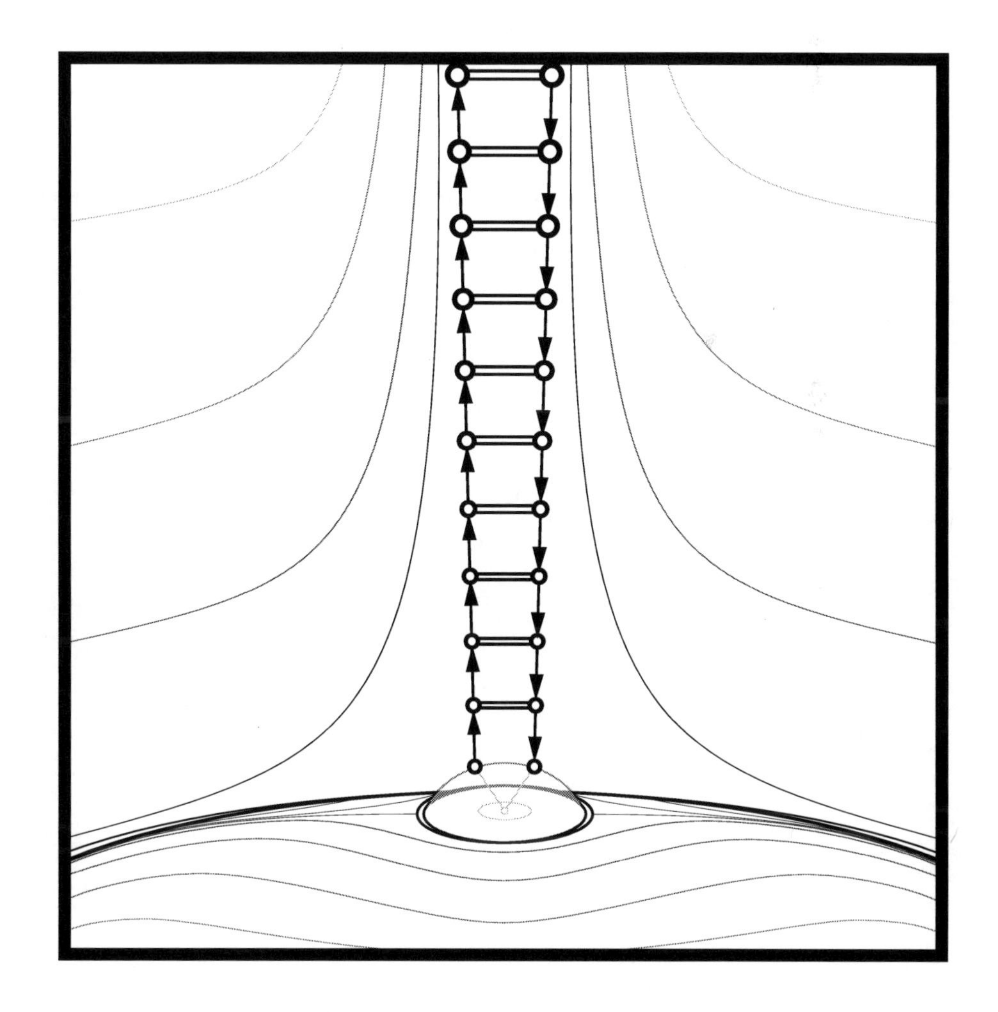

Homology takes as its input a chain complex – a hierarchical assembly line of parts – and returns its global features. The elemental tools for analyzing homology are likewise linelike devices: **exact sequences**. When outfitted with functoriality, these sequences assemble into the homological engines of inference.

## 5.1 Homotopy invariance

Homology begins by replacing topological spaces with complexes of algebraic objects (vector spaces, abelian groups, or modules, depending on one's preferences). Other topological notions – continuous functions, homeomorphisms, homotopies, *etc.* – also have analogues at the level of chain complexes. Generalizing from the pattern set in §4.10, define a **chain map** to be any graded homomorphism $\varphi_\bullet\colon \mathcal{C} \to \mathcal{C}'$ between chain complexes that respects the grading and commutes with the boundary maps. This is best expressed in the form of a commutative diagram as per (4.10):

$$\begin{array}{ccccccccc} \cdots \longrightarrow & C_{n+1} & \xrightarrow{\partial} & C_n & \xrightarrow{\partial} & C_{n-1} & \xrightarrow{\partial} & \cdots \\ & \downarrow{\scriptstyle \varphi_\bullet} & & \downarrow{\scriptstyle \varphi_\bullet} & & \downarrow{\scriptstyle \varphi_\bullet} & & \\ \cdots \longrightarrow & C'_{n+1} & \xrightarrow{\partial'} & C'_n & \xrightarrow{\partial'} & C'_{n-1} & \xrightarrow{\partial'} & \cdots \end{array} \tag{5.1}$$

Commutativity means that homomorphisms are path-independent in the diagram: $\varphi_\bullet \circ \partial = \partial' \circ \varphi_\bullet$. Chain maps are the analogues of continuous maps, since, via respect for the boundary operators, neighbors are sent to neighbors. The appropriate generalization of a homeomorphism to chain complexes is therefore an invertible chain map – one which is an isomorphism for all $C_n \to C'_n$. Clearly, a homeomorphism $f\colon X \to Y$ induces a chain map $f_\bullet\colon \mathcal{C}^{\text{sing}}_X \to \mathcal{C}^{\text{sing}}_Y$ which is an isomorphism. As such, $H^{\text{sing}}_\bullet(X) \cong H^{\text{sing}}_\bullet(Y)$.

The extension to homotopy is more subtle. Recall that $f, g\colon X \to Y$ are homotopic if there is a map $F\colon X \times [0,1] \to Y$ which restricts to $f$ on $X \times \{0\}$ and to $g$ on $X \times \{1\}$. A **chain homotopy** between chain maps $\varphi_\bullet, \psi_\bullet\colon \mathcal{C} \to \mathcal{C}'$ is a homomorphism $F\colon \mathcal{C} \to \mathcal{C}'$ sending $n$-chains to $(n+1)$-chains so that $\partial' F - F\partial = \varphi_\bullet - \psi_\bullet$:

$$\begin{array}{ccccccccc} \cdots \longrightarrow & C_{n+1} & \xrightarrow{\partial} & C_n & \xrightarrow{\partial} & C_{n-1} & \xrightarrow{\partial} & \cdots \\ {\scriptstyle F} \swarrow & \psi_\bullet \downarrow \varphi_\bullet & {\scriptstyle F} \swarrow & \psi_\bullet \downarrow \varphi_\bullet & {\scriptstyle F} \swarrow & \psi_\bullet \downarrow \varphi_\bullet & {\scriptstyle F} \swarrow & \\ \cdots \longrightarrow & C'_{n+1} & \xrightarrow[\partial']{} & C'_n & \xrightarrow[\partial']{} & C'_{n-1} & \xrightarrow[\partial']{} & \cdots \end{array} \tag{5.2}$$

One calls $F$ a map of **degree** $+1$, indicating the upshift in the grading.[1] Note the morphological resemblance to homotopy of maps: a chain homotopy maps each $n$-chain to a $n+1$-chain, the algebraic analogue of a 1-parameter family. The difference between the ends of the homotopy, $\partial' F - F\partial$, gives the difference between the chain maps.

**Lemma 5.1.** *Chain homotopic maps induce the same homomorphisms on homology.*

[1] The overuse of the term *degree* in graphs, maps of spheres, and chain complexes is unfortunate.

**Proof.** Consider $[\alpha] \in H_\bullet(\mathcal{C})$. Assuming $\varphi_\bullet$ and $\psi_\bullet$ are chain homotopic maps from $\mathcal{C}$ to $\mathcal{C}'$,

$$\begin{aligned} H(\varphi_\bullet)[\alpha] - H(\psi_\bullet)[\alpha] &= [(\varphi_\bullet - \psi_\bullet)\alpha] \\ &= [(\partial' F + F\partial)\alpha] \\ &= [\partial'(F\alpha)] + [F(\partial\alpha)] = 0, \end{aligned}$$

since $\alpha$ is a cycle and $\partial'(F\alpha)$ is a boundary. ⊙

The following theorem is proved by constructing an explicit chain homotopy [176]:

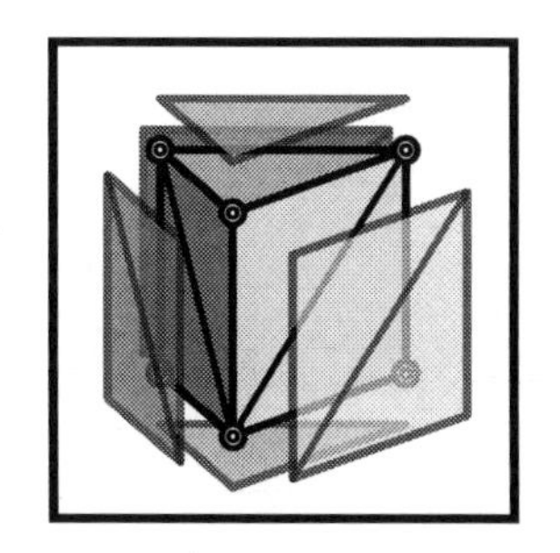

**Theorem 5.2 (Homotopy Invariance of Homology).** *Homotopic maps $f, g\colon X \to Y$ induce chain homotopic maps $f_\bullet, g_\bullet$ from $\mathcal{C}_X^{\mathrm{sing}}$ to $\mathcal{C}_Y^{\mathrm{sing}}$.*

The idea behind the proof is simple. For each singular $n$-simplex $\sigma$, one considers the $F$-image of $\sigma \times [0,1]$ as a family of singular $n$-simplices parameterized by the homotopy. This prism is then triangulated into singular $(n+1)$-simplices that encode the homotopy. It is shown that this chain map $\mathcal{P}$ (called a **prism operator**) is a chain homotopy.

**Corollary 5.3.** *Singular homology is a homotopy invariant of topological spaces.*

## 5.2 Exact sequences

In the analogy between topological spaces and algebraic complexes, there is a special class of complexes that are elementary building blocks. A complex $\mathcal{C} = (C_\bullet, \partial)$ is **exact** when its homology vanishes: $\ker \partial_n = \operatorname{im} \partial_{n+1}$ for all $n$. Exact sequences are as useful and primal as the nullhomologous spaces they mirror.

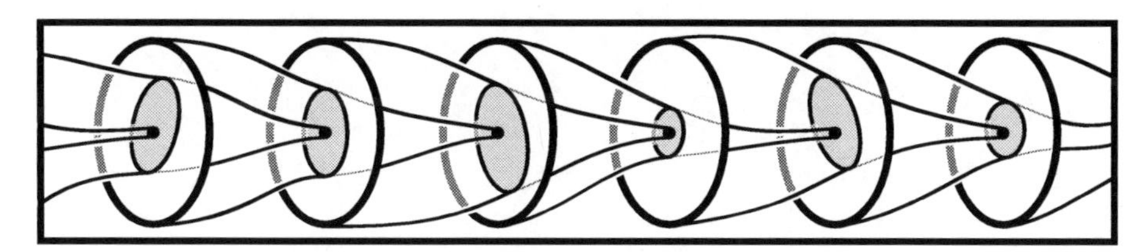

**Example 5.4 (Exact sequence)** ⊚

The following simple examples of exact sequences help build intuition:

1. Two groups are isomorphic, $\mathbf{G} \cong \mathbf{H}$, iff there is an exact sequence of the form:

$$0 \longrightarrow \mathbf{G} \longrightarrow \mathbf{H} \longrightarrow 0$$

2. The 1st Isomorphism Theorem says that for a homomorphism $\varphi$ of $\mathbf{G}$, the following sequence is exact:

$$0 \longrightarrow \ker\varphi \longrightarrow \mathbf{G} \xrightarrow{\varphi} \operatorname{im}\varphi \longrightarrow 0$$

   Such a 5-term sequence framed by zeroes is called a **short exact sequence**. In any such short exact sequence, the second map is injective; the penultimate, surjective.
3. More generally, the kernel and cokernel of a homomorphism $\varphi\colon \mathbf{G} \to \mathbf{H}$ fit into an exact sequence:

$$0 \longrightarrow \ker\varphi \longrightarrow \mathbf{G} \xrightarrow{\varphi} \mathbf{H} \longrightarrow \operatorname{coker}\varphi \longrightarrow 0$$

4. Consider $C = C^\infty(\mathbb{R}^3)$, the vector space of differentiable functions and $\mathcal{X} = \mathcal{X}(\mathbb{R}^3)$, the vector space of $C^\infty$ vector fields on $\mathbb{R}^3$. These fit together into an exact sequence,

$$0 \longrightarrow \mathbb{R} \longrightarrow C \xrightarrow{\nabla} \mathcal{X} \xrightarrow{\nabla\times} \mathcal{X} \xrightarrow{\nabla\cdot} C \longrightarrow 0\,, \tag{5.3}$$

   where $\nabla$ is the gradient differential operator from vector calculus, and the initial $\mathbb{R}$ term in the sequence represents the constant functions on $\mathbb{R}^3$. The exactness of this sequence encodes the fact that *curl-of-grad* and *div-of-curl* vanish, as well as the fact that, on $\mathbb{R}^3$, all curl-free fields are gradients and all div-free fields are curls. This one exact sequence compactly encodes many of the relations of vector calculus.

⊙

The most important examples of exact sequences are those relating homologies of various spaces and subspaces. The critical technical tool for the generation of such weaves an exact thread through a loom of chain complexes.

**Lemma 5.5 (Snake Lemma).** *If $\mathcal{A} = (A_\bullet, \partial)$, $\mathcal{B} = (B_\bullet, \partial)$, and $\mathcal{C} = (C_\bullet, \partial)$ form a short exact sequence of chain complexes,*

$$0 \longrightarrow A_\bullet \xrightarrow{i_\bullet} B_\bullet \xrightarrow{j_\bullet} C_\bullet \longrightarrow 0\,,$$

*then this induces the* **long exact sequence***:*

$$\longrightarrow H_n(\mathcal{A}) \xrightarrow{H(i)} H_n(\mathcal{B}) \xrightarrow{H(j)} H_n(\mathcal{C}) \xrightarrow{\delta} H_{n-1}(\mathcal{A}) \xrightarrow{H(i)}\,. \tag{5.4}$$

*Moreover, the long exact sequence is* **natural***: a commutative diagram of short exact sequences and chain maps*

$$\begin{array}{ccccccccc} 0 & \longrightarrow & A_\bullet & \longrightarrow & B_\bullet & \longrightarrow & C_\bullet & \longrightarrow & 0 \\ & & \downarrow{\scriptstyle f_\bullet} & & \downarrow{\scriptstyle g_\bullet} & & \downarrow{\scriptstyle h_\bullet} & & \\ 0 & \longrightarrow & \tilde{A}_\bullet & \longrightarrow & \tilde{B}_\bullet & \longrightarrow & \tilde{C}_\bullet & \longrightarrow & 0 \end{array}$$

*induces a commutative diagram of long exact sequences*

$$\begin{array}{ccccccccc} \longrightarrow H_n(\mathcal{A}) & \longrightarrow & H_n(\mathcal{B}) & \longrightarrow & H_n(\mathcal{C}) & \xrightarrow{\delta} & H_{n-1}(\mathcal{A}) & \longrightarrow \\ \downarrow{\scriptstyle H(f)} & & \downarrow{\scriptstyle H(g)} & & \downarrow{\scriptstyle H(h)} & & \downarrow{\scriptstyle H(f)} & \\ \longrightarrow H_n(\tilde{\mathcal{A}}) & \longrightarrow & H_n(\tilde{\mathcal{B}}) & \longrightarrow & H_n(\tilde{\mathcal{C}}) & \xrightarrow{\delta} & H_{n-1}(\tilde{\mathcal{A}}) & \longrightarrow \end{array} \quad (5.5)$$

An *exact sequence of chain complexes* means that there is a short exact sequence in each grading, and these short exact sequences fit into a commutative diagram with respect to the boundary operators. The induced **connecting homomorphism** $\delta\colon H_n(\mathcal{C}) \to H_{n-1}(\mathcal{A})$ comes from the boundary map in $\mathcal{C}$ as follows:

1. Fix $[\gamma] \in H_n(\mathcal{C})$; thus, $\gamma \in C_n$.
2. By exactness, $\gamma = j(\beta)$ for some $\beta \in B_n$.
3. By commutativity, $j(\partial\beta) = \partial(j\beta) = \partial\gamma = 0$.
4. By exactness, $\partial\beta = i\alpha$ for some $\alpha \in A_{n-1}$.
5. Set $\delta[\gamma] := [\alpha] \in H_{n-1}(\mathcal{A})$.

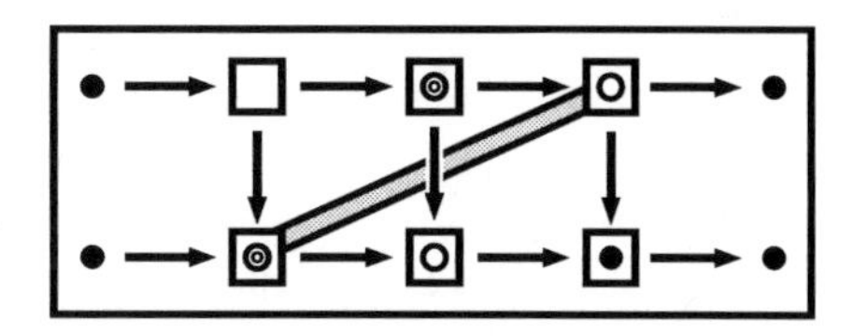

Every topologist, no matter how wedded to geometric intuition, must possess a thorough understanding of the connecting homomorphism. This is perhaps best grasped via animation; a static illustration is a poor substitute, but nevertheless conveys the critical shift in grading that exactness and weaving enacts. The reader should demonstrate that $\delta$ is well-defined and that the resulting long exact sequence is indeed exact. Doing so solidifies the invaluable technique of diagrammatic argument.

## 5.3 Pairs and Mayer-Vietoris

Of the many exact sequences that a topologist must master, two are central: the long exact sequence of a pair, and the Mayer-Vietoris sequence. The former unwraps the relative homology of §4.7. Given $A \subset X$ (a subset in the singular case, or a subcomplex in the cellular), the following sequence of chain complexes is exact:

$$0 \longrightarrow C_\bullet(A) \xrightarrow{i_\bullet} C_\bullet(X) \xrightarrow{j_\bullet} C_\bullet(X, A) \longrightarrow 0\,,$$

where $i\colon A \hookrightarrow X$ is an inclusion and $j\colon (X, \varnothing) \hookrightarrow (X, A)$ is an inclusion of pairs. This yields the **long exact sequence of the pair** $(X, A)$:

$$\longrightarrow H_n(A) \xrightarrow{H(i)} H_n(X) \xrightarrow{H(j)} H_n(X, A) \xrightarrow{\delta} H_{n-1}(A) \longrightarrow\,, \quad (5.6)$$

The connecting homomorphism $\delta$ takes a relative homology class $[\alpha] \in H_n(X, A)$ to the homology class $[\partial\alpha] \in H_{n-1}(A)$. This is an excellent sequence for decomposing homology of quotient spaces in terms of the homological analogues of images and kernels of a projection.

**Example 5.6 (Spheres)** ◎

Computing the homology of the sphere $\mathbb{S}^k$ is a simple application of the long exact sequence of the pair $(\mathbb{D}^k, \partial\mathbb{D}^k)$, made simpler still by using reduced homology:

$$\longrightarrow \tilde{H}_n(\mathbb{D}^k) \xrightarrow{H(j)} H_n(\mathbb{D}^k, \partial\mathbb{D}^k) \xrightarrow{\delta} \tilde{H}_{n-1}(\partial\mathbb{D}^k) \xrightarrow{H(i)} \tilde{H}_{n-1}(\mathbb{D}^k) \longrightarrow .$$

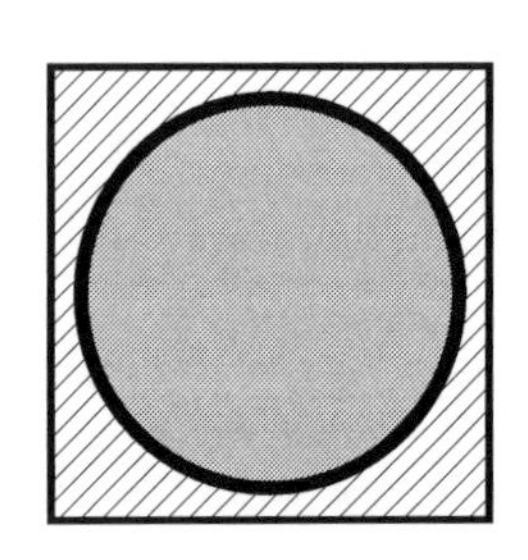

By definition, $\partial\mathbb{D}^k \cong \mathbb{S}^{k-1}$. By excision, $H_\bullet(\mathbb{D}^k, \partial\mathbb{D}^k) \cong \tilde{H}_\bullet(\mathbb{D}^k/\partial\mathbb{D}^k) \cong \tilde{H}_\bullet(\mathbb{S}^k)$. For all $n$, $\tilde{H}_n(\mathbb{D}^k) = 0$. As the first and last terms in the sequence above vanish, exactness yields the recurrence relation

$$\tilde{H}_n(\mathbb{S}^k) \cong \tilde{H}_{n-1}(\mathbb{S}^{k-1}),$$

for all $n$. Beginning with the explicit and trivial computation of $\tilde{H}_n(\mathbb{S}^0) \cong 0$ for $n > 0$ and $\cong \mathbb{Z}$ for $n = 0$, one inducts via the above to show that $\tilde{H}_n(\mathbb{S}^k) \cong \mathbb{Z}$ for $n = k$, and $\cong 0$ else.

◎

The second key sequence is derived from a decomposition of $X$ into subsets (or subcomplexes) $A$ and $B$. In the singular setting, one requires $X = \text{int}(A) \cup \text{int}(B)$; in the cellular case, subcomplexes suffice. Consider the short exact sequence

$$0 \longrightarrow C_\bullet(A \cap B) \xrightarrow{\phi_\bullet} C_\bullet(A) \oplus C_\bullet(B) \xrightarrow{\psi_\bullet} C_\bullet(A + B) \longrightarrow 0,$$

with chain maps $\phi_\bullet \colon c \mapsto (c, -c)$, and $\psi_\bullet \colon (a, b) \mapsto a + b$. The term on the right, $C_\bullet(A + B)$, consists of those chains which can be expressed as a sum of chains on $A$ and chains on $B$. In cellular homology with $A, B$ subcomplexes, $C_\bullet(A + B) \cong C_\bullet(X)$; in the singular category, one shows (via the techniques of Čech homology) that $H_\bullet(A + B) \cong H_\bullet(X)$. In both settings, the resulting long exact sequence yields the **Mayer-Vietoris sequence**:

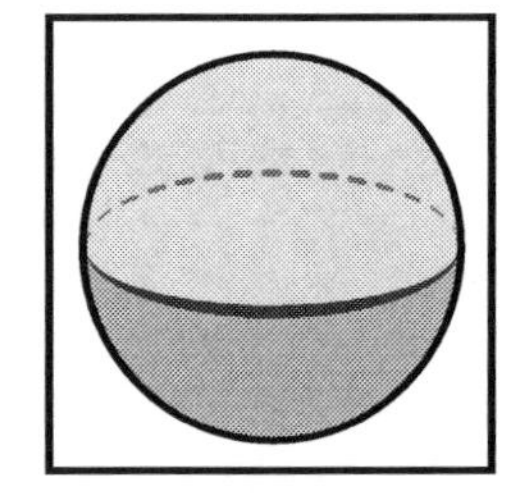

$$\longrightarrow H_n(A \cap B) \xrightarrow{H(\phi)} H_n(A) \oplus H_n(B) \xrightarrow{H(\psi)} H_n(X) \xrightarrow{\delta} H_{n-1}(A \cap B) \longrightarrow$$

The connecting homomorphism decomposes a cycle in $X$ into a sum of chains in $A$ and $B$, then takes the boundary of one of these chains in $A \cap B$. This sequence captures the additivity of homology, *cf.* Lemma 3.8.

**Example 5.7 (Spheres, redux)** ◎

The computation of $H_\bullet(\mathbb{S}^k)$ can be carried out via Mayer-Vietoris as follows. Let $A$ and $B$ be upper and lower hemispheres, homeomorphic to $\mathbb{D}^k$, intersecting at an equatorial $\mathbb{S}^{k-1}$.

$$\longrightarrow \tilde{H}_n(\mathbb{D}^k)\oplus\tilde{H}_n(\mathbb{D}^k) \xrightarrow{H(\psi)} \tilde{H}_n(\mathbb{S}^k) \xrightarrow{\delta} \tilde{H}_{n-1}(\mathbb{S}^{k-1}) \xrightarrow{H(\phi)} \tilde{H}_{n-1}(\mathbb{D}^k)\oplus\tilde{H}_{n-1}(\mathbb{D}^k) \longrightarrow$$

As $\tilde{H}_\bullet(\mathbb{D}^k)\cong 0$, one obtains by exactness that $\delta\colon \tilde{H}_n(\mathbb{S}^k)\cong\tilde{H}_{n-1}(\mathbb{S}^{k-1})$ for all $n$ and all $k$. Thus, again, via $\tilde{H}_\bullet\mathbb{S}^0$, one has immediately that $\tilde{H}_n(\mathbb{S}^k)\cong 0$ unless $n=k$, where it equals $\mathbb{Z}$. ⊚

There are many other exact sequences, only a few of which this chapter will unfurl. For future use, note the existence of relative versions of the two sequences above. The long exact sequence of a **triple** $(X, A, B)$, where $B\subset A\subset X$, is derived from the short exact sequence of chains:

$$0 \longrightarrow C_\bullet(A,B) \xrightarrow{i_\bullet} C_\bullet(X,B) \xrightarrow{j_\bullet} C_\bullet(X,A) \longrightarrow 0\,,$$

where, as before, $i_\bullet$ and $j_\bullet$ are induced by inclusions on pairs. In the case of Mayer-Vietoris, if one decomposes the pair $(X,Y)$ for $Y\subset X$ as $X=A\cup B$ and $Y=C\cup D$, with $C\subset A$ and $D\subset B$, then the following is exact:

$$0 \longrightarrow C_\bullet(A\cap B, C\cap D) \xrightarrow{\phi_\bullet} C_\bullet(A,C)\oplus C_\bullet(B,D) \xrightarrow{\psi_\bullet} C_\bullet(A+B, C+D) \longrightarrow 0,$$

As before, when $X$ lies in the union of interiors of $A$ and $B$ (likewise with $Y$ in the union of $C$ and $D$), then the penultimate term in the sequence becomes $H_\bullet(X,Y)$ when passing to homology. Section 5.12 will put this relative tool to use.

## 5.4 Equivalence of homology theories

This chapter slowly builds the argument that even elementary homological algebra is a powerful upgrade to the basic linear algebra so useful in applied mathematics. Exactness and commutativity are two such simple tools: much more is available. Consider the following diagrammatic lemma from homological algebra.

**Lemma 5.8 (The 5-Lemma).** *Given a commutative diagram of abelian groups of the form*

$$\begin{array}{ccccccccc} \bullet & \longrightarrow & \bullet & \longrightarrow & \bullet & \longrightarrow & \bullet & \longrightarrow & \bullet \\ \cong\downarrow & & \cong\downarrow & & \downarrow & & \downarrow\cong & & \downarrow\cong \\ \bullet & \longrightarrow & \bullet & \longrightarrow & \bullet & \longrightarrow & \bullet & \longrightarrow & \bullet \end{array} \qquad (5.7)$$

*whose top and bottom rows are exact, and whose four* outer *vertical maps are isomorphisms; the middle vertical map is an isomorphism as well.*

This lemma is extremely useful: one example suffices. This text handles competing homology theories glibly, with the justification that they usually agree.[2] This almost always can be shown using the 5-Lemma and induction.

[2]Exotic spaces on which they disagree are not of primary importance in most applications.

**Theorem 5.9.** *On simplicial complexes, singular and simplicial homology are isomorphic.*

**Proof.** Choose a filtration $X_i \subset X_{i+1} \subset \cdots$ for $X$ that adds one simplex per step and induct on this sequence order. On the 0-skeleton of $X$, the isomorphism is clear. Assume that $\sigma$ is the $k$-simplex which when added to $X_i$ yields $X_{i+1}$. This $\sigma$ is glued to $X_i$ along its boundary $\partial\sigma$, homeomorphic to the sphere $\mathbb{S}^{k-1}$. The Mayer-Vietoris sequences for $(X_i, \sigma)$ in simplicial and singular homology fit together in a commutative diagram,

$$\begin{array}{ccccccccc}
H_n^{\text{simp}}\partial\sigma & \longrightarrow & H_n^{\text{simp}}X_i \oplus H_n^{\text{simp}}\sigma & \longrightarrow & H_n^{\text{simp}}X_{i+1} & \longrightarrow & H_{n-1}^{\text{simp}}\partial\sigma & \longrightarrow & H_{n-1}^{\text{simp}}X_i \oplus H_{n-1}^{\text{simp}}\sigma \\
\downarrow \cong & & \downarrow \cong & & \downarrow & & \downarrow \cong & & \downarrow \cong \\
H_n^{\text{sing}}\partial\sigma & \longrightarrow & H_n^{\text{sing}}X_i \oplus H_n^{\text{sing}}\sigma & \longrightarrow & H_n^{\text{sing}}X_{i+1} & \longrightarrow & H_{n-1}^{\text{sing}}\partial\sigma & \longrightarrow & H_{n-1}^{\text{sing}}X_i \oplus H_{n-1}^{\text{sing}}\sigma
\end{array}$$

with vertical arrows induced from the map interpreting a simplicial chain as a singular chain. By induction and previous computations of the homologies of balls and spheres, four of the five vertical maps are isomorphisms. The 5-Lemma completes the induction step and the proof. ⊙

The same result holds for other homology theories, such as Čech and cellular, assuming the appropriate defining structures (covers, cell structures, *etc.*) exist. The proof of Theorem 5.9 works for cellular homology of a regular cell complex. In the general case, more care concerning the definitions is required. In all these cases, the isomorphism is *natural*, meaning that, as in the case of the long exact sequences of §5.2, a map $f\colon X \to Y$ between spaces induce the same homomorphisms on homology, independent of which theory (singular, cellular, *etc.*) is used, so long as these homologies are well-defined. This sameness is expressed as a commutative diagram.

$$\begin{array}{ccc}
H_\bullet^{\text{cell}}X & \xrightarrow{H(f)} & H_\bullet^{\text{cell}}Y \\
\downarrow \cong & & \cong \downarrow \\
H_\bullet^{\text{sing}}X & \xrightarrow[H(f)]{} & H_\bullet^{\text{sing}}Y
\end{array}$$

## 5.5 Cellular homology, redux

The treatment of cellular homology in Chapter 4 was incomplete, especially in regards to defining the incidence number $[\sigma\colon \tau]$ for a pair of cells $\sigma \trianglelefteq \tau$, and showing that the boundary operator in Equation (4.6) yields a chain complex. Moreover, the boundary operator was only defined for *regular* cell complexes, and a formal definition of induced orientation was never given at all outside of the simplicial setting. These issues can now be rectified.

Let $X$ be a finite-dimensional cell complex. Choose an orientation for each cell $\tau$: recall from Example 4.18 that this is a choice of generator for $H_n(\tau, \partial\tau; \mathbb{Z})$ for $n = \dim \tau$. To define the induced orientation on the boundary of the cell (without using one's right hand), use the connecting homomorphism $\delta\colon H_n(\tau, \partial\tau) \to H_{n-1}(\partial\tau)$ from the long exact sequence on the pair $(\tau, \partial\tau)$. Recall from §4.3 that the incidence number $[\sigma\colon \tau]$ of a face $\sigma \trianglelefteq \tau$ records (for a *regular* cell complex) whether the orientations on $\partial\tau$ and $\sigma$ agree $(+1)$ or disagree $(-1)$. This can now be both interpreted and extended to the general cellular setting as a degree:

$$[\sigma\colon \tau] := \deg\left(H_n(\tau, \partial\tau) \xrightarrow{\delta} H_{n-1}(\partial\tau) \xrightarrow{H(q)} H_{n-1}(\sigma/\partial\sigma)\right), \tag{5.8}$$

where for $\tau$ an $n$-cell, $q\colon \partial\tau \to \sigma/\partial\sigma$ is the map that quotients out the complement of $\sigma$ in $X^{(n-1)}$. In short, *incidence number is the degree of the attaching map on the boundary.* For a non-regular cell complex, this number may be nonzero: witness the cell structure on $\mathbb{P}^2$ that has one cell in each dimension $0, 1, 2$, and an attaching map from the 2-cell to the 1-skeleton of degree 2.

What remains is to show that the incidence numbers cancel to give the cellular boundary operator as per Equation (4.7). This requires a deeper look at what the quotient map in (5.8) is. Consider $X^{(n)} \subset X$ the $n$-skeleton comprised of all cells of dimension less than or equal to $n$. To analyse how the $n$-cells are glued onto the $(n-1)$-skeleton, one focuses on the long exact sequence on the pair $(X^{(n)}, X^{(n-1)})$ in *singular* homology near grading $n$:

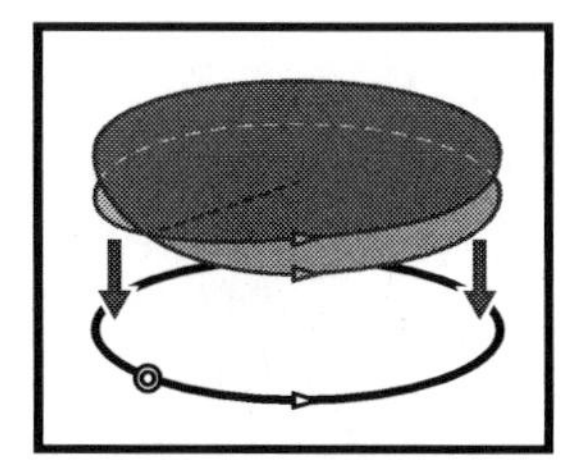

$$\longrightarrow H_n X^{(n)} \xrightarrow{H(j)} H_n(X^{(n)}, X^{(n-1)}) \xrightarrow{\delta} H_{n-1}X^{(n-1)} \xrightarrow{H(i)} H_{n-1}X^{(n)} \longrightarrow .$$

The critical observation is this: $H_n(X^{(n)}, X^{(n-1)}) \cong H_n(X^{(n)}/X^{(n-1)}) \cong C_n^{\mathrm{cell}}(X)$, since the quotient is a wedge of spheres of dimension $n$, one for each $n$-cell of $X$. This allows for a definition of the cellular chain complex via weaving together the long exact sequences of incident skeletal pairs:

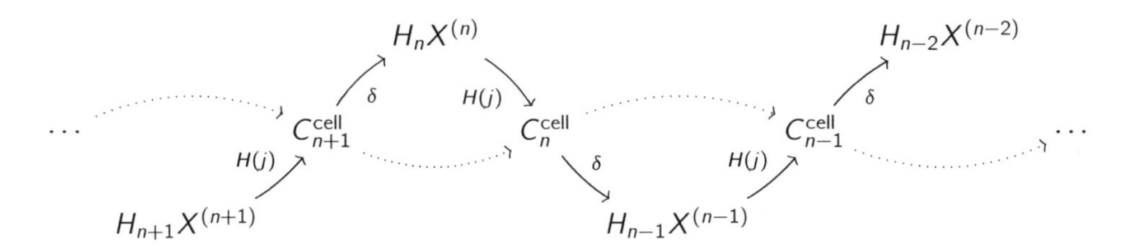


One defines the cellular boundary operator $\partial^{\mathrm{cell}}$ from this diagram as the composition $\partial^{\mathrm{cell}} := H(j) \circ \delta$. This gives an immediate proof that $\partial^2 = 0$ in cellular homology, since $\partial^2 = H(j) \circ (\delta \circ H(j)) \circ \delta$, and the middle two terms vanish due to exactness. That this definition of cellular homology agrees with that of §4.3 follows from a close examination of $\partial^{\mathrm{cell}}$, which takes $X^{(n)}/X^{(n-1)}$ to $X^{(n-1)}/X^{(n-2)}$ by means of $\delta$ and quotients. On an $n$-cell $\tau$, the $n$-sphere $\tau/\partial\tau$ is sent to the wedge of $(n-1)$-spheres $\sigma/\partial\sigma$ for each $\sigma \trianglelefteq \tau$. This map, coming from $\delta$, yields precisely the incidence number $[\sigma\colon \tau]$ on each face. Compared to juggling the combinatorics of incidence number cancellation, this exploitation of exactness is incisive.

## 5.6 Coverage in sensor networks

Sensors – devices which return data tied to a location – are ubiquitous. The problem of collating distributed pieces of sensor data over a communications network is an engineering challenge for which the tools of topology and homological algebra seem strangely fitting.

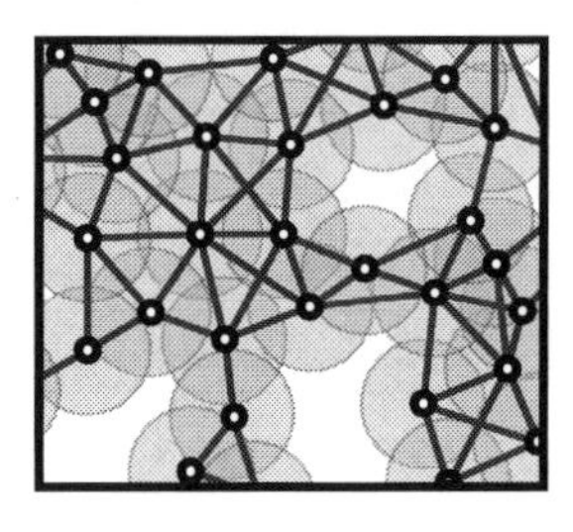

One simple-to-state problem is that of **coverage**. Fix a domain $\mathcal{D} \subset \mathbb{R}^2$ and consider a finite collection $\mathcal{Q}$ of sensors nodes in the plane with two tasks: they (1) sense a neighborhood of their locale in $\mathcal{D}$; and (2) communicate with other sensors. Both of these actions are assumed to be local in the sense that individual nodes cannot extract sensing data from or communicate data over all of $\mathcal{D}$. The problem of coverage, or more precisely, **blanket coverage**, is the question of whether there are holes in the sensor network – are there any regions in $\mathcal{D}$ which are not sensed? Other important coverage problems include **barrier coverage**, in which one wants to determine whether a sensor network separates $\mathcal{D}$ or surrounds a critical region, and **sweeping coverage**, the time-dependent problem familiar to users of robotic vacuum sweepers.

When the coordinates of the sensors are known, computational geometry suffices to determine coverage. For non-localized sensors, the following simple application of homology gives effective criteria. Specific assumptions are kept to a minimum, for clarity and ease of proofs:

1. Sensors are modeled as a finite collection of nodes $\mathcal{Q} \subset \mathbb{R}^2$.
2. Each sensor is assumed to have a unique identity which it broadcasts; nearby neighbors detect the transmission and establish a communication link.
3. Communication is symmetric and generates a communications graph, $X$, on $\mathcal{Q}$ with corresponding flag complex $\mathsf{F} = \mathsf{F}(X)$.
4. Sensor coverage regions are based on communication proximity: for any subset of sensors $S \subset \mathcal{Q}$ which pairwise communicate, the union of coverage regions of $S$ contains the convex hull of $S$ in $\mathbb{R}^2$.
5. One fixes a 'fence' cycle $\mathsf{C} \subset X$ whose image in $\mathbb{R}^2$ is a simple closed curve bounding a domain $\mathcal{D} \subset \mathbb{R}^2$ of interest.

One wants to know whether $\mathcal{D}$ is contained in the coverage region of the network. The critical assumption is the fourth, connecting the communications and sensing of the network by means of the flag complex $\mathsf{F}$. It is satisfied by systems with radially-symmetric communications networks (or *unit disc graphs*) and radially symmetric sensing regions with the proper ratio between sensing and communication [87, 88], but asymmetric systems are permissible in this framework.

**Theorem 5.10 ([87]).** *Given $\mathcal{Q}$, $X$, $\mathsf{F}$, $\mathsf{C}$, and $\mathcal{D}$ as above, then all of $\mathcal{D}$ is contained in the sensor-covered region if, equivalently:*

*1.* $[\mathsf{C}] = 0 \in H_1(\mathsf{F})$.

2. *There exists* $[\alpha] \in H_2(\mathsf{F}, \mathsf{C})$ *with* $\partial\alpha = \mathsf{C}$.

**Proof.** Equivalence of the two conditions comes from the long exact sequence of the pair $(\mathsf{F}, \mathsf{C})$ induced by the inclusion $i\colon \mathsf{C} \hookrightarrow \mathsf{F}$,

$$\cdots \longrightarrow H_2(\mathsf{F}, \mathsf{C}) \xrightarrow{\delta} H_1(\mathsf{C}) \xrightarrow{H(i)} H_1(\mathsf{F}) \longrightarrow \cdots,$$

since $\ker H(i) = \operatorname{im} \delta$ by exactness. Consider (from §2.2) the shadow map $\mathcal{S}\colon \mathsf{F} \to \mathbb{R}^2$ which sends vertices of the flag complex $\mathsf{F}$ to the physical sensor locations $\mathcal{Q} \subset \mathcal{D}$ and which sends a $k$-simplex of $\mathsf{F}$ to the (potentially singular) $k$-simplex given by the convex hull of the vertices implicated. By construction, $\mathcal{S}$ takes the pair $(\mathsf{F}, \mathsf{C})$ to $(\mathbb{R}^2, \partial\mathcal{D})$, acting as a homeomorphism on the second terms of the pair. This map induces the following diagram on the long exact sequences of the pairs:

$$\begin{array}{ccccccc} \cdots \longrightarrow & H_2(\mathsf{F}, \mathsf{C}) & \xrightarrow{\delta} & H_1(\mathsf{C}) & \longrightarrow \cdots . \\ & \downarrow{\scriptstyle H(\mathcal{S})} & & \cong \downarrow{\scriptstyle H(\mathcal{S})} & \\ \cdots \longrightarrow & H_2(\mathbb{R}^2, \partial\mathcal{D}) & \xrightarrow{\delta} & H_1(\partial\mathcal{D}) & \longrightarrow \cdots \end{array} \tag{5.9}$$

By assumption, $\partial\alpha = \mathsf{C} \neq 0$; hence, $H(\mathcal{S})\delta[\alpha] = H(\mathcal{S})[\partial\alpha] \neq 0$. Naturality implies that the diagram is commutative: $\delta H(\mathcal{S}) = H(\mathcal{S})\delta$. Commutativity implies that $\delta H(\mathcal{S})[\alpha] \neq 0$, and thus $H(\mathcal{S})[\alpha] \neq 0$. If the sensors do not cover some point $p \in \mathcal{D}$, then $p$ does not lie in the image of $\mathcal{S}$; thus, the map $\mathcal{S}\colon \mathsf{F} \to \mathbb{R}^2$ is a composition of maps $\mathsf{F} \to \mathbb{R}^2{-}p \hookrightarrow \mathbb{R}^2$. Diagram (5.9) is restructured as:

$$\begin{array}{ccccc} & & H_2(\mathsf{F}, \mathsf{C}) & \xrightarrow{\delta} & H_1(\mathsf{C}) \\ & \swarrow & \downarrow{\scriptstyle H(\mathcal{S})} & & \cong \downarrow{\scriptstyle H(\mathcal{S})} \\ H_2(\mathbb{R}^2{-}p, \partial\mathcal{D}) & \searrow & H_2(\mathbb{R}^2, \partial\mathcal{D}) & \xrightarrow{\delta} & H_1(\partial\mathcal{D}) \end{array} \tag{5.10}$$

However, $H_2(\mathbb{R}^2{-}p, \partial\mathcal{D}) = 0$, as the long exact sequence of the pair $(\mathbb{R}^2{-}p, \partial\mathcal{D})$ reveals:

$$\longrightarrow H_2(\mathbb{R}^2{-}p) \longrightarrow H_2(\mathbb{R}^2{-}p, \partial\mathcal{D}) \xrightarrow{\delta} H_1(\partial\mathcal{D}) \xrightarrow{H(i)} H_1(\mathbb{R}^2{-}p) \longrightarrow$$

The first term in this sequence is zero since $\mathbb{R}^2{-}p \simeq \mathbb{S}^1$, which has vanishing $H_2$. The two last terms are $H_1(\partial\mathcal{D}) \cong H_1(\mathbb{R}^2{-}p)$, and, moreover, $H(i)\colon H_1(\partial\mathcal{D}) \to H_1(\mathbb{R}^2{-}p)$ gives the winding number (§4.12) of $\partial\mathcal{D}$ about $p \in \mathbb{R}^2$. Since $p$ lies in the interior of $\mathcal{D}$, the winding number is $\pm 1$, and $H(i)$ is an isomorphism. Exactness then implies that $H_2(\mathbb{R}^2{-}p, \partial\mathcal{D}) = 0$. Commutativity of (5.10) completes the proof. ⊙

The assumption on sensor coverage specifies that certain regions are guaranteed to be covered while passing no information about lack of coverage elsewhere. As such,

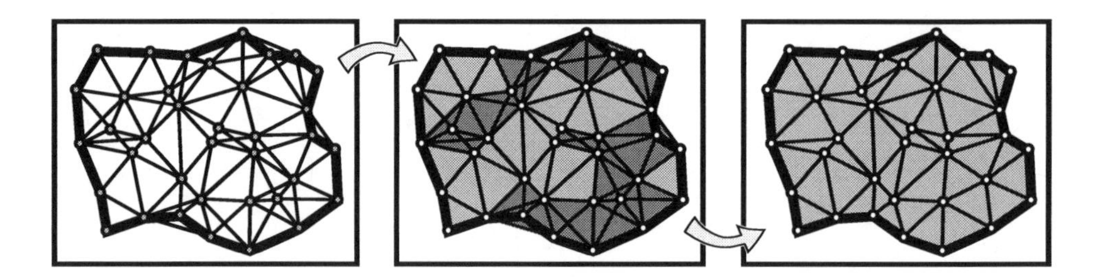

the homological coverage criterion cannot be if-and-only-if. It, like the assumptions on which it is built, is a conservative criterion. However, when the homological coverage criterion fails, choosing a basis for $H_1(\mathsf{F})$ which is *sparse* (implicating few nodes and edges) gives information about where the coverage holes may reside. Finally, *any* relative cycle $\alpha \in Z_2(\mathsf{F}, \mathsf{C})$ with $\partial\alpha = \mathsf{C}$ suffices to cover $\mathcal{D}$: only those nodes implicated in $\alpha$ are required to be actively sensing/communicating. This allows one to conserve power or establish a sleep-wake cycle by homological means.

## 5.7 Degree and computation

The crucial step in the proof of Theorem 5.10 used a winding number (degree) whose computation was, fortunately, obvious. More difficult degree computations are often possible by means of local formulae. For example, given a point $p \in \mathbb{R}^2$ and a closed curve $\gamma\colon \mathbb{S}^1 \to \mathbb{R}^2 - p$, the winding number of $\gamma$ about $p$ is easily computed as follows. Draw a ray from $p$ and perturb it to intersect the image of $\gamma$ in a finite set of points $Q = \{q_i\}_1^K$. At each intersection point $q_i$, the curve *kisses* or *crosses* the ray; either the curve traverses (left-to-right or right-to-left, since both are oriented) or it osculates, touching the ray and immediately turning back. Each action contributes a *local* degree: $\pm 1$ if crossing and 0 if kissing. The net winding number of $\gamma$ about $p$ is the sum of these local contributions to degree.

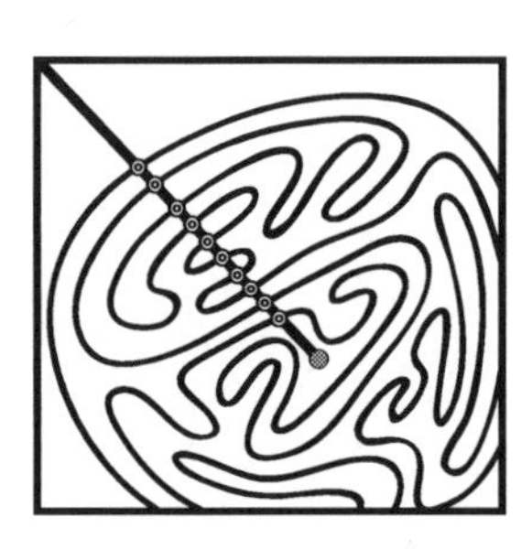

This simple example of a local computation rewards rumination. What happens if, instead of a curve in the plane, one needs to know whether a cycle in an ad hoc non-localized network surrounds a node (as in, *e.g.*, a network of security cameras [151])? To know whether a curve surrounds a point in the plane, it suffices to know the local behavior of the curve at a (small) finite number of points. What the curve does elsewhere is irrelevant. This has the pattern and stamp of topology. This intuitively simple procedure has a rigorous footing in transversality [169]; better still is the use of local homology. Assume $f\colon \mathbb{S}^n \to \mathbb{S}^n$ and $q \in f^{-1}(p)$ is an isolated point in the inverse image. Define the **local degree** of $f$ at $q$ to be

$$\deg(f; q) := \deg H(f)\colon H_n(U, U - q) \to H_n(V, V - p), \tag{5.11}$$

for $U$ and $V$ sufficiently small neighborhoods of $q$ and $p$ satisfying $f(U) \subset V$. The local degree is an integer since $H_n(\mathbb{S}^n, \mathbb{S}^n - p) \cong \mathbb{Z}$ for any point $p \in \mathbb{S}^n$. The validity of local computation can be shown using basic tools:

**Proposition 5.11.** *Assume that for $f\colon \mathbb{S}^n \to \mathbb{S}^n$, $p$ has discrete inverse image $f^{-1}(p) = Q = \{q_i\}_i$. Then*

$$\deg f = \sum_i \deg(f, q_i). \tag{5.12}$$

**Proof.** From Lemma 5.5, the long exact sequences of the pairs $(\mathbb{S}^n, \mathbb{S}^n - Q)$ and $(\mathbb{S}^n, \mathbb{S}^n - p)$ form a commutative diagram:

$$\begin{array}{ccccccc}
H_n(\mathbb{S}^n - Q) & \longrightarrow & H_n(\mathbb{S}^n) & \xrightarrow{H(j)} & H_n(\mathbb{S}^n, \mathbb{S}^n - Q) & \xrightarrow{\delta} & H_{n-1}(\mathbb{S}^n - Q) \\
\downarrow{\scriptstyle H(f)} & & \downarrow{\scriptstyle H(f)} & & \downarrow{\scriptstyle H(f)} & & \downarrow{\scriptstyle H(f)} \\
H_n(\mathbb{S}^n - p) & \longrightarrow & H_n(\mathbb{S}^n) & \xrightarrow[H(j)]{\cong} & H_n(\mathbb{S}^n, \mathbb{S}^n - p) & \xrightarrow{\delta} & H_{n-1}(\mathbb{S}^n - p)
\end{array}$$

The first terms in both rows and the last term in the bottom row vanish. By exactness, the lower map $H(j)$ is an isomorphism. Thus, by commutativity

$$\deg f = \deg \left( H_n(\mathbb{S}^n) \xrightarrow{H(j)} H_n(\mathbb{S}^n, \mathbb{S}^n - Q) \xrightarrow{H(f)} H_n(\mathbb{S}^n, \mathbb{S}^n - p) \right).$$

Choose a small neighborhood $V$ of $p$ so that $f^{-1}(V) = \sqcup_i U_i$ is a disjoint collection of neighborhoods of the $q_i$. It follows that

$$H_n(\mathbb{S}^n, \mathbb{S}^n - Q) \cong^{(1)} H_n\left(f^{-1}(V), f^{-1}(V) - Q\right) \cong^{(2)} \bigoplus_i H_n(U_i, U_i - q_i),$$

where the isomorphisms come from (1) excision, and (2) additivity. The local degree at $q_i$, $\deg(f, q_i)$, is by definition the degree of the induced map $H(f)\colon H_n(U_i, q_i) \to H_n(V, p)$. Thus, by additivity,

$$\deg f = \deg \left( H_n(\mathbb{S}^n) \to \bigoplus_i H_n(U_i, U_i - q_i) \to H_n(V, V - p) \right) = \sum_i \deg(f, q_i).$$

⊙

**Corollary 5.12.** *If $f\colon \mathbb{S}^n \to \mathbb{S}^n$ is not surjective, then* $\deg(f) = 0$.

**Proof.** If not surjective, apply (5.12) to an empty inverse image. ⊙

The long exact sequence of the pair also permits computation of degree in slightly different settings. For example, consider the case of a map of the form $f\colon (\mathbb{D}^n, \partial\mathbb{D}^n) \to (\mathbb{D}^n, \partial\mathbb{D}^n)$ that maps a closed disc to itself, restricting to a map on the boundary sphere. It is sensible to speak of the degree of $f$ by using relative homology: $\deg(f) := \deg(H_n(\mathbb{D}^n, \partial\mathbb{D}^n) \to H_n(\mathbb{D}^n, \partial\mathbb{D}^n))$. This is well-defined since $H_n(\mathbb{D}^n, \partial\mathbb{D}^n) \cong \mathbb{Z}$. Moreover, it is easily computed in terms of what $f$ does either on the boundary or on the interior, as follows. There is both an induced map $\tilde{f}\colon \mathbb{S}^n \to \mathbb{S}^n$ on the quotient sphere, given by collapsing the boundary to a point, and a restriction map $\overline{f}\colon \mathbb{S}^{n-1} \to \mathbb{S}^{n-1}$ given by restriction to the boundary.

**Lemma 5.13.** *For the above,* $\deg(f) = \deg(\tilde{f}) = \deg(\overline{f})$.

**Proof.** That $\deg(f) = \deg(\tilde{f})$ follows from naturality. The equivalence of this to $\deg(\overline{f})$ comes from the long exact sequence applied to $f\colon (\mathbb{D}^n, \partial\mathbb{D}^n) \to (\mathbb{D}^n, \partial\mathbb{D}^n)$:

$$\begin{array}{ccccccc} 0 = H_n(\mathbb{D}^n) & \longrightarrow & H_n(\mathbb{D}^n, \partial\mathbb{D}^n) & \xrightarrow{\delta} & H_{n-1}(\partial\mathbb{D}^n) & \longrightarrow & H_{n-1}(\mathbb{D}^n) = 0\,, \\ & & \downarrow{\scriptstyle H(f)} & & \downarrow{\scriptstyle H(\overline{f})} & & \\ 0 = H_n(\mathbb{D}^n) & \longrightarrow & H_n(\mathbb{D}^n, \partial\mathbb{D}^n) & \xrightarrow{\delta} & H_{n-1}(\partial\mathbb{D}^n) & \longrightarrow & H_{n-1}(\mathbb{D}^n) = 0 \end{array}$$

By exactness, the connecting homomorphisms $\delta$ on top and bottom are isomorphisms. Thus, by commutativity, $\deg(f) = \deg(\overline{f})$. ⊙

**Example 5.14 (Colorings)** ⊚

There are numerous results in combinatorics that are inherently topological, several of which involve **colorings**. The following is a classical example implicating degree. Consider a 2-simplex $T$ with vertices labeled by $\{0, 1, 2\}$. Let $T'$ be a subdivision of $T$. Label all the new vertices of $T'$ using any label $\{0, 1, 2\}$ subject to the following *boundary condition*: on $\partial T'$, each vertex must *not* be labeled by the label of $T$ on the vertex opposite that edge. Hence, on the portion of the boundary of $T'$ connecting the $\{0\}$ and $\{1\}$ vertices of $T$, the label $\{2\}$ must not be used.

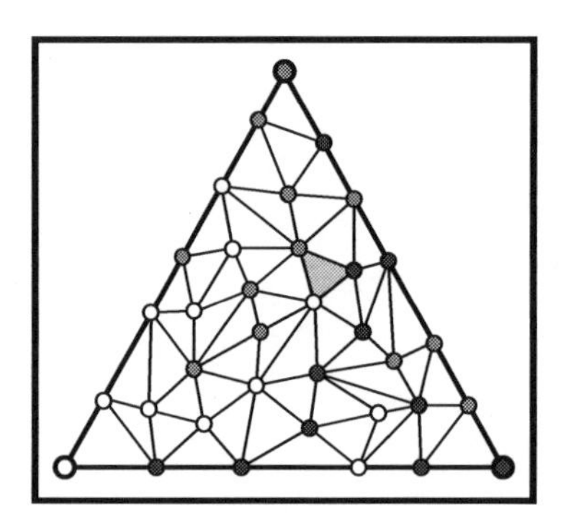

**Lemma 5.15 (Sperner's Lemma).** *Any* $\{0, 1, 2\}$-*labeling of the vertices of* $T'$ *obeying the boundary condition must possess at least one triangle with all three labels at vertices.*

**Proof.** Consider the piecewise-linear simplicial map $f\colon T' \to T$ that sends each simplex in $T'$ to the unique simplex in $T$ determined by the convex hull of the labels in $T'$. Note that the conclusion of the theorem holds if and only if $f$ is surjective. The boundary coloring condition implies that $f\colon \partial T' \to \partial T$ with degree $+1$. Thus, by Corollary 5.12 and Lemma 5.13, $f$ is surjective. ⊙

From this proof, one sees clearly that the boundary condition can be relaxed to allow for any coloring of the vertices on the boundary that imparts a nonzero degree.

Furthermore, it is clear that the proof holds for a subdivided $n$-simplex with $n+1$ distinct colors. With the appropriate homological restrictions, other cellular spaces are likewise admissible: everything hinges on homology. Sperner's Lemma can be viewed as a discretized version of the Brouwer fixed point theorem of §4.13; as such, it is useful in discrete versions of fair-division and consensus problems [283]. ◎

Sperner's Lemma can also be restated as saying that any $(n+1)$-coloring of the vertices of a simplicial $\mathbb{S}^n$ must have an *even* number of $(n+1)$-colored simplices. In this setting, the proof above reduces to an examination of a simplicial map $\mathbb{S}^n \to \Delta^n$. The topological features of such maps are covered by another classical theorem from algebraic topology implicating degrees and exact sequences.

## 5.8 Borsuk-Ulam theorems

There are a number of results which fall under the names Borsuk-Ulam; all orbit about spheres and the antipodal map $a\colon \mathbb{S}^n \to \mathbb{S}^n$ that sends $x \mapsto -x$. The following is the key step in these results, written in the language of degree theory.

**Theorem 5.16 (Borsuk-Ulam Theorem).** *Odd maps of $\mathbb{S}^n$ have odd degree; even maps of $\mathbb{S}^n$ have even degree.*

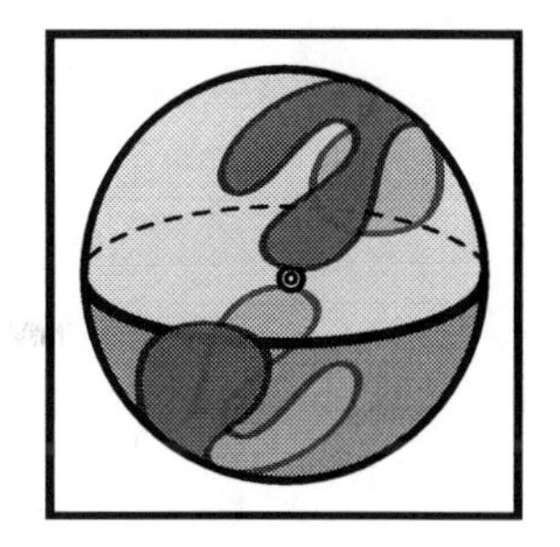

**Proof.** The proof in the (harder) odd case is sketched, the key to which is a long exact sequence to track antipodes. Assume that $f$ is odd, so that $f \circ a = a \circ f$. Recall from Example 1.2, the quotient space $\mathbb{S}^n/a$ consisting of equivalence classes of antipodal points is one definition of the real projective space $\mathbb{P}^n$. Let $\pi\colon \mathbb{S}^n \to \mathbb{P}^n$ denote the quotient projection map; this map is a 2-to-1 local homeomorphism. Let $\overline{f}\colon \mathbb{P}^2 \to \mathbb{P}^2$ be the induced map. There is a commutative diagram of short exact sequences of chain complexes in $\mathbb{F}_2$ coefficients,

$$\begin{array}{ccccccccc} 0 & \longrightarrow & C_\bullet(\mathbb{P}^n) & \xrightarrow{\tau} & C_\bullet(\mathbb{S}^n) & \xrightarrow{\pi_\bullet} & C_\bullet(\mathbb{P}^n) & \longrightarrow & 0\,, \\ & & \downarrow{\scriptstyle \overline{f}_\bullet} & & \downarrow{\scriptstyle f_\bullet} & & \downarrow{\scriptstyle \overline{f}_\bullet} & & \\ 0 & \longrightarrow & C_\bullet(\mathbb{P}^n) & \xrightarrow{\tau} & C_\bullet(\mathbb{S}^n) & \xrightarrow{\pi_\bullet} & C_\bullet(\mathbb{P}^n) & \longrightarrow & 0 \end{array}$$

where $\tau$ is the **transfer map**, lifting a chain in $\mathbb{P}^n$ via $a$ to an antipodal pair of chains in $\mathbb{S}^n$. The $\mathbb{F}_2$ coefficients ensures that the sequence is exact. The corresponding long exact sequences yield:

$$\begin{array}{ccccccccccc} 0 & \longrightarrow & H_n(\mathbb{P}^n) & \xrightarrow{H(\tau)} & H_n(\mathbb{S}^n) & \xrightarrow{H(\pi)} & H_n(\mathbb{P}^n) & \xrightarrow{\delta} & H_{n-1}(\mathbb{P}^n) & \longrightarrow & 0\,, \\ & & \downarrow{\scriptstyle H(\overline{f})} & & \downarrow{\scriptstyle H(f)} & & \downarrow{\scriptstyle H(\overline{f})} & & \downarrow{\scriptstyle H(\overline{f})} & & \\ 0 & \longrightarrow & H_n(\mathbb{P}^n) & \xrightarrow{H(\tau)} & H_n(\mathbb{S}^n) & \xrightarrow{H(\pi)} & H_n(\mathbb{P}^n) & \xrightarrow{\delta} & H_{n-1}(\mathbb{P}^n) & \longrightarrow & 0 \end{array}$$

By exactness and knowledge of $H_\bullet(\mathbb{P}^n)$, one argues that in the above diagram, $\delta$ and $H(\tau)$ are isomorphisms, while $H(\pi) = 0$. By inducting on the dimension $n$ and using the commutative square penultimate to the right, one shows that $H(\overline{f})$ is an isomorphism. By commutativity (or the 5-Lemma), this implies that $H(f)\colon H_n(\mathbb{S}^n; \mathbb{F}_2) \to H_n(\mathbb{S}^n; \mathbb{F}_2)$ is an isomorphism. This, in turn, is the mod 2 reduction of $H(f)$ from $H_n(\mathbb{S}^n; \mathbb{Z}) \to H_n(\mathbb{S}^n; \mathbb{Z})$; *i.e.*, $\deg(f) \bmod 2 = 1$. ⊙

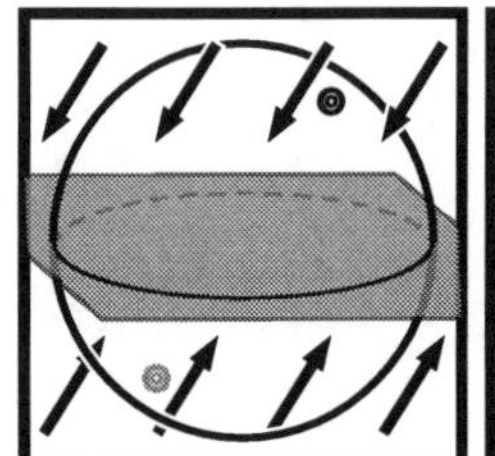 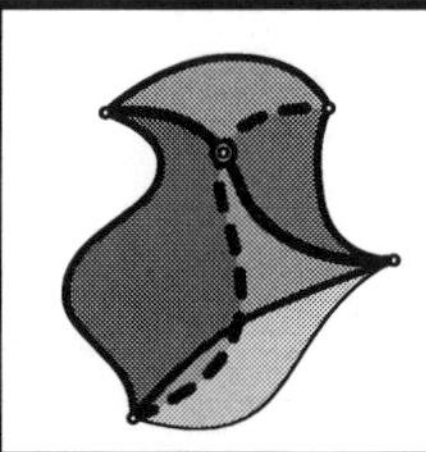 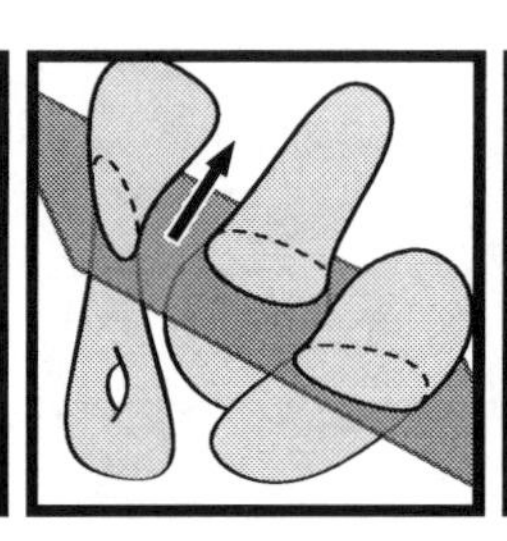 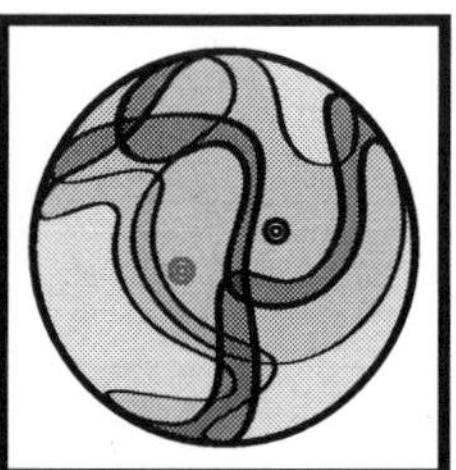

There are a number of famous corollaries of Theorem 5.16:

1. **Borsuk-Ulam:** Any map $\mathbb{S}^n \to \mathbb{R}^n$ must identify some pair of antipodal points.
2. **Radon:** Any map $\Delta^{n+1} \to \mathbb{R}^n$ of an $n+1$-simplex has a pair of disjoint closed faces (simplices of $\partial\Delta^{n+1}$) whose images in $\mathbb{R}^n$ intersect.
3. **Stone-Tukey:** Given a collection of $n$ Lebesgue-measurable bodies in $\mathbb{R}^n$, there is a hyperplane which bisects evenly the volume of each body.
4. **Lusternik-Schnirelmann:** Any cover $\mathcal{U} = \{U_i\}_0^n$ of $\mathbb{S}^n$ by $n+1$ open sets must have at least one element $U_j$ containing an antipodal pair of points.

Several of these theorems have physical interpretations. For example, it is common to express the first corollary above as saying that (assuming meteorological continuity) some antipodal pair of points on the earth have the same temperature and barometric pressure. Applications more relevant to this text lie in economics and *fair division* problems; for applications to combinatorics, see [219].

## 5.9 Euler characteristic

Sequences solve the mystery of the topological invariance of the Euler characteristic. In what follows, use field coefficients. The subtle step, following the theme of §5.1, is to lift the notion of Euler characteristic from a cell complex to an arbitrary (finite, finite-dimensional) sequence $C_\bullet$ of vector spaces:

$$\chi(C_\bullet) := \sum_k (-1)^k \dim C_k. \tag{5.13}$$

The alternating sum is a binary variant of exactness. A short exact sequence of vector spaces $0 \to A \to B \to C \to 0$ has $\chi = 0$, since $C \cong B/A$. By applying this to individual rows of a short exact sequence of (finite, finite-dimensional) chain complexes,

$$0 \longrightarrow A_\bullet \longrightarrow B_\bullet \longrightarrow C_\bullet \longrightarrow 0$$

one sees that $\chi$ of this sequence also vanishes: $\chi(A_\bullet) - \chi(B_\bullet) + \chi(C_\bullet) = 0$. This, then, provides the key to understanding why the Euler characteristic is a homological invariant. The following lemma is the homological version of the Rank-Nullity Theorem (Lemma 1.10):

**Lemma 5.17.** *The Euler characteristic of a chain complex $C_\bullet$ and its homology $H_\bullet$ are identical, when both are defined.*

**Proof.** From the definitions of homology and chain complexes, one has two short exact sequences of chain complexes, arranged like so:

$$\begin{array}{ccccccccc}
0 & \longrightarrow & B_\bullet & \longrightarrow & Z_\bullet & \longrightarrow & H_\bullet & \longrightarrow & 0\,. \\
0 & \longrightarrow & Z_\bullet & \longrightarrow & C_\bullet & \longrightarrow & B_\bullet^- & \longrightarrow & 0 \\
 & & + & & - & & + & &
\end{array}$$

Here, $B_\bullet^-$ is the shifted boundary complex: $B_k^- := B_{k-1}$. By exactness, the Euler characteristic of the sum of these two sequences is zero. The $Z$ terms cancel. The $B$ terms cancel, since $\chi(B_\bullet^-) = -\chi(B_\bullet)$. This leaves $\chi(H_\bullet) - \chi(C_\bullet) = 0$. ⊙

**Corollary 5.18.** *For a compact cell complex $X$ with subcomplexes $A$ and $B$,*

$$\chi(X) = \sum_k (-1)^k \dim H_k(X)$$
$$\chi(A \cup B) = \chi(A) + \chi(B) - \chi(A \cap B)$$
$$\chi(X - A) = \chi(X) - \chi(A)$$

*Furthermore, $\chi$ is a homotopy invariant among this class of spaces.*

These results follow from applications of Lemma 5.17 to (1) the chain complex for cellular homology; (2) the Mayer-Vietoris sequence; and (3) the long exact sequence of the pair $(X, A)$ respectively, the last requiring a little excisive effort to relate $C_\bullet(X, A)$ to $X - A$.

## 5.10 Lefschetz index

There is a generalization of Euler characteristic from spaces to self-maps. For any chain map $\varphi_\bullet \colon C_\bullet \to C_\bullet$ on a finite-dimensional chain complex $\mathcal{C}$ over a field $\mathbb{F}$, define the **Lefschetz index** as the graded alternating sum of the traces of chain maps $\tau(\varphi_\bullet) := \sum_k (-1)^k \mathrm{trace}\,(\varphi_\bullet \colon C_k \to C_k)$.

The analogue of Lemma 5.17 holds: the alternating sum of traces of $H(f)$ equals the alternating sum of traces of $f_\bullet$ via the same argument. This index, like the Euler characteristic which it mimics, is intimately connected to the question of fixed points, not of vector fields, but of self-maps in general. For a self-map $f \colon X \to X$ of a space

$X$, one defines its Lefschetz index as, equivalently:

$$\tau_f := \tau(f_\bullet) = \sum_k (-1)^k \text{trace}\,(H(f) \colon H_k X \to H_k X)\,. \tag{5.14}$$

**Theorem 5.19 (Lefschetz Fixed Point Theorem).** *For $X$ a finite (thus, compact) cell complex, any map $f \colon X \to X$ must have a fixed point if $\tau_f \neq 0$.*

**Proof.** The technical portion of the proof (omitted) is to show that $f$ may be approximated by a cellular map (also labeled $f$) for a suitably subdivided cell structure on $X$ in such a way that the approximation also has empty fixed point set. By compactness and the lack of fixed points, this subdivided cellular map sends each $n$-cell to a *different* cell. Thus, the trace of the chain map $f_\bullet \colon C_\bullet \to C_\bullet$ vanishes and $\tau_f = 0$. The approximation step does not change the action on homology or (therefore) Lefschetz index. ⊙

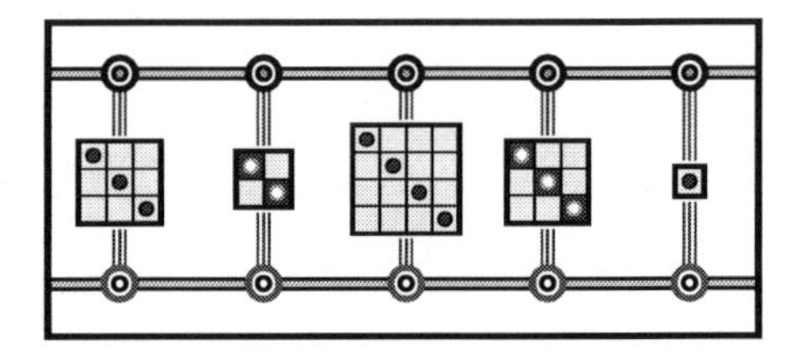

The theorem can be extended greatly. Assuming that the image $f(X)$ is contained in a compact subset of $X$, the theorem holds for any $X$ homeomorphic to a neighborhood retract in some Euclidean space. [3] Moreover, the theorem holds not only for maps but for multivalued maps $F : X \rightrightarrows X$ with the appropriate modifications: see §5.11 and §7.7. Finally, the computation of the Lefschetz index can be reduced to a sum over the fixed point set of $f$: see §7.7. This hints at a different proof of Theorem 5.19, since the sum over the empty set vanishes.

The Lefschetz theorem provides a simple proof of Theorem 3.3, that any vector field on a manifold $M$ with $\chi \neq 0$ has at least one fixed point. If a nonvanishing vector field existed, the time-$\epsilon$ map of the flow (for $\epsilon$ sufficiently small) would be a map of $M$ without fixed points that is a small perturbation of the identity map. The Lefschetz index of this map is $\tau(\mathrm{Id}) = \chi(M)$, since the identity map on $H_\bullet$ has trace equal to the dimension of $H_\bullet$. Indeed, it follows that for any Euclidean neighborhood retract $X$ with $\chi \neq 0$, any map $f \colon X \to X$ homotopic to the identity has a fixed point.

## 5.11 Nash equilibria

Among the many applications of fixed point theorems, few are as celebrated as that of the existence of **Nash equilibria** in multiplayer games. The following is a terse rendition.[4] Consider a collection of $N$ players, each of whom can choose from a finite set $A_i$, of *pure* strategies in, say, a game or an economy. The **payoff function**, $f : \prod_i A_i \to \mathbb{R}^N$, records the numerical return to each of the $N$ players upon executing

[3]Specifically, $Y \subset \mathbb{R}^n$ is a neighborhood retract if there is a retraction $r \colon U \to Y$ for $U$ an open neighborhood. Any $X$ homeomorphic to such a $Y$ is called an **ENR**, or Euclidean neighborhood retract.

[4]Nash' 1950 article – less than a page – is terser still.

the strategies chosen in each $A_i$. Naturally, each player wishes to maximize payoff. A **Nash equilibrium** for this game is a choice of strategies $x^*$ such that for each $i$, $f_i(x^*)$ is maximal with respect to varying all inputs except $x_i^*$ – that is, each player has chosen an optimal strategy with respect to fixing everyone else's [known] strategies. The existence of a Nash equilibrium means, in principle, that *everyone is content.* Such an equilibrium may or may not exist.

The insightful step is to allow for *mixed* or probabilistic strategies. Each player chooses from his set of strategies $A_i$ according to some fixed probability distribution, playing repeatedly. This has the effect of generating a **strategy space**, $X_i$, for each player, corresponding to the space of probability distributions on $A_i$: each $X_i$ is a simplex. The payoff function extends by linearity to a continuous map $f : \prod_i X_i \to \mathbb{R}^N$, recording the *expected* return to each of the $N$ players. Nash showed that for mixed strategies, an equilibrium always exists [239]. His initial proof used the Brouwer fixed point theorem (Theorem 4.26) in a manner not unlike that of Example 4.27; he quickly converged to a simpler proof via a better fixed point theorem for multivalued functions. The *Kakutani fixed point theorem* [187] states that for $F\colon \mathbb{D}^n \rightrightarrows \mathbb{D}^n$ a multivalued map whose graph is closed in $\mathbb{D}^n \times \mathbb{D}^n$ and has $F(x)$ convex for each $x$, then there exists a fixed point – some $x$ satisfying $x \in F(x)$.

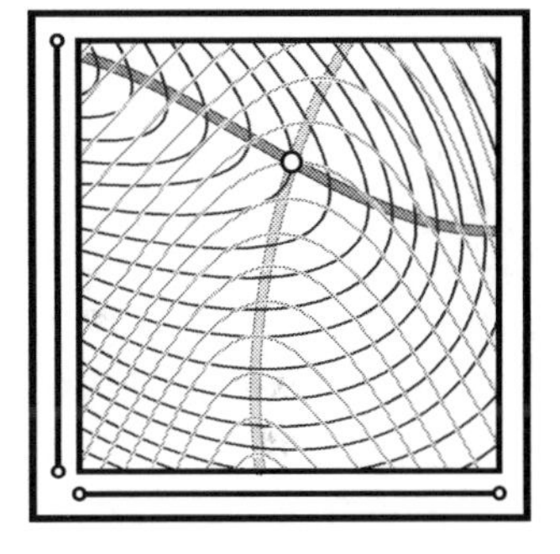

Nash's proof follows. For mixed strategies $x, x' \in \prod_i X_i$, one says that $x'$ *counters* $x$ if, for each $i$, the $i^{\text{th}}$ strategy of $x'$ is $f$-optimal with respect to all the not-$i$ strategies of $x$: that is, the player-$i$ payoff,

$$f_i(\ldots, x_{i-1}, x_i', x_{i+1}, \ldots),$$

is maximal with respect to varying only $x_i'$. Consider the multivalued map $F\colon \prod_i X_i \rightrightarrows \prod_i X_i$ which sends $x$ to the set of countering strategies $\{x'\}$. This satisfies the criteria for the Kakutani theorem, since the domain is homeomorphic to some $\mathbb{D}^n$, the images of $F$ are convex, and the graph of $F$ is closed (via continuity of $f$). The existence of a self-countering strategy – a Nash equilibrium by definition – follows via Kakutani.

The classical Nash theorem has been extended in numerous ways. One such extension is to more general strategy spaces $X_i$. It is possible to handle this setting with the Lefschetz Theorem [289], relaxing the conditions on the strategy space and on images of $F$. Let $X = \prod_i X_i$ be a product of reasonable spaces such as cell complexes.[5] Assume $F\colon X \rightrightarrows X$ a multivalued map whose graph is closed in $X \times X$ and whose images $F(x)$ are acyclic (have $\tilde{H}_\bullet = 0$). It follows that the two factor projection maps $p_1, p_2\colon X \times X \to X$ can be used to define a Lefschetz index for $F$ via $\tau_f = \tau(H(p_2 \circ p_1^{-1}))$, since $H(p_1)$ is an isomorphism. A multi-valued extension of the Lefschetz Theorem [106, 164] states that $F$ must have a fixed point when $\tau_f \neq 0$. This implies the existence of the classical Nash equilibrium (in that case, $\tau_f = 1$), but it also allows for strategy spaces which are noncontractible, if the conditions on $\tau_f$ are met.

[5]The technical conditions can be relaxed to an ANR – absolute neighborhood retract – and can be made to work even in infinite-dimensional settings.

The lesson here – not for the last time – is that a qualitative theorem which requests convexity or piecewise-linearity can be persuaded to relax given the appropriate invocation of homology.

## 5.12 The game of Hex

Nash equilibria arise frequently in the context of mathematical games, itself a fruitful field of topological intricacies. The **game of Hex** is a classical game of topological type. The traditional version of the game is played on a rhombus-shaped board with a uniform hexagonal tiling. The two players (traditionally *black* and *white*, representing colors of markers used in the game) alternate laying down one marker of their color on an open hex cell on the board, filling that cell.

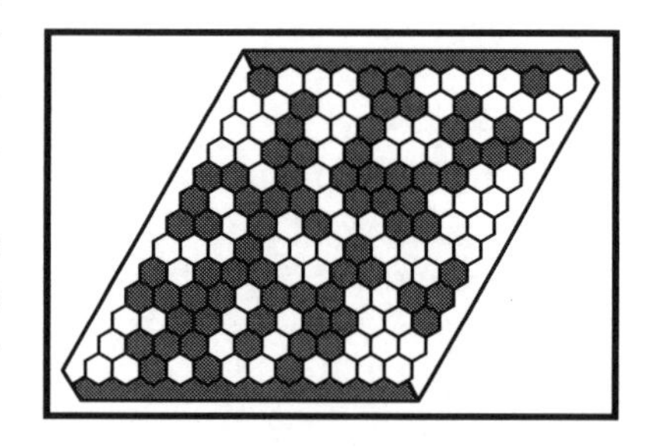

The player goals are, respectively, to build a connection from the top-to-bottom (black) or left-to-right (white) of the board, each player trying to win while blocking the other. Note that the corner pieces of the board border two sides. It is a classical result that this game always ends with one and only one winner. While it is obvious that this is a topological result, a clear topological proof is a worthy exercise. The Brouwer fixed point theorem was initially used by Nash and then Gale [138] to prove existence of a winner (Gale gave also a converse proof of the fixed point theorem via the Hex Theorem). The following is a different approach that may help the beginner learn to work with diagrams and exact sequences. It has the virtue of permitting very general (though not arbitrary) playing boards.

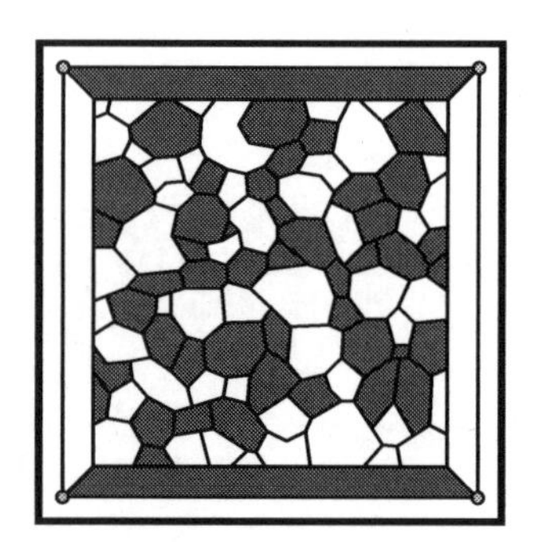

Let $D = [0, 1]^2$ be a Euclidean square, outfitted with a definable cell decomposition satisfying the following: (1) there are a disjoint pair of (black) 2-cells, $B_0$, containing the top and bottom edges, and another (white) disjoint pair, $W_0$, containing the left and right edges; (2) each vertex in the cell structure has degree exactly three – there are three edges that terminate at each vertex. Note that the four corners of the square satisfy this condition. Players alternate choosing 2-cells which are then colored white-or-black depending on the player. Since the cell decomposition is finite, each game must terminate, and it suffices to consider what happens when all 2-cells are thus colored. Let $B$ denote the union of the closure of each black 2-cell (including boundary cells $B_0$), and let $W$ be the corresponding white subset. The classical Hex Theorem translated into this setting states that *either* there is a path in $B$ connecting the two components of $B_0$ *or* a path in $W$ connecting the two components of $W_0$, but *not both*.

**Theorem 5.20 (Hex Theorem).** *This game of Hex has one and only one winner.*

**Proof.** Clearly, $B \cup W$ covers $D$ and $S = B_0 \cup W_0$ deformation retracts to $\partial D$. The pairs $(B, B_0)$, $(W, W_0)$, and $(D, S)$ fit together via the relative Mayer-Vietoris sequence and the long exact sequences of pairs (§5.3) into a commutative diagram:

$$\begin{array}{ccccccc}
H_2(D,S) & \longrightarrow & H_1(B\cap W, B_0\cap W_0) & \longrightarrow & H_1(B,B_0)\oplus H_1(W,W_0) & \longrightarrow & H_1(D,S) \\
\uparrow & & \uparrow & & \uparrow \quad \uparrow & & \uparrow \\
H_2(D) & \longrightarrow & H_1(B\cap W) & \longrightarrow & H_1(B)\oplus H_1(W) & \longrightarrow & H_1(D) \\
 & & \uparrow & & \uparrow \quad \uparrow & & \\
 & & 0 = H_1(B_0\cap W_0) & & 0 = H_1(B_0)\oplus H_1(W_0) = 0 & &
\end{array}$$

Horizontal rows are Mayer-Vietoris; vertical columns are sequences of pairs; field $\mathbb{F}$ coefficients are used. Only the relevant portions are displayed. The entire right column is zero, as is $H_2(D)$ in the lower-left entry. All vertical maps are injective by exactness. Note the lone copy of $\mathbb{F}$ in the upper-left corner, due to $H_2(D, S)$: this *is* the source of the unique solution to the game. To explain: the goal is to prove the existence of either a path in $B$ connecting $B_0$ or a path in $W$ connecting $W_0$. Such a path is *precisely* a relative homology class in one of the factors of $H_1(B, B_0) \oplus H_1(W, W_0)$ which is *not* in the image of the absolute homology $H_1(B) \oplus H_1(W)$ (necessarily injective, by exactness of columns). From exactness of the middle row, $H_1(B \cap W) \cong H_1(B) \oplus H_1(W)$.

The crucial observation is this: $B \cap W$ is a 1-manifold with boundary, since each point of $B \cap W$ is either (1) along a 1-cell of $D$; (2) at a degree-three vertex of $D$ with two-out-of-three cells of one color and the third another; or (3) one of the four corners of $D$. The *only* boundary points of $B \cap W$ are these four points. Therefore, $B \cap W$ consists of $N$ circles and exactly two compact intervals, with endpoints the four corners of $D$. The homology $H_1(B \cap W) \cong \mathbb{F}^N$ injects into $H_1(B \cap W, B_0 \cap W_0)$, which must have dimension $N + 2$, since the two intervals become relative cycles. By reducing the above diagram to dimensions of the homologies and adding one more (vanishing) term to the left of the top row, one obtains:

$$\begin{array}{ccccccccc}
0 & \longrightarrow & 1 & \longrightarrow & N+2 & \longrightarrow & M & \longrightarrow & 0\,, \\
 & & \uparrow & & \uparrow & & \Uparrow & & \uparrow \\
 & & 0 & \longrightarrow & N & \longrightarrow & N & \xrightarrow{\cong} & 0
\end{array}$$

with vertical maps injective. The top row is short-exact and implies that $M+1 = N+2$; thus $M = N + 1$, *i.e.*,

$$\dim\left(H_1(B, B_0) \oplus H_1(W, W_0)\right) = \dim\left(H_1(B) \oplus H_1(W)\right) + 1.$$

There exists exactly one relative homology class for *either* $H_1(B, B_0)$ *or* $H_1(W, W_0)$ which is not an absolute homology class: the winner. $\odot$

## 5.13 Barcodes and persistent homology

The capstone applications of this chapter comprise a short survey of the exciting work being done in **topological data analysis** using sequences and homologies. The motivation is that, for a parameterized family of spaces (*e.g.*, Vietoris-Rips complexes) modeling a point-cloud data set, qualitative features which persist over a larger parameter range have greater statistical significance. This branch of applied topology is advancing very rapidly; see [105, 306] for initial primary works, [55, 103, 144] for surveys, and [104, 305] for texts. The work described here spans contributions of Carlsson, de Silva, Edelsbrunner, Harer, Zomorodian, and many others, viewed from a representation-theoretic aspect.

Consider a sequence of spaces $X_i$ with maps:

$$X_0 \longrightarrow X_1 \longrightarrow \cdots \longrightarrow X_{N-1} \longrightarrow X_N. \tag{5.15}$$

The sequence may be finite, as shown, or infinite. This is motivated by a sequence of Vietoris-Rips complexes of a set of data points with an increasing sequence of radii $(\epsilon_i)_{i=0}^N$, in which case the maps are inclusions. This topological sequence is converted to an algebraic sequence by passing to homology and invoking functoriality:

$$H_k(X_0) \longrightarrow H_k(X_1) \longrightarrow \cdots \longrightarrow H_k(X_{N-1}) \longrightarrow H_k(X_N). \tag{5.16}$$

The induced homomorphisms on homology encode local topological changes in the $X_i$: the question is, what are the *global* changes? A homology class in $H_\bullet(X_i)$ is said to *persist* if its image in $H_\bullet(X_{i+1})$ is also nonzero; otherwise it is said to *die*. A homology class in $H_\bullet(X_j)$ is said to be *born* when it is not in the image of $H_\bullet(X_{j-1})$. This is most easily seen in the context of $H_0$, classifying connected components of a space. In the context of increasing Vietoris-Rips complexes of a point cloud this sequence of homologies in grading zero gives information about *clustering* of points.

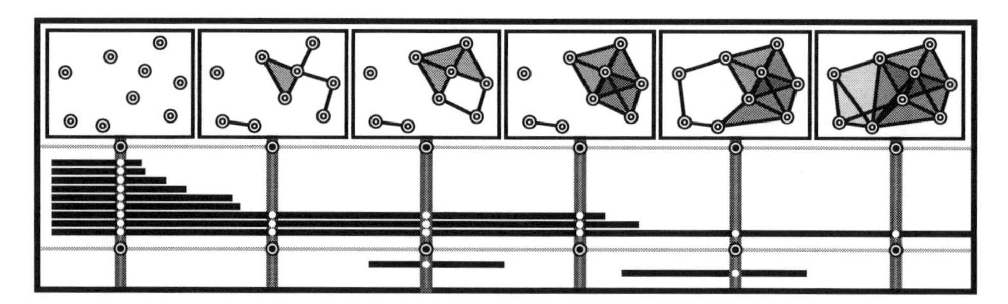

This language of birth, death, and everything-in-between is convenient, but suspiciously informal for mathematics. The injection of a little representation theory yields a principled approach that fulfills intuition in a manner consistent with the themes of this chapter. Consider the generalization of (5.16) to an arbitrary $\mathbb{Z}$-graded sequence of $\mathbb{F}$-vector spaces and linear transformations:

$$V_\bullet = \quad \cdots \longrightarrow V_{i-1} \longrightarrow V_i \longrightarrow V_{i+1} \longrightarrow \cdots . \tag{5.17}$$

Two such sequences $V_\bullet$ and $V'_\bullet$ are said to be *isomorphic* if there are isomorphisms $V_k \cong V'_k$ which commute with the linear transformations in $V_\bullet$ and $V'_\bullet$ as in Equation (5.1). The simplest such sequence is an **interval indecomposable** of the form

$$I_\bullet = \quad \cdots \longrightarrow 0 \longrightarrow 0 \longrightarrow \mathbb{F} \xrightarrow{\text{Id}} \mathbb{F} \xrightarrow{\text{Id}} \cdots \xrightarrow{\text{Id}} \mathbb{F} \longrightarrow 0 \longrightarrow 0 \longrightarrow \cdots$$

where the *length* of the interval equals the number of nonzero terms minus 1, so that an interval of length zero consists of $0 \to \mathbb{F} \to 0$ alone. Infinite or bi-infinite intervals are also included. Intervals – indeed, arbitrary linear sequences – can be formally added by taking the direct sum, $\oplus$, term-by-term and map-by-map. Interval indecomposables are precisely *indecomposable* with respect to $\oplus$ and cannot be expressed as a sum of simpler sequences.

The following result is a simple version of a deeper decomposition theorem from representation theory:

**Theorem 5.21 (Structure Theorem for finite linear sequences).** *Any linear sequence of finite-dimensional vector spaces and linear transformations decomposes as a direct sum of interval indecomposables, unique up to reordering.*

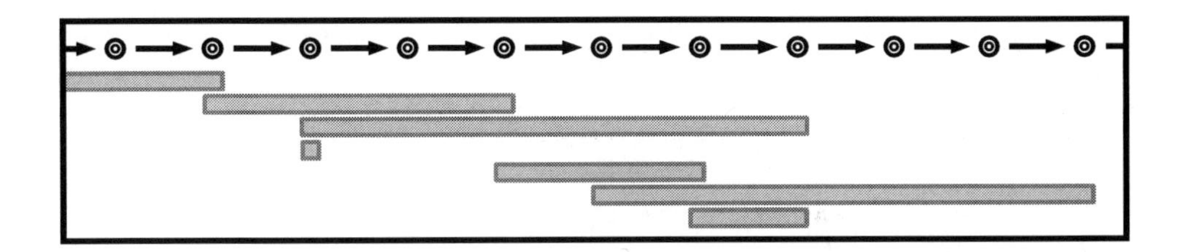

This prompts a graphical language for interpreting basic linear algebra. For example, any linear transformation $\mathbb{F}^n \xrightarrow{A} \mathbb{F}^m$ extends to a linear sequence with two nonzero entries. According to the Structure Theorem, such a sequence may be decomposed in precisely $\min(m, n) + 1$ different ways, depending on the number of interval indecomposables of length 1. The student of linear algebra knows this number of indecomposable intervals of length 1 under the guise of $\text{rank}(A)$.

Applying the Structure Theorem to a sequence of homologies (as in Equation (5.16)) in field coefficients yields a pictograph that is called a homology **barcode**. The phenomena of homology class *birth*, *persistence*, and *death* corresponds precisely to the *beginning*, *middle*, and *end* of an interval indecomposable. The barcode provides a simple descriptor for topological evolution: the shorter an interval, the more ephemeral the hole; long bars indicate robust topological features with respect to the parameter. This is salient in the context of point clouds $\mathcal{Q}$ and Vietoris-Rips complexes $\text{VR}_\epsilon(\mathcal{Q})$ using an increasing sequence $\{\epsilon_i\}$ as parameter. For $\epsilon$ too small or too large, the homology of $\text{VR}_\epsilon(\mathcal{Q})$ is unhelpful. Instead of trying to choose an *optimal* $\epsilon$, choose them *all*: the barcode reveals significant features.

There is no need to restrict to the case of metric-based simplicial complexes or spaces at all. One can begin with a **persistence complex**: a sequence of chain complexes $\mathcal{P} = (\mathcal{C}_i)$, $i \in \mathbb{Z}$, together with chain maps $x\colon \mathcal{C}_i \longrightarrow \mathcal{C}_{i+1}$. (For notational simplicity, suppress the index subscript on the chain maps $x$.) Note that each $\mathcal{C}_i = (C_{\bullet,i}, \partial)$ is itself a complex. The **persistent homology** of a persistence complex $\mathcal{P}$ on the interval $[i, j]$, denoted $H_\bullet(\mathcal{P}[i, j])$, is defined to be the image of the homomorphism $H(x^{j-i})\colon H_\bullet(\mathcal{C}_i) \to H_\bullet(\mathcal{C}_j)$ induced by $x^{j-i}$. The homology barcode is an infographic of persistent homology: dim $H_k(\mathcal{P}[i, j])$ is equal to the number of intervals in the barcode of $H_k(\mathcal{P})$ containing the parameter interval $[i, j]$.

## 5.14 The space of natural images

The number, scope, and impact of examples of persistent homology and barcodes are too many to encapsulate: to date, example applications include computer vision [120, 173], Gaussian random fields [4], genetic markers [93], hypothesis testing [43], materials science [199, 213], molecular compression [139], sensor coverage [88], signal processing [107], and much more. One of the first examples of discovering subtle topological structure in a high-dimensional data set came from an examination of *natural images*. A collection of 4167 digital photographs of random outdoor scenes was assembled in the late 1990s by van Hateren and van der Schaaf [294, 236]. Mumford, Lee, and Pederson [237] sampled this data by choosing at random 5000 three-pixel by three-pixel squares within each digital image and retaining the top 20% of these with respect to contrast. The full data set consisted of roughly 8,000,000 vectors in $\mathbb{R}^9$ whose components represent grey-scale intensities. By normalizing with respect to mean intensity and high-contrast images (those away from the origin), and by utilizing a certain norm for contrast, the data set $\mathcal{M}$ was fit on a topological seven-sphere $\mathbb{S}^7 \subset \mathbb{R}^8$.

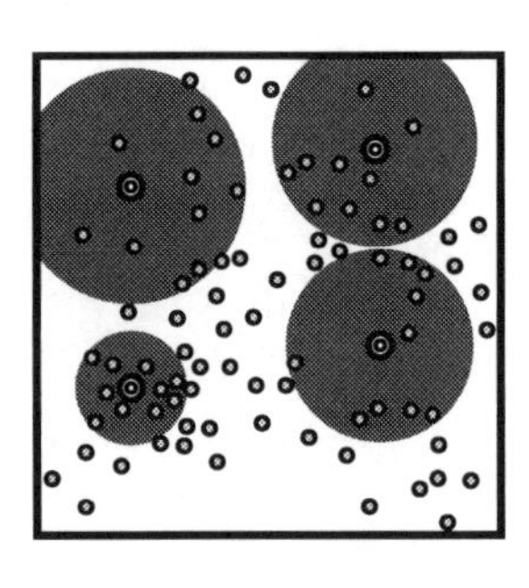

A cursory visualization reveals points distributed seemingly densely over the entire $\mathbb{S}^7$, prompting judicious use of density filtrations. A **codensity** function is used in [59] as follows. Fix a positive integer $k > 0$. For any point $x_\alpha$ in the data set, define $\delta_k(x_\alpha)$ as the distance in $\mathbb{R}^n$ from $x_\alpha$ to $k^{\text{th}}$ nearest neighbor of $x_\alpha$ in the data set. For a fixed value of $k$, $\delta_k$ is a positive distribution over the point cloud which measures the radius of the ball needed to enclose $k$ neighbors. Values of $\delta_k$ are thus inversely related to the point cloud density. The larger a value of $k$ used, the more averaging occurs among neighbors, blurring finer variations. Denote by $\mathcal{M}[k, T]$ the subset of $\mathcal{M}$ in the upper $T$-percent of density as measured by $\delta_k$. This is a two-parameter subset of the point cloud which, for reasonable values of $k$ and $T$, represents an appropriate core.

The first interesting persistent homology computation on this data set occurs at the level of $H_1$. Taking a density threshold of $T = 25$ at neighbor parameter $k = 300$, with 5000 points sampled at random from $\mathcal{M}[k, T]$, computing the barcode for the first homology $H_1$ reveals a unique persistent generator. This indicates that the data set is diffused about a primary circle in the 7-sphere. The structure of the barcode is robust with respect to the random sampling of the points in $\mathcal{M}[k, T]$. In practice, witness complexes from §2.3 provide small enough spaces for computations to be done quickly.

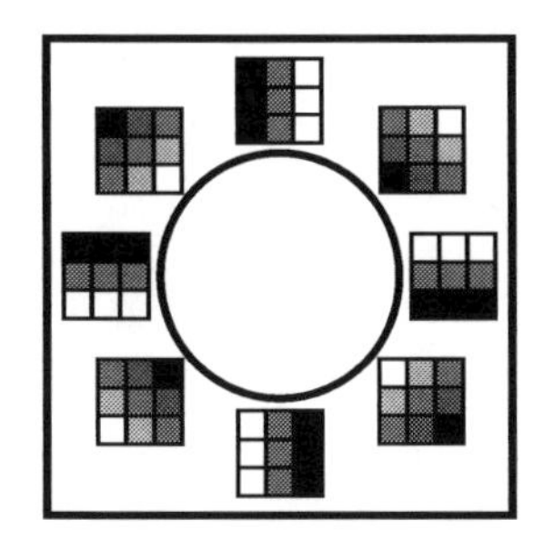

A closer examination of the data corresponding to this primary circle reveals a pattern of 3-by-3 patches with one light region and one dark region separated by a linear transition. This curve between light and dark is linear and appears in a circular family parameterized by the angle of the transition line. As seen from the barcode, this generator is dominant at the threshold and codensity parameters chosen. An examination of the barcodes for the first homology group $H_1$ of the data set filtered by codensity parameter $k = 15$ and threshold $T = 25$ reveals a different persistent $H_1$. The reduction in $k$ leads to less averaging and more localized density sensitivity.

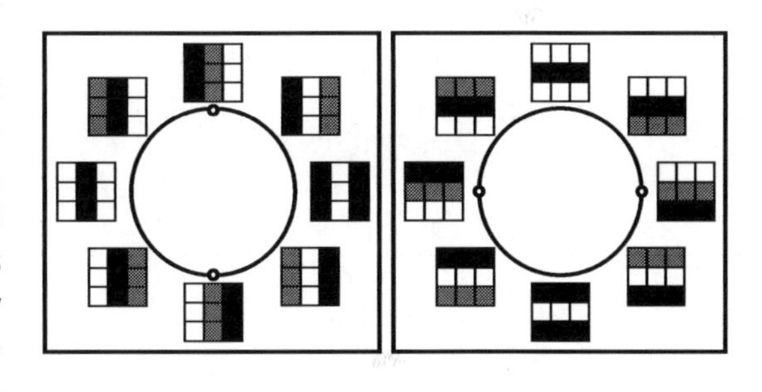

The barcode reveals that the persistent $H_1$ of samples from $\mathcal{M}[k, T]$ has dimension five. This does not connote the presence of five disjoint circles in the data set. Rather, by focusing on the generators and computing the barcode for $H_0$, it is observed that, besides the primary circle from the high-$k$ $H_1$ computation, there are two secondary circles which come into view at the lower density parameter. A close examination of these three circles reveals that each intersects the primary circle twice, yet the two secondary circles are disjoint. As noted in [59], each secondary circle regulates images with three contrasting regions and interpolates between these states and the primary circle. The difference between the two secondary circles lies in their bias for horizontal and vertical stratification respectively.

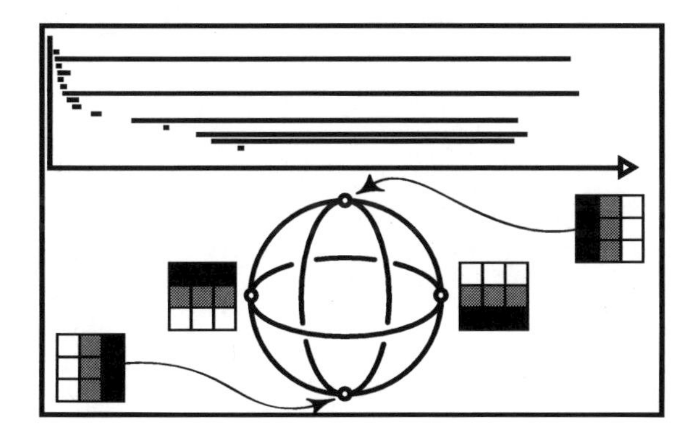

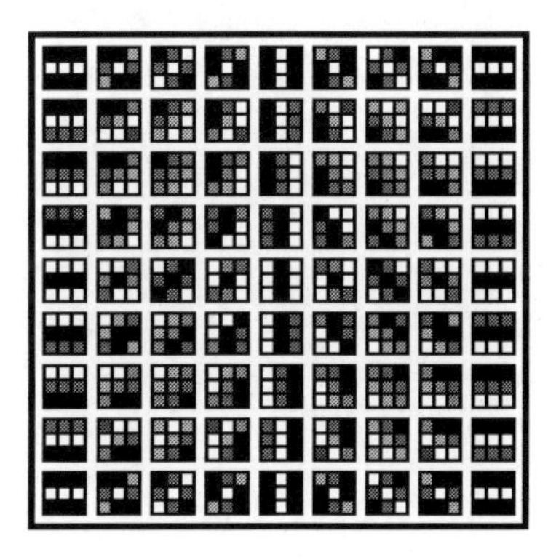

The barcodes for the second persistent homology $H_2$ are more volatile with respect to changes in density and thresholding. This is not surprising: the lowest order terms in any series expansion are always most easily perceived. However, there is indication of a persistent $H_2$ generator (in $\mathbb{F}_2$ coefficients) at certain settings of $k$ and $T$. Combined with the basis of $H_1$ generators, one obtains predictive insight to the structure of the space of high-contrast patches. At certain density thresholds, the $H_2$ barcode suggests a two-dimensional completion of the low-$k$ persistent $H_1$ basis into a Klein bottle $K^2$. The primary and secondary circles appear with the appropriate intersection properties. A comparison of homology computations in $\mathbb{F}_2$ and $\mathbb{F}_3$ coefficients resolves the ambiguity that $H_2(K^2; \mathbb{F}_2) \cong H_2(\mathbb{T}^2; \mathbb{F}_2)$ and verifies that the persistent surface found is $K^2$ and not $\mathbb{T}^2$.

## 5.15 Zigzag persistence

The icon of persistence is the monotone sequence, $\cdots \longrightarrow \bullet \longrightarrow \bullet \longrightarrow \bullet \longrightarrow \cdots$, where arrows connote maps of spaces or chains or the induced homomorphisms on homology. However, other non-monotone sequences are possible and relevant to data management [56]. For example, sequences of the form $\cdots \longrightarrow \bullet \longleftarrow \bullet \longrightarrow \bullet \longleftarrow \cdots$, arise in consistency tests for sampling point clouds as follows.

Given a large set of nodes $\mathcal{Q}$, as in §5.14, it may be infeasible to construct the full Vietoris-Rips complexes and compute persistent homology. A small sample of points is taken (perhaps at random) and a witness complex is constructed as in §2.3: this is indeed the method used for the natural images example in the previous section. To check for accuracy of the sampling, one could compare the homologies resulting from a pair of samples. However, it is not the dimensions of the homology that matter, but the correspondence. Assume that several samples of the data all detect the presence of a single hole. Is it the same hole? Or are there many holes in the true data set, each sampling detecting a different one? Given an ordered sequence of small subsamples, build simplicial complexes $X_i$ based on them. One means of performing comparisons is to build an alternating sequence of spaces and inclusions:

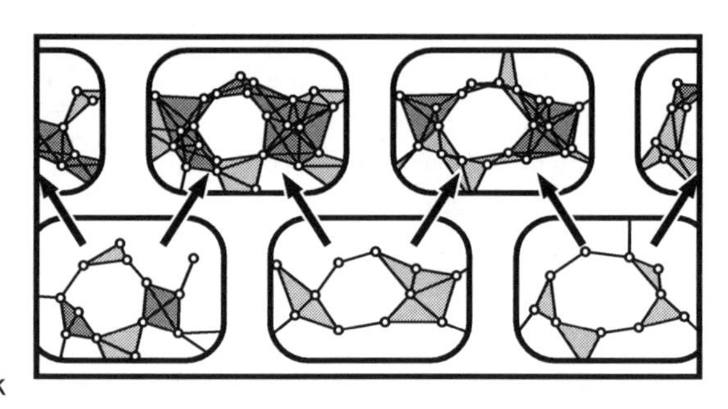

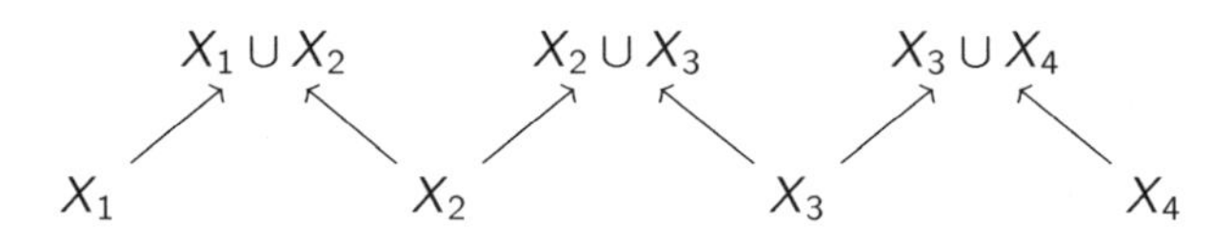


An increase in the complexity of a diagram prompts a concomitant increase in the sophistication of algebra brought to bear. Note that, although not monotone, this sequence is *linear*. A little deeper into representation theory (the classification theorem of Gabriel [136] for quivers of type $A_n$) reveals that the Structure

Theorem classifies indecomposables of *any* linear sequence,

where each $\bullet$ is an $\mathbb{F}$-vector space and each $\longleftrightarrow$ is a linear transformation going either to the left or to the right. An indecomposable interval, as before, is a sequence of copies of $\mathbb{F}$ with identity maps, in this setting respecting the directions of arrows in the original sequence. In classical persistence, all arrows go to the right. In consistency tests, the arrows alternate. The resulting **zigzag persistence** and associated barcodes exist, are computable [61], and are extremely useful. In the context of consistency checks, a long bar in the barcode means that a homological feature is sampled consistently over the sequence.

The moral of these latter sections and of this chapter is that the homology of a sequence is worth more than a sequence of homologies.

# Notes

1. The Snake Lemma and 5-Lemma are two of many wonderfully useful general results in homological algebra: for others, see [142, 193, 300]. The subject of homological algebra does not exactly make for colorful reading, but [142], combined with the mantra that complexes are algebraic representations of spaces, is a good place to start.
2. This chapter on exact sequences is just the beginning of a diagrammatic calculus. The next step is to build a **double complex**, a bi-graded 2-d array of chains $\mathcal{C}_{\bullet,\bullet}$ with horizontal $\partial$ and vertical $\partial'$ chain maps satisfying $\partial\partial' + \partial'\partial = 0$. Such techniques lead quickly to a **spectral sequence**, a structure reminiscent of a book whose pages are double complexes, outfitted with homomorphisms that *turn the pages*. Such structures, though notationally intricate, are quite powerful.
3. The Radon Theorem (in the corollaries of Theorem 5.16) is usually stated in terms of convex hulls of points in Euclidean space, and its proof is often by means of convex geometry. The deeper meaning of the Radon Theorem is, like that of the Helly Theorem in §6.6, topological in nature. All the Borsuk-Ulam type results are greatly generalizable, with the *Colored Tverberg Theorem* being one of the most general [41, 219]. It is to be suspected that such generalizations of the Borsuk-Ulam Theorem are useful – perhaps to economics most readily [33, 180].
4. One should not underestimate the utility of local degree computations as in §5.7; the ability to infer global features from a small number of local measurements is greatly desirable in the sciences.
5. The Cellular Approximation Theorem casually alluded to in the proof of Theorem 5.19 is deep and significant: any map between cell complexes can be homotoped to a cellular map relative to a subcomplex on which the map is already cellular [176].
6. Every proof of the Hex Theorem the author has seen uses two different strategies for the existence and uniqueness of the winner. The (new) proof presented here – though not the simplest or most direct – is pleasant in that a single diagram yields both.
7. The extension of the coverage criterion to higher-dimensional networks is possible, but less clean than the 2-d case presented here. The difficulty resides in specifying the boundary of the domain in question. For regions in the plane, it is not difficult to imagine choosing a cycle or even establishing a *fence* of sensors. In $\mathbb{R}^n$, one must specify a triangulated $n-1$ cycle: this seems awkward to the author, but some applications may permit this condition, in which case a simple modification of the existing proofs

suffices. A persistent homology approach is given in [88].

8. Why was the Čech complex of the sensor cover not used to obtain a coverage criterion? Determining the depths of overlaps of coverage sets requires explicit distances, hence, coordinates; *cf.* the use of Vietoris-Rips complexes versus Čech complexes in point cloud data.
9. As stated, the criterion for computing homological coverage in sensor networks is *centralized*, in the sense that nodes must upload connectivity data to a central computer. More desirable is a *decentralized* or *distributed* computation, performable by nodes communicating with neighbors. Algorithms for decentralized computation of homology have just recently emerged [30, 97, 209, 210, 267].
10. The available perspectives on persistent homology for an author to choose from are daunting, and the treatment in this chapter is necessarily elementary. See the books by Edelsbrunner-Harer [104] and Zomorodian [305] (and, eventually, the book by Blumberg, Carlsson, and Vejdemo-Johansson) for more details from several perspectives. In particular, there are different conventions for representing persistence information beyond the barcodes of §5.13.
11. Since the core idea of persistent homology is little more that iterated functoriality, it is difficult to declare when it was discovered. Early formulations of the notion of persistent homology appear independently in the work of Frosini and Ferri [53, 132, 133], the thesis of Robins [251], and the paper of Delfinado and Edelsbrunner [91]. The subsequent history is one of simultaneous crystallization of theorems, algorithms, and applications about this notion of persistence.
12. From the classification theorem of Gabriel [136], it is shown that not only linear sequences, but certain Dynkin diagrams (of types A, D, and E) have a nice structure theorem classifying indecoposables. Other results from representation theory are poised to impact persistence (such as the Auslander-Reiten quiver [110]). This is especially salient in the context of multi-dimensional persistence – an algebraically challenging scenario [62].
13. Besides being a clear and convenient descriptor for topological data analysis, barcodes possess a very useful **stability** in the context of point-cloud data – nearby point-clouds return barcodes which are close in a certain (*interleaving*) distance: see §7.2 for an example and §10.6 for details. It is this stability that makes barcodes useful in describing noisy point-cloud approximations.

## Chapter 6

# Cohomology

Cohomology is the mirror-image of homology, flipping geometric intuition for algebraic dexterity. The duality implicit in this theory is a subtle and easily underestimated tool in algebraic topology.

## 6.1 Duals

The first form of duality one encounters in Mathematics is combinatorial.[1]

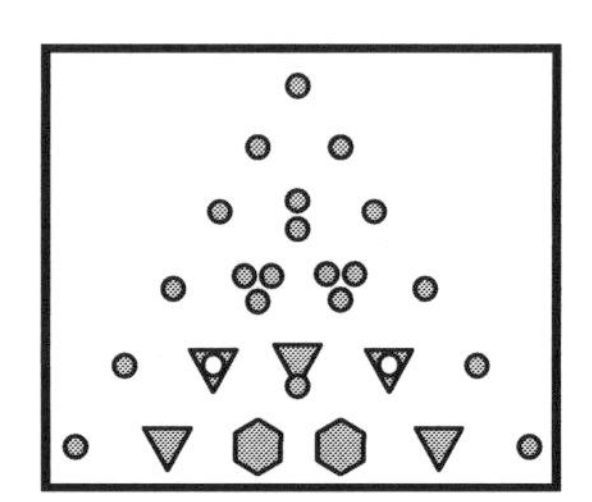

The number of ways to choose $k$ items from $n \geq k$ is exactly the same as the number of ways to (not) choose $n - k$ items. This symmetry in counting manifests itself in numerous numerical miracles, from the reflection symmetry of Pascal's triangle, to the fact that $\chi(\mathbb{S}^{2n+1}) = 0$. The manner in which duality presents itself in Topology is best discovered through the familiar forms of linear algebra and calculus.

The **dual space** of a real vector space $V$ is $V^\vee$, the vector space of all linear functionals $V \to \mathbb{R}$. The dual space satisfies $\dim(V^\vee) = \dim(V)$, and $(V^\vee)^\vee \cong V$ for $V$ finite-dimensional. There is a corresponding notion of duality for linear transformations. If $f\colon V \to W$ is linear, the **dual map** or **adjoint** of $f$ is $f^\vee\colon W^\vee \to V^\vee$ given by $(f^\vee(\eta))(v) = \eta(f(v))$. Note how the dual transformation reverses the direction of the map: it is the archetypal construct of this chapter.

**Example 6.1 (Gradients)** ⊚

Dual vector spaces play an important role in calculus on manifolds. The **cotangent space** to a manifold $M$ at $p \in M$ is the vector space dual $T_p^*M = (T_pM)^\vee$ to the tangent space. The cotangent spaces, like their tangent space duals, fit together to form a bundle of vector spaces over $M$, the **cotangent bundle** $T^*M$. The analogue of a vector field is a **1-form**: a choice of $T_p^*M$ continuous in $p$. For example, given a real-valued function $f\colon M \to \mathbb{R}$, the **gradient** of $f$ is the 1-form $df$ which, in local coordinates $\{x_i\}_1^n$, evaluates to

$$df = \sum_{i=1}^{n} \frac{\partial f}{\partial x_i} dx_i\,,$$

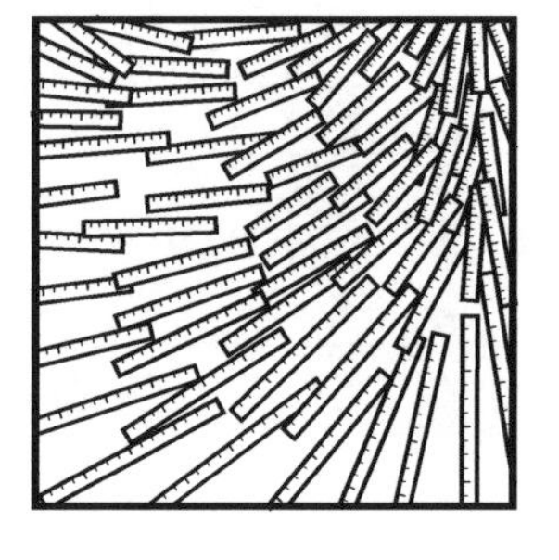

where $dx_i$ is the dual to the $x_i$ unit tangent vector. The *Chain Rule* implies that $df$ is independent of the local coordinates used to express it. It is a common mistake to conflate the gradient of a function with a vector field $\nabla f$: this is permissible in Euclidean space, but the pairing is not canonical. It is best to imagine a gradient $df$ not as a vector field, but as a *ruler field* – a field of rulers with direction, orientation, and scale – along which tangent vectors are measured. ⊚

[1]That grammatical dualities cannot fail to precede even these is not a false statement.

## 6.2 Cochain complexes

A **cochain complex** is a sequence $\mathcal{C} = (C^\bullet, d)$ of **R**-modules $C^k$ (**cochains**) and module homomorphisms $d^k \colon C^k \to C^{k+1}$ (**coboundary maps**) with the property that $d^{k+1} \circ d^k = 0$ for all $k$. The **cohomology** of a cochain complex is,

$$H^k(\mathcal{C}) = \ker d^k / \operatorname{im} d^{k-1}.$$

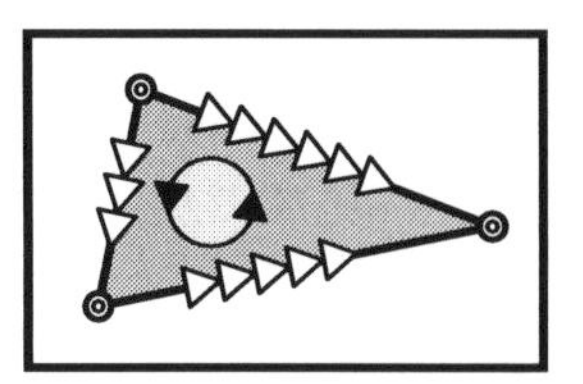

Cohomology classes are equivalence classes of **cocycles** in $\ker d$. Two cocylces are **cohomologous** if they differ by a **coboundary** in $\operatorname{im} d$. The simplest means of constructing cochain complexes is to dualize a chain complex $(C_\bullet, \partial)$. Given such a complex (with coefficients in, say, **R**), define $C^k = C_k^\vee$, the module of homomorphisms $C_k \to \mathbf{R}$. The coboundary $d$ is the adjoint of the boundary $\partial$, so that

$$d \circ d = \partial^\vee \circ \partial^\vee = (\partial \circ \partial)^\vee = 0^\vee = 0.$$

The coboundary operator $d$ is explicit: $(df)(\tau) = f(\partial\tau)$. For $\tau$ a $k$-cell, $d$ implicates the **cofaces** – those $(k+1)$-cells having $\tau$ as a face.

**Example 6.2 (Simplicial cochains)** ⊚

Examples in the simplicial category are illustrative. Consider a triangulated disc with a 1-cocycle on edges using $\mathbb{F}_2$ coefficients.

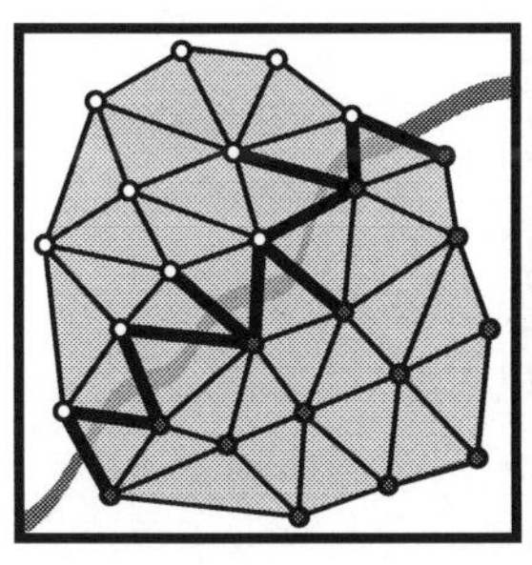

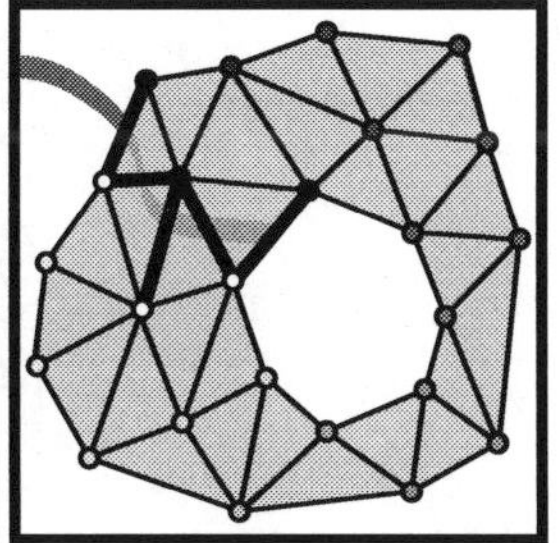

Any such 1-cocycle is the coboundary of a 0-cochain which labels vertices with 0 and 1 *on the left* and *on the right* of the 1-cocycle, so to speak: this is what a trivial class in $H^1(\mathbb{D}^2)$ looks like. On the other hand, if one considers a surface with some nontrivial $H_1$ – say, an annulus – then one can construct a similar 1-cocycle that is nonvanishing in $H^1$. The astute reader will notice the implicit relationship between such cocycles and gradients of a local *potential* over the vertices, with cohomology class in $H^1$ differentiating between those which are or are not globally expressible as a gradient of a potential. ⊚

**Example 6.3 (Integration)** ⊚

Consider a chain complex $\mathcal{C}$ with **R**-coefficients, freely generated by (oriented) simplices $\{\sigma\}$ in a simplicial complex $X$. The dual basis cochains can be thought of as characteristic functions $\{\mathbb{1}_\sigma\}$. The obvious pairing between $\mathcal{C}$ and its dual $\mathcal{C}^\vee$ in these bases permits an integral interpretation: for basis elements $\sigma$ and $\tau$, define $\int_\sigma \mathbb{1}_\tau$ to be the evaluation taking the value 1 if and only if $\tau = \sigma$ and 0 else.

This notation is illustrative. By definition of the coboundary $d$ as the dual of the boundary $\partial$, one has $(d\mathbb{1}_\sigma)(\tau) = \mathbb{1}_\sigma(\partial\tau)$. Using linearity and the integral notation, one has for all cochains $\alpha \in C^p$ and chains $c \in C_{p+1}$,

$$\int_c d\alpha = \int_{\partial c} \alpha. \tag{6.1}$$

This notation should appeal to scientists and engineers who know from early education the utility of Stokes' Theorem. The hint of differentials is no accident: calculus and cohomology are integrally bound, see §6.9. ⊚

**Example 6.4 (Kirchhoff's voltage rule)** ⊚

Consider an electric circuit as a 1-dimensional cell complex, with circuit elements (resistors, capacitors, etc.) located in the interiors of edges. **Kirchhoff's voltage rule** states that the sum of the voltage potential differences across any loop in the circuit is zero. In the language of this chapter: *voltage is a 1-coboundary.* ⊚

**Example 6.5 (Cuts and flows)** ⊚

Consider a **directed graph** $X$: a 1-d cell complex with each edge oriented. A classical set of problems in combinatorial optimization concerns flows on $X$. Choose two nodes of $X$ to be the source (s) and target (t) nodes, and assume that the graph is connected from s $\rightarrow$ t (respecting direction). A **flow** on $X$ is an assignment of coefficients (in,

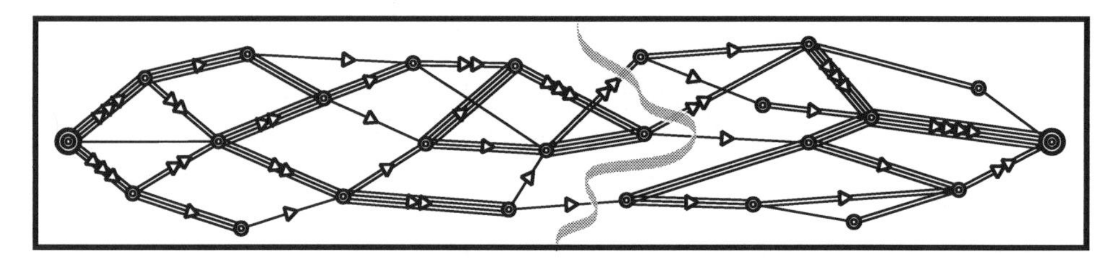

say, $\mathbb{N}$) to edges of $X$ so that, for each node except s and t, the sum of the in-pointing edge flow values equals the sum of the out-pointing edge flow values. Motivated by problems in transportation and railway shipping, the classical **max flow problem** seeks to maximize the **flow value** (the net amount flowing from s or, equivalently, into t) over all flows. The problem is constrained in that each edge is assigned a **capacity** that dominates the possible flow value on that edge.

The **max-flow-min-cut theorem** states that the maximum possible flow value equals the minimal **cut value** as follows, where a **cut** is a subset of edges of $X$ whose removal disconnects s and t (there are no longer directed paths from source to target) and its value is the sum of the edge capacities over the cut. This theorem is a minimax-type theorem in which a duality between flows and cuts is prominent.

In the language of this chapter, given $X$, s, and t, cuts and flows correspond to various chains and cochains. For clarity, assume that $X$ has been augmented to have a single *feedback* edge from t to s with infinite capacity. In this case, any flow is an element of $H_1(X;\mathbb{Z})$, where the sign of the $\mathbb{Z}$-coefficients is consistent with the edge orientations: the condition for being a 1-cycle is *precisely* that of a conservative flow. A cut gives a 1-cocycle which is nontrivial in $H^1(X;\mathbb{Z})$, since, with the feedback edge, there is no way to separate vertices of $X$ on either *side* of the cut. The reader rightly suspects that flow/cut duality is at heart a co/homological duality: see Example 10.25. ◎

**Example 6.6 (Arbitrage)** ◎

Certain simple examples from economics and social networks are expressible in the language of chains and cochains: the following is from [185]. Consider a static exchange market on a set of $N$ commodities, $V = \{v_i\}_1^N$. Assume the existence of **exchange rates** – coefficients $\epsilon_{ij} > 0$ such that one unit of $v_i$ is worth $\epsilon_{ij}$ units of $v_j$, with the natural symmetry that $\epsilon_{ji} = \epsilon_{ij}^{-1}$. Build a graph $X$, with nodes $V$ and edge set equal to all $\{v_i, v_j\}$ such that exchange rates $\epsilon_{ij}$ are known.

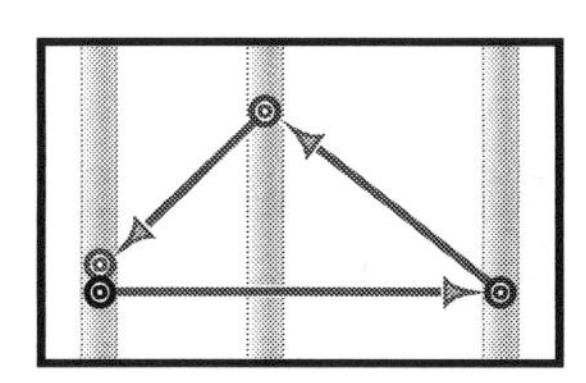

There is a natural $\mathbb{R}$-valued 1-cochain given by exchange rates as follows. Orient the 1-cells of $X$ in an arbitrary but fixed fashion, and let the 1-cochain $\xi$ evaluate on an oriented edge $(v_i, v_j)$ to $\ln \epsilon_{ij}$. In a perfect exchange system, $\xi$ is a cocycle: any cyclic sequence of trades from commodity $v_i$ back to $v_i$ returns the same amount (assuming zero transaction costs). The failure of $\xi$ to be a cocycle indicates an arbitrage – a sequence of trades resulting in net gain. It is fascinating to consider what happens when the goods at issue are more complex financial items, whose restrictions on availability and convertibility yield richer topological objects both noisy (edge values are uncertain) and dynamic (time-varying). The reader may enjoy contemplating whether higher-order algebraic coefficients and/or higher-dimensional simplicial models (say, the flag complex) yield a useful approach to characterizing arbitrage in complex systems. ◎

## 6.3 Cohomology

The definition of cohomology in terms of dualizing chain complexes, though less than intuitive, is efficient. By dualizing results of the previous two chapters, one immediately obtains the following:

1. Cohomology theories: cellular, singular, local, Čech, relative, reduced; all with arbitrary coefficients;
2. Functoriality and the homotopy invariance of cohomology;
3. Cohomological long exact sequences of pairs; the Mayer-Vietoris sequence; excision.

To correctly interpret and use these results *arrows must be reversed.* For example, the induced homomorphisms on cohomology twist composition: $H(f \circ g) = H(g) \circ H(f)$, in keeping with the way that duals of linear transformations behave. The Snake Lemma

(Lemma 5.5) holds, but when dualizing a short exact sequence of chains to cochains, the sequence flips. This manifests itself in, *e.g.*, the long exact sequence of a pair $(X, A)$ as follows. The short exact sequence of cochain complexes,

$$0 \longrightarrow C^\bullet(X,A) \xrightarrow{j^\bullet} C^\bullet(X) \xrightarrow{i^\bullet} C^\bullet(A) \longrightarrow 0 ,$$

becomes the long exact sequence,

$$\longrightarrow H^{n-1}(A) \xrightarrow{\delta} H^n(X,A) \xrightarrow{H(j)} H^n(X) \xrightarrow{H(i)} H^n(A) \longrightarrow ,$$

where, in this text, the induced and connecting homomorphisms have the same notation in homology and cohomology and are distinguishable via context and direction. The novice reader should as an exercise write out the Mayer-Vietoris sequence on cohomology.

The beginner may be deflated at learning that cohomology seems to reveal no new data, at least as far as linear algebra can count.

**Theorem 6.7 (Universal Coefficient Theorem).** *For $X$ a space with finite-dimensional homology over a field $\mathbb{F}$, $H^n(X;\mathbb{F}) \cong H_n(X;\mathbb{F})^\vee$.*

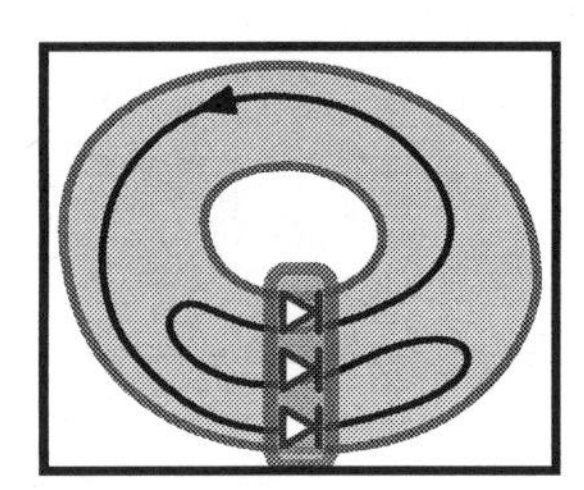

The situation is much more delicate in $\mathbb{Z}$-coefficients, but it is nevertheless true that cohomology is determined by knowing the homology and certain algebraic properties of the coefficient ring. Students may wonder why they should bother with this seemingly supererogatory cohomology, as it alike to homology in every respect except intuition. In the beginning, one should repeat the mantra that duality often simplifies algebraic difficulties.

**Example 6.8 (Connectivity)** ⊚

Note that the dimension of $H^0$, like that of $H_0$, yields connectivity data. The proof in the case of cohomology is *simpler* than for homology, since, by definition, $H^0 = \ker d^0$. There is nothing to quotient out, due to arrow reversal. In the cellular case, elements of $H^0(X)$ are functions on vertices whose oriented differences along edges (a 'discrete derivative') is everywhere zero – these equate to the singular interpretation of locally-constant functions on $X$. In either case, a suitable basis consists of characteristic functions of connected components of $X$. Note the differences. In homology, $H_0$ determines the number of path-connected components (homologous 0-cycles are connected by 1-chain paths) while cohomology $H^0$ measures connected components (as seen by functionals). ⊚

**Example 6.9 (1-Cocycles)** ⊚

One cartoon for understanding this distinction between local and global coboundaries is the popular optical illusion of the *impossible tribar*. When one looks at the tribar, the drawn perspective is locally realizable – one can construct a local depth function. However, a global depth function cannot be defined. The impossible tribar is a cartoon of a non-zero class in $H^1$ (properly speaking, $H^1(\mathbb{S}^1; \mathbb{R}^+)$, using the multiplicative reals for coordinates as a way to encode projective geometry [243]).

An even more cartoonish example that evokes nontrivial 1-cocycles is the popular game of *Rock, Paper, Scissors*, for which there are local but not global ranking functions. A local gradient of *rock-beats-scissors* does not extend to a global gradient. Perhaps this is why customers are asked to conduct rankings (*e.g.*, *Netflix* movie rankings or *Amazon* book rankings) as a 0-cochain (*"how many stars?"*), and not as a 1-cochain (*"which-of-these-two-is-better?"*): nontrivial $H^1$ is, in this setting, undesirable. The *Condorcet paradox* – that locally consistent comparative rankings can lead to global inconsistencies – is a favorite topic in voting theory. Its best explanation, cohomology, is less popular. ⊚

## 6.4 Poincaré duality

Homology and cohomology of manifolds express duality as a dimensional symmetry. Based on data from spheres, compact orientable surfaces, tori, and the Künneth formula in §4.2, one might guess that for $M$ a compact orientable $n$-manifold and coefficients in a field, $\dim H_k M = \dim H_{n-k} M$, and likewise with cohomology. This is true and is a version of duality due to Poincaré.

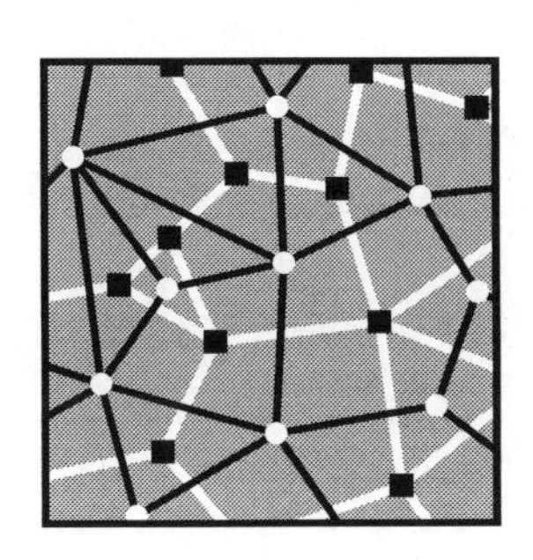

At the level of cellular homology, this duality has a geometric interpretation. Consider, *e.g.*, a compact surface with a polyhedral cell structure, and let $\mathcal{C}$ be the cellular chain complex with $\mathbb{F}_2$ coefficients. There is a dual polyhedral cell structure, yielding a chain complex $\overline{\mathcal{C}}$, where the dual cell structure places a vertex in the center of each original 2-cell, has 1-cells transverse to each original 1-cell, and, necessarily, has as its 2-cells neighborhoods of the original vertices. Each dual 2-cell is a polyhedral $n$-gon, where $n$ is the degree of the original 0-cell dual. Note that these cell decompositions are truly dual and have the effect of reversing the dimensions of cells: $k$-cells generating $C_k$ are in bijective correspondence with $(2-k)$-cells generating a modified cellular chain group $\overline{C}_{2-k}$. The dual complex consisting of $\overline{C}^k := \overline{C}_k^\vee$ and $\overline{d} = \overline{\partial}^\vee$ entwines with $C_\bullet$ in a diagram:

$$\begin{array}{ccccccccc} 0 & \longrightarrow & C_2 & \xrightarrow{\partial} & C_1 & \xrightarrow{\partial} & C_0 & \longrightarrow & 0 \\ & & \downarrow\cong & & \downarrow\cong & & \downarrow\cong & & \\ 0 & \longrightarrow & \overline{C}^0 & \xrightarrow{\overline{d}} & \overline{C}^1 & \xrightarrow{\overline{d}} & \overline{C}^2 & \longrightarrow & 0 \end{array}$$

The reader should check that the vertical maps are isomorphisms and, crucially, that the diagram is commutative. The equivalence of singular and cellular co/homology, along with Theorem 6.7, implies that, for a compact surface with $\mathbb{F}_2$ coefficients, $H_k \cong H^{2-k} \cong (H_{2-k})^\vee \cong H_{2-k}$. Though this style of proof generalizes to higher-dimensional manifolds, a better explanation for the symmetries present in co/homology is more algebraic, and proceeds in a manner that adapts to non-compact manifolds as well. To unwrap this, a modified cohomology theory is helpful.

Given a (singular, simplicial, cellular) cochain complex $\mathcal{C}^\bullet$ on a space $X$, consider the subcomplex $\mathcal{C}^\bullet_c$ of cochains which are compactly supported: each cochain is zero outside some compact subset of $X$. (In the simplicial or cellular setting, this is equivalent to building cochains from a finite number of basis cochains.) The coboundary map restricts to $d\colon C^k_c \to C^{k+1}_c$ and $d^2 = 0$, yielding a well-defined **cohomology with compact supports**, $H^\bullet_c(X)$. This cohomology satisfies the following:

1. $H^\bullet_c$ is not a homotopy invariant, but is a proper-homotopy (and hence a homeomorphism) invariant.
2. $H^k_c(\mathbb{R}^n) = 0$ for all $k$ except $k = n$, in which case it is of rank one.
3. $H^\bullet_c(X) \cong H^\bullet(X)$ for $X$ compact.
4. $H^\bullet_c(X) \cong H^\bullet(X, X{-}K)$ for $K$ a sufficiently large compact set.

This can be subtle: in brief, one orders compact sets by inclusion and uses induced maps to take a limit. See, *e.g.*, [176, 3.3]. Compactly supported cohomology cleanly expresses manifold duality:

**Theorem 6.10 (Poincaré Duality).** *For $M$ an $n$-manifold, there is a natural isomorphism* $\mathrm{PD}\colon H_k(M;\mathbb{F}_2) \xrightarrow{\cong} H^{n-k}_c(M;\mathbb{F}_2)$.

For $M$ a compact $n$-manifold, $H_k(M;\mathbb{F}_2) \cong H^{n-k}(M;\mathbb{F}_2)$. For orientable manifolds, the theorem holds with any field coefficients. Integer coefficients are more problematic: torsional elements lag in duality [46, 176]. With a slight change of perspective (in §6.9), a more precise form of PD will be given.

## 6.5 Alexander duality

Poincaré duality can be adapted to several related settings involving manifolds. Among the most useful is **Alexander duality**.

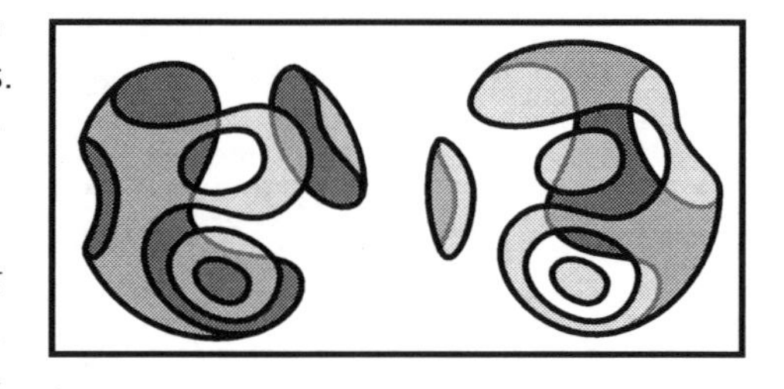

**Theorem 6.11 (Alexander Duality).** *Let $A \subset \mathbb{S}^n$ be compact, nonempty, proper, and locally-contractible. There is an isomorphism*

$$\mathrm{AD}\colon \tilde{H}_k(\mathbb{S}^n{-}A) \xrightarrow{\cong} \tilde{H}^{n-k-1}(A).$$

**Proof.** The condition on $A$ is a form of tameness and is crucial, allowing one to choose a small open neighborhood $U$ of $A$ that deformation-retracts onto $A$. For $k > 0$,

$$\begin{aligned} H_k(\mathbb{S}^n - A) &\cong H_c^{n-k}(\mathbb{S}^n - A) && \text{[Poincaré duality]} \\ &\cong H^{n-k}(\mathbb{S}^n - A, (\mathbb{S}^n - A) - (\mathbb{S}^n - U)) && \text{[compact supports]} \\ &= H^{n-k}(\mathbb{S}^n - A, U - A) && \\ &\cong H^{n-k}(\mathbb{S}^n, U) && \text{[excision]} \\ &\cong \tilde{H}^{n-k-1}(U) && \text{[long exact sequence of pair]} \\ &\cong \tilde{H}^{n-k-1}(A). && \text{[deformation retraction]} \end{aligned}$$

Slight modifications are required for $k = 0$. The theorem holds for all coefficients. ⊙

## 6.6 Helly's Theorem

The following theorem is a classic result in convex geometry and geometric combinatorics. Using co/homology yields a transparent proof.

**Theorem 6.12 (Helly's Theorem).** *Let $\mathcal{U} = \{U_\alpha\}$ be a collection of $M > n+1$ compact convex subsets of $\mathbb{R}^n$ such that every $(n+1)$-tuple of distinct elements of $\mathcal{U}$ has a point in common. Then all elements of $\mathcal{U}$ have a point in common.*

**Proof.** Induct on $M$, beginning at $M = n+2$. Consider the nerve $\mathcal{N}(\mathcal{U})$ of $\mathcal{U}$. It is a subcomplex of the $(n+1)$-simplex which, by hypothesis, contains all faces. If the common intersection is empty, then $\mathcal{N}(\mathcal{U}) = \partial\Delta^{n+1} \simeq \mathbb{S}^n$. As the cover $\mathcal{U}$ is by convex sets, it is an acyclic cover (all nonempty intersections are homologically acyclic) and, via Theorem 2.4,

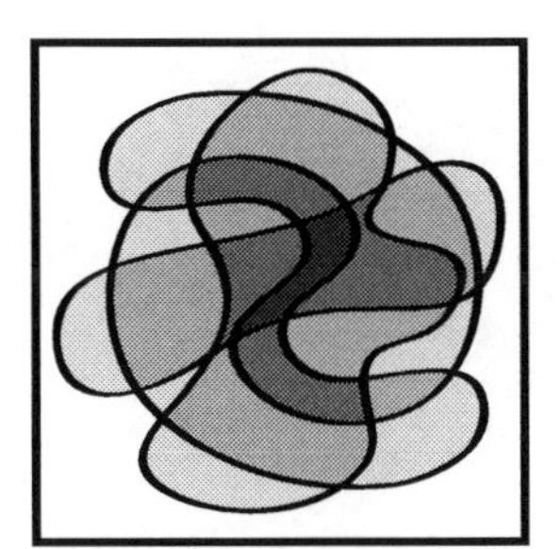

$$\check{H}_\bullet(\mathcal{U}) \cong H_\bullet(\mathcal{N}(\mathcal{U})) \cong H_\bullet\left(\cup_\alpha U_\alpha\right).$$

In other words, $\mathcal{N}(\mathcal{U}) \simeq \mathbb{S}^n$ has the homology of a subset of $\mathbb{R}^n$. However, it is impossible for a subset $A \subset \mathbb{R}^n$ to have the homology type of $\mathbb{S}^n$, thanks to Alexander duality, as follows. Note that the hypotheses for Theorem 6.11 are satisfied, and therefore,

$$H_n(A) \cong \tilde{H}^{n-n-1}(\mathbb{R}^n - A) = \tilde{H}^{-1}(\mathbb{R}^n - A) = 0,$$

since $\tilde{H}^{-1} = 0$ for all nonempty spaces. Thus, $\mathcal{N}(\mathcal{U}) \not\simeq \mathbb{S}^n$, and the cover $\mathcal{U}$ must have a common intersection point. The induction step is a simple modification of this proof. ⊙

The reader may have seen proofs of Helly's Theorem based on convex geometry or functional analysis. One benefit of the topological approach is that extensions to the non-compact and non-convex world are natural and easily discerned. The critical ingredient is that the resulting cover $\mathcal{U}$ is acyclic.

## 6.7 Numerical Euler integration

Alexander duality is a key step in developing a highly effective numerical method for performing integration with respect to Euler characteristic over a non-localized planar network. Recall from §3.7 that certain problems in data aggregation over a network are expressible as an integral with respect to $d\chi$ over a tame space $X$.

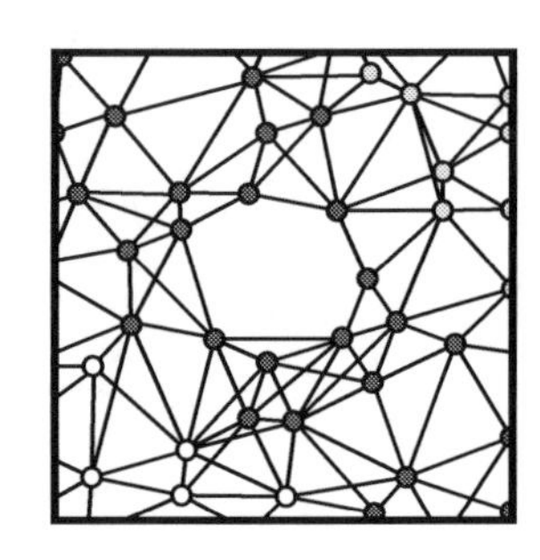

This assumption of integrating sensors over a continuous domain is highly unrealistic; however, expressing the true answer as an integral over a continuum provides a clue as to how to approximate the result over a discretely sampled domain. Given an integrand $h \in \mathrm{CF}(\mathbb{R}^n)$ sampled over a discrete set, computational formulae such as Equation (3.10) suggest that the estimation of the Euler characteristics of the upper excursion sets is an effective approach. However, if the sampling occurs over a network with communication links, then it is potentially difficult to approximate those Euler characteristics. Taking the flag complex of the network can lead to the existence of *fake holes* – higher-dimensional spheres (*cf.* §2.2) that ruin an Euler characteristic approximation.

**Proposition 6.13 ([24]).** *For $h\colon \mathbb{R}^2 \to \mathbb{N}$ constructible and upper semi-continuous,*

$$\int_{\mathbb{R}^2} h\, d\chi = \sum_{s=0}^{\infty} \left(\beta_0\{h > s\} - \beta_0\{h \leq s\} + 1\right), \tag{6.2}$$

*where the zero*$^{\text{th}}$ *Betti number* $\beta_0 = \dim H_0$ *is the number of connected components.*

**Proof.** Let $A$ be a compact nonempty tame subset of $\mathbb{R}^2$. From the homological definition of the Euler characteristic and compactness of $A$,

$$\chi(A) = \sum_{k=0}^{\infty} (-1)^k \dim H_k(A),$$

where $H_\bullet$ denotes singular homology in field coefficients. Since $A \subset \mathbb{R}^2$, $H_k(A) = 0$ for all $k > 1$, and it suffices to compute $\chi(A) = \dim H_0(A) - \dim H_1(A)$. By a slight modification of Alexander duality,

$$\dim H_1(A) = \dim \tilde{H}^0(\mathbb{S}^2 - A) = \dim H_0(\mathbb{R}^2 - A) - 1,$$

where the last equality (if not obvious) follows from a long exact sequence on the pair $(\mathbb{S}^2 - A, \mathbb{R}^2 - A)$. Since $h$ is upper semi-continuous, each of the upper excursion sets $A = \{h > s\}$ is compact. Noting that $\mathbb{R}^2 - A = \{h \leq s\}$, one has:

$$\int h\, d\chi = \sum_{s=0}^{\infty} \chi\{h > s\} = \sum_{s=0}^{\infty} \dim H_0(\{h > s\}) - (\dim H_0(\{h \leq s\}) - 1).$$

$\odot$

**Corollary 6.14.** *A sufficient sampling condition to ensure exact computation of $\int h\, d\chi$ over a planar network is that the network correctly samples the connectivity of all the upper and lower excursion sets of h.*

This formulation is extremely important to numerical implementation of this integration theory to planar sensor networks, which, in practice, may be both non-localized and not dense enough to sufficiently cover regions [79].

## 6.8 Forms and Calculus

In the setting of smooth manifolds, cohomology flows naturally from multivariable calculus [46, 169]. One begins with multilinear algebra. Given a $\mathbb{R}$-vector space $V$, let $\mathbf{\Lambda}(V)$ denote the algebra of **forms** on $V$ – alternating multilinear maps from products of $V$ to $\mathbb{R}$. Given a basis $\{x_i\}_1^n$ for $V$, explicit generators for $\mathbf{\Lambda}(V)$ are given by the dual 1-forms $dx_i \in V^\vee$, where $dx_i \colon V \to \mathbb{R}$ returns the $x_i^{\text{th}}$ coordinate of a vector in $V$. Scalar multiplication in $\mathbf{\Lambda}(V)$ is over $\mathbb{R}$ and the sum is induced by that on $V^\vee$. The product in the algebra $\mathbf{\Lambda}(V)$ is called the **wedge product** and denoted $\wedge$. It is alternating, meaning, in particular, that $dx_i \wedge dx_j = -dx_j \wedge dx_i$ for all $i$ and $j$. The algebra $\mathbf{\Lambda}(V)$ is graded:

$$\mathbf{\Lambda}(V) := \bigoplus_{p=0}^{\infty} \mathbf{\Lambda}^p(V),$$

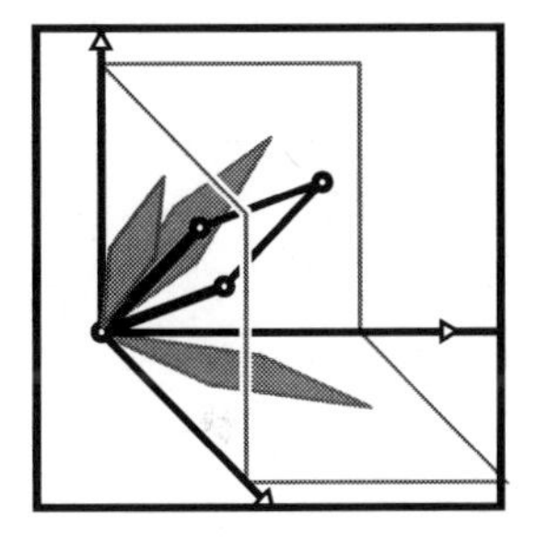

where $\mathbf{\Lambda}^p(V)$ is the vector space of $p$-forms, with basis $dx_{i_1} \wedge \cdots \wedge dx_{i_p}$ for $1 \leq i_1 < \cdots < i_p \leq n = \dim V$. A $p$-form takes as its argument an ordered $p$-tuple of vectors in $V$ and returns a real number in a manner that is multilinear and alternating, *cf.* determinants. The uniqueness of the determinant implies that $\dim \mathbf{\Lambda}^n = 1$; the alternating property of $\wedge$ implies that $\mathbf{\Lambda}^p = 0$ for all $p > n = \dim V$.

These pointwise constructions pass to manifolds, yielding **differential forms**. Recall from §1.3 that for an $n$-manifold $M$, the tangent bundle $T_*M$ is a collection of $n$-dimensional vector spaces, parameterized by points in $M$. A (smooth) vector field $V$ is a choice of elements of $T_xM$ varying (smoothly) in $x$, or, more precisely, a **section** taking $x \mapsto V(x) \in T_xM$. In like manner, the vector spaces of $p$-forms, $\mathbf{\Lambda}^p(T_xM)$, collectively form a "bundle" varying smoothly with $x \in M$, of which the cotangent bundle $T^*M$ is the case $p = 1$. As with the gradient 1-forms of Example 6.1, a ***p*-form field** (shortened to *p-form* in practice) is a section $\alpha \colon M \to \mathbf{\Lambda}^p(T_*M)$ giving $\alpha_x \in \mathbf{\Lambda}^p(T_xM)$ varying smoothly in $x$. The space of all such sections – the $p$-form fields on $M$ – is denoted $\Omega^p = \Omega^p(M)$. On all manifolds, $\Omega^0(M) = C^\infty(M; \mathbb{R})$, since $\dim \mathbf{\Lambda}^0 = 1$. Likewise, $\Omega^p(M) = 0$ for all $p > \dim M$. Algebraic operations on $\mathbf{\Lambda}$ pass to operations on $\Omega := \oplus_p \Omega^p$ operating pointwise. For example, the wedge product extends to $\wedge \colon \Omega^p \times \Omega^q \to \Omega^{p+q}$.

In the passage from $\mathbf{\Lambda}$ to $\Omega$, form fields change from point-to-point: such changes are measured by a derivative. The appropriate differential operator for forms is the **exterior derivative** $d\colon \Omega^p \to \Omega^{p+1}$, which has the following properties:

1. $d$ is linear with respect to addition of forms and scalar multiplication;
2. $d$ satisfies a Leibniz rule: $d(\alpha \wedge \beta) = d\alpha \wedge \beta + d\beta \wedge \alpha$;
3. on 0-forms, $d$ is the usual differential operator $d\colon f \mapsto df$; and
4. $d(df) = 0$ for all 0-forms $f$.

**Example 6.15 (Vector calculus on $\mathbb{R}^3$)** ⊙

On Euclidean $\mathbb{R}^3$, 1-forms $\alpha \in \Omega^1$ are representable as $\alpha = f_x dx + f_y dy + f_z dz$, where the coefficient functions $f_i$ are smooth. A 2-form, $\beta \in \Omega^2$, is representable as $\beta = g_x dy \wedge dz + g_y dz \wedge dx + g_z dx \wedge dy$. Every 3-form on $\mathbb{R}^3$ is of the form $h\, dx \wedge dy \wedge dz$ for some $h$. The differential operator $d$ is familiar to students of vector calculus. Recall from Equation (5.3) how the gradient, curl, and divergence operators tie together functions $C = C^\infty(\mathbb{R}^3)$ and vector fields $\mathcal{X} = \mathcal{X}(\mathbb{R}^3)$. This is conveyed via a commutative diagram,

$$\begin{array}{ccccccc} C & \xrightarrow{\nabla} & \mathcal{X} & \xrightarrow{\nabla\times} & \mathcal{X} & \xrightarrow{\nabla\cdot} & C \\ \downarrow\cong & & \downarrow\cong & & \downarrow\cong & & \downarrow\cong \\ \Omega^0 & \xrightarrow{d} & \Omega^1 & \xrightarrow{d} & \Omega^2 & \xrightarrow{d} & \Omega^3 \end{array} . \qquad (6.3)$$

On $\Omega^0(\mathbb{R}^3)$, $d$ is the gradient operator, taking a function $f$ not to its gradient vector field, but to the more natural gradient 1-form $df$. On $\Omega^1(\mathbb{R}^3)$, $d$ is the curl operator; on $\Omega^2(\mathbb{R}^3)$, $d$ acts as the divergence operator. Here, the vertical arrows identify 0- and 3-forms with functions, and identify vector fields with 1- and 2-forms in the obvious ways:

$$\begin{array}{lcl} \vec{F} = F_x\hat{\imath} + F_y\hat{\jmath} + F_z\hat{k} & \longrightarrow & F_x dx + F_y dy + F_z dz = \alpha_{\vec{F}} \\ \vec{F} = F_x\hat{\imath} + F_y\hat{\jmath} + F_z\hat{k} & \longrightarrow & F_x dy \wedge dz + F_y dz \wedge dx + F_z dx \wedge dy = \beta_{\vec{F}} \end{array}$$

The reader should return to simplicial examples of cochains and coboundaries and be convinced that what is measured on the algebraic level is indeed a discrete analogue of gradients, curls, and divergences.

**Example 6.16 (Maxwell's equations)** ⊙

The language of differential forms is ubiquitous in mathematical physics. Maxwell's equations admit a particularly simple interpretation. The calculus version of Maxwell's equations (on Euclidean $\mathbb{R}^3$, in a vacuum) are as follows:

$$-\frac{1}{c}\frac{\partial \vec{B}}{\partial t} = \nabla \times \vec{E} \qquad\qquad \nabla \cdot \vec{B} = 0$$
$$\frac{1}{c}\frac{\partial \vec{E}}{\partial t} = \nabla \times \vec{B} - \frac{4\pi}{c}\vec{J} \qquad\qquad \nabla \cdot \vec{E} = 4\pi\rho,$$

where $\vec{E}$, $\vec{B}$, $\vec{J}$, $\rho$, and $c$ are the electric field, magnetic field, current, charge density, and speed of light, respectively. Using the Euclidean structure to convert a field $\vec{F}$

into a 1-form $\alpha_{\vec{F}}$ or a 2-form $\beta_{\vec{F}}$ as in Example 6.15, one obtains:

$$d(c\alpha_{\vec{E}} \wedge dt + \beta_{\vec{B}}) = 0$$
$$d(c\alpha_{\vec{B}} \wedge dt - \beta_{\vec{E}}) = 4\pi\beta_{\vec{J}} \wedge dt - \rho\, dx \wedge dy \wedge dz.$$

These equations can be made more compact still and extended to arbitrary geometric manifolds using some of the constructions in §6.12. ⊚

**Example 6.17 (Ideal fluids)** ⊚

The equations of motion for a perfect inviscid fluid are likewise lifted to forms. Let $\vec{V}(t)$ be a time-dependent volume-preserving vector field representing the instantaneous motion of a fluid on Euclidean $\mathbb{R}^n$. Volume-preserving means that for the chosen volume form $\mu \in \Omega^n$ with $\mu \neq 0$ nowhere vanishing, the derivative $d\mu(\vec{V}, \cdot) = 0$ vanishes. Let $\alpha_{\vec{V}}$ be the (time-dependent) 1-form dual to $\vec{V}$ as per Example 6.15. Then, the Euler equations become

$$\frac{\partial \alpha_{\vec{V}}}{\partial t} + d\alpha_{\vec{V}}(\vec{V}, \cdot) = -dH, \tag{6.4}$$

where $H \colon M \to \mathbb{R}$ is a function (sometimes known as the *head* or *Bernoulli* function) combining pressure and kinetic energy terms. This formulation permits doing fluid dynamics on an arbitrary **Riemannian manifold** – a manifold $M$ with a smoothly varying inner product $g(\cdot, \cdot)$ on tangent spaces. The dual 1-form $\alpha_{\vec{V}}$ to $\vec{V}$ is given by contraction into the metric: $\alpha_{\vec{V}} := g(\vec{V}, \cdot)$ [16]. The **vorticity** of such a fluid is usually presented as a vector field representing the curl of the velocity field: in fact, it is more properly defined to be the 2-form $d\alpha_{\vec{V}}$, the derivative of the 1-form dual to velocity. ⊚

The language of differential forms is designed for integration: a $p$-form is perhaps best thought of as an object that can be integrated over a $p$-dimensional domain. Specifically, given an oriented $p$-dimensional submanifold with corners, $S$, there is an integral operator $\int_S \colon \Omega^p \to \mathbb{R}$ defined by evaluating the $p$-form pointwise on oriented $p$-tuples of tangent vectors to $S$ and integrating on coordinate charts using the standard Lebesgue integral. As is the case in the more familiar setting of line integrals in vector calculus, the Chain Rule implies an invariance of the integral with respect to local orientation-preserving coordinate representations. Modulo details about induced orientations on boundaries, the fundamental theorem of calculus-with-forms is transparent:

**Theorem 6.18 (Stokes' Theorem).** *For $\alpha$ a compactly-supported $p$-form and $S$ an oriented $(p+1)$-dimensional manifold (with boundary and/or corners),*

$$\int_S d\alpha = \int_{\partial S} \alpha.$$

## 6.9 De Rham cohomology

The antisymmetry property of forms is reminiscent of the cancelations via judicious choice of signs at the heart of all co/chain complexes. Thanks to this antisymmetry and the commutativity of mixed partial derivatives, $d^2 = 0$ always, as presaged by Example 6.15. This prompts the interpretation of $\Omega$ as a complex: the **de Rham complex** of a manifold $M$ is the cochain complex ${}_{dR}\mathcal{C} = (\Omega^\bullet, d)$ of forms with coboundary the exterior derivative $d$. The **de Rham cohomology** of $M$, ${}_{dR}H^\bullet(M)$, is the cohomology of $\Omega^\bullet$. In this theory, it is traditional to call the cocycles **closed** forms and the coboundaries **exact** forms. *A de Rham cohomology class is an equivalence class of closed forms modulo exact forms.*

**Example 6.19 (Winding numbers and de Rham cohomology)** ⊚

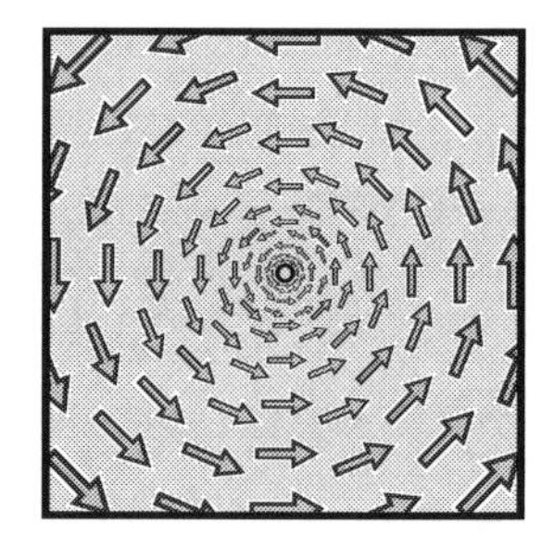

The reduced de Rham cohomology ${}_{dR}\tilde{H}^\bullet(\mathbb{R}^n)$ of $\mathbb{R}^n$ is trivial for all $n$, thanks to the Fundamental Theorem of Integral Calculus: closed forms are exact. On the punctured plane $\mathbb{R}^2{-}0$, the closed 1-form,

$$d\theta = \frac{x\,dy - y\,dx}{x^2 + y^2},$$

is *not* exact. Despite being denoted $d\theta$, there is no single-valued 0-form $\theta$ whose gradient 1-form is $d\theta$; thus, $d\theta$ defines a non-trivial cohomology class and generator of ${}_{dR}H^1(\mathbb{R}^2{-}0) \cong \mathbb{R}$. The integral of $d\theta$ over an oriented piecewise-smooth closed curve $\gamma$ in the punctured plane yields an integer, and this is, precisely, the winding number of $\gamma$ about 0: *cf.* Equation (3.4). ⊚

**Example 6.20 (Wedge product)** ⊚

In de Rham cohomology, the wedge product for forms descends to a product on cohomology. By defining $[\alpha] \wedge [\beta] := [\alpha \wedge \beta]$, one notes that since $\alpha$ and $\beta$ are closed, $d(\alpha \wedge \beta) = d\alpha \wedge \beta + d\beta \wedge \alpha = 0$; furthermore, if $\alpha$ or $\beta$ is exact, then so is $\alpha \wedge \beta$. Thus, $\wedge\colon {}_{dR}H^p \times {}_{dR}H^q \to {}_{dR}H^{p+q}$ turns ${}_{dR}H^\bullet$ into a ring. Since, locally, a basis Euclidean $k$-form measures oriented projected $k$-dimensional volumes, the wedge product inherits a volumetric interpretation. On the torus $\mathbb{T}^n$ with angular coordinates $\theta_i$, the 1-forms $d\theta_i$, $i = 1 \ldots n$ generate the cohomology ring ${}_{dR}H^\bullet(\mathbb{T}^n)$. For example, the generator for ${}_{dR}H^n(\mathbb{T}^n) \cong \mathbb{R}$ is the volume form $d\theta_1 \wedge \cdots \wedge d\theta_n$. ⊚

It is no coincidence that the de Rham cohomology of $\mathbb{R}^n$ and $\mathbb{T}^n$ have the same dimensions as in the singular theory.

**Theorem 6.21 (de Rham Isomorphism Theorem).** *For $M$ a manifold, ${}_{dR}H^\bullet(M) \cong H^\bullet(M;\mathbb{R})$.*

A little more effort yields a calculus-based version of Poincaré duality. Let $\Omega^\bullet_c(M)$ denote the complex of *compactly supported* forms on $M$, complete with differential

$d$. This yields a well-defined cohomology ${}_{dR}H^\bullet_c$ with compact supports, which, as an extension of Theorem 6.21, is isomorphic to the singular $H^\bullet_c$.

**Theorem 6.22 (Poincaré Duality, de Rham version).** *For $M$ an oriented manifold of dimension $n$, wedge and integration of forms yield isomorphisms*

$$\int_M \blacksquare \wedge \square : {}_{dR}H^p(M) \xrightarrow{\cong} {}_{dR}H^{n-p}_c(M)^\vee$$

$$\int_{\blacksquare} \square : H_p(M;\mathbb{R}) \xrightarrow{\cong} {}_{dR}H^p(M)^\vee$$

It is an instructive exercise to show that integration descends to homology and cohomology. For $[\alpha] \in {}_{dR}H^p$, $\beta \in \Omega^{p-1}$, $S$ a boundaryless $p$-dimensional submanifold and $T$ a $(p+1)$-dimensional submanifold with boundary, then

$$\int_{S+\partial T} \alpha + d\beta = \int_S \alpha + \int_{\partial S} \beta + \int_T d\alpha + \int_{\partial T} d\beta = \int_S \alpha,$$

via Stokes' Theorem. Thus, only $[S]$ and $[\alpha]$ matter. The isomorphisms of Theorem 6.22 effect the Poincaré Duality isomorphism PD from §6.4 (in the oriented case, with $\mathbb{R}$ coefficients).

## 6.10 Cup products

The de Rham isomorphism of Theorem 6.21 allows one to import calculus-based intuition into cohomology theory. Several of the constructs that are natural and clear in the setting of manifolds lift to the more general (cellular, singular, algebraic) cohomology theories. This section explores the algebraic generalization of the wedge product. Recall from Example 6.20 that the wedge product on forms, $\wedge$, so implicitly familiar to students of multivariable calculus, descends to a product on de Rham cohomology classes, giving ${}_{dR}H^\bullet$ the structure of a ring. What is the singular analogue?

Just as one defines the wedge $\wedge$ on forms and then passes to cohomology, one defines a product on cochains. Let $\alpha \in C^p(X;\mathbf{R})$ be a $p$-cochain and $\beta \in C^q(X;\mathbf{R})$ a $q$-cochain. Define the **cup product**, $\alpha \smile \beta \in C^{p+q}(X;\mathbf{R})$ to be the cochain whose value on a singular $(p+q)$-simplex $\sigma\colon \Delta^{p+q} \to X$ is given by restriction of the canonical simplex $\Delta^{p+q} = [v_0, v_1, \ldots, v_{p+q}]$ to the 'first' $p$-simplex and the 'last' $q$-simplex:

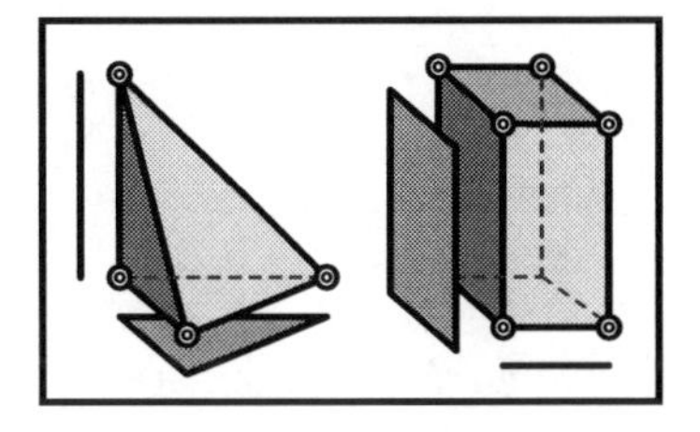

$$(\alpha \smile \beta)\sigma := \alpha\left(\sigma|[v_0, v_1, \ldots, v_p]\right) \cdot \beta\left(\sigma|[v_p, v_{p+1}, \ldots, v_{p+q}]\right),$$

where the product is in the ring structure of the coefficients $\mathbf{R}$. One shows that for $\alpha$ and $\beta$ cocycles, the cup product is a cocycle as well, therefore inducing an operation on $H^\bullet$ which gives it the structure of a ring. The cup product on $H^\bullet(X;\mathbf{R})$ for $\mathbf{R}$ a commutative ring is, like the wedge product, graded anti-commutative: $\beta \smile \alpha = (-1)^{pq}\alpha \smile \beta$.

If the reader finds this definition confusing, it is perhaps advisable to think of cubes rather than of simplices. For a cubical complex with $\alpha$ and $\beta$ cochains on $p$-cubes and $q$-cubes respectively, then the product chain $\alpha \smile \beta$ evaluated on a *topological product* $(p+q)$-cube is the *algebraic product* of the cochain values on the factor $p$-cube and $q$-cube. In this setting, the parallel to the de Rham wedge product is clearest.

**Example 6.23 (Projective space cohomology rings)** ⊚

The ring structure on $\mathbb{P}^n$ in $\mathbb{F}_2$ coefficients is particularly satisfying: it is the ring of polynomials in one variable, $x$, modulo the ideal generated by $x^{n+1}$:

$$H^\bullet(\mathbb{P}^n;\mathbb{F}_2) \cong \mathbb{F}_2[x]/\left(x^{n+1}\right), \tag{6.5}$$

where $x \in H^1(\mathbb{P}^n;\mathbb{F}_2)$. This computation is not elementary (see, *e.g.*, [176]), but it has important consequences. For example, the ring structure reveals that $\mathbb{P}^3$ is not homotopic to $\mathbb{P}^2 \vee \mathbb{S}^3$, even though they have isomorphic $\mathbb{F}_2$ cohomology *groups*. Though both spaces have $H^k \cong \mathbb{F}_2$ for $k \leq 3$ and 0 otherwise, their ring structures differ since the generator $x \in H^1$ satisfies $x^3 = 0$ for $\mathbb{P}^2 \vee \mathbb{S}^3$ but $x^3 \neq 0$ for $\mathbb{P}^3$. ⊚

## 6.11 Currents

On smooth manifolds, calculus provides the convenient language of forms for cohomology. Duals of forms provide an extremely flexible interpolation between smooth and discrete homological structures on manifolds that allow one to talk about, *inter alia*, the homology class of a vector field. This section touches on analytic tools based on geometric measure theory [119, 134, 234]. To avoid the numerous technicalities involving regularity and rectifiability, let the reader assume (via restriction to the o-minimal structure of globally subanalytic sets) sufficient (piecewise) smoothness where needed.

Fix $M$ an oriented manifold of dimension $n$. Let $\Omega_p(M) := \left(\Omega^p_c(M)\right)^\vee$ be the space of $p$-**currents** – real-valued functionals on compactly supported $p$-forms. Currents have a homological nature. Given any $p$-current $T \in \Omega_p$, the boundary of $T$, $\partial T \in \Omega_{p-1}$, is defined via the adjoint to the exterior derivative: $\partial T(\alpha) = T(d\alpha)$. A **cycle** is a current $T$ with $\partial T = 0$. Clearly, $\partial^2 = 0$, and there is a resulting chain complex $(\Omega_\bullet(M), \partial)$ with ensuing **de Rham homology** ${}_{dR}H_\bullet(M)$. The analogue of Theorem 6.21 holds: ${}_{dR}H_\bullet(M) \cong H_\bullet(M;\mathbb{R})$.

The chief advantage in using currents is, as with all things homological, visualizability. An oriented $p$-dimensional submanifold in $M$ is a $p$-current, since one can integrate a $p$-form over it: its de Rham homology class coincides with its singular homology class. For example, any (piecewise-smooth) oriented knot or link is a 1-current, since one can integrate a 1-form over oriented curves. Upon fixing a volume form on $M$, a piecewise-smooth vector field $V$ is likewise a 1-current, since any 1-form

$\alpha$ pairs with $V$ pointwise as $\alpha(V)$, which may then be integrated over $M$ (to a finite value, thanks to compact support of $\alpha$). The 2-currents on a manifold with volume form can range in shape from oriented surfaces to pairs of vector fields to a pair of tangent vectors at a single point.

**Example 6.24 (Volume preserving links)** ⊚

One beauty of the language of currents is that it allows one to compare both knots and vector fields on a manifold. Recall the definitions of links from Example 4.24 as disjoint embedded loops in $\mathbb{S}^3$. A vector field is an entirely different class of objects; one notes that the flowlines of a vector field have the potential to close up into periodic orbits – a link. The similarity seems to end there.

However, it follows from a result of Sullivan [284, Prop. II.25] that any closed nullhomologous 1-current, such as a volume-preserving vector field on an oriented manifold, can be realized as a limit of 1-currents supported on a compact 1-dimensional submanifold: an oriented link. This implies that any volume-preserving flow on $\mathbb{S}^3$ is the limit (in the sense of 1-currents) of a sequence of ever-lengthening, ever-coiling links. This suggests a reformulation of knot/link theory in terms of volume-preserving vector fields on $\mathbb{S}^3$. ⊚

**Example 6.25 (Helicity and fluids)** ⊚

It has been known for a long time what is the appropriate asymptotic analogue of linking number for volume-preserving vector fields on $\mathbb{S}^3$ [14, 16, 273]. The construction is as follows: given any two points $x, y \in \mathbb{S}^3$, evolve them forward under the flow of the vector field for times $s$ and $t$ respectively, until the flowlines come close to their starting points (that this happens for almost-every $x$ and $y$ infinitely often follows from the Poincaré Recurrence Theorem for volume-preserving flows [258]). Close these curves with short paths and compute the linking number (well-defined for almost all $x$ and $y$). The limit of this linking number, normalized by $st$, converges as $s, t \to +\infty$ to a function $\ell k(x, y)$, which, when integrated over $\mathbb{S}^3 \times \mathbb{S}^3$ with the conserved volume form, yields the **asymptotic linking number** of the flow,

$$\ell k(V) := \int_{\mathbb{S}^3}\int_{\mathbb{S}^3} \ell k(x, y)\, \mathrm{dvol}_x\, \mathrm{dvol}_y.$$

The techniques of forms and currents makes the computation of this seemingly-intractable quantity elementary. A volume-preserving vector field $V$ on $\mathbb{S}^3$ is closed and nullhomologous as a 1-current; this implies that the vector field contracted into the volume form $\mu$ yields an exact 2-form $\mu(V, \cdot, \cdot) = d\alpha$ for some $\alpha$. The **helicity** of $V$ is the integral of the wedge product of $\alpha$ with its derivative:

$$\mathcal{H}(V) := \int_{\mathbb{S}^3} \alpha \wedge d\alpha,$$

One shows well-definedness with respect to choice of $\alpha$ via Stokes' Theorem. Arnol'd [14] (following Moffat [232] (following Calŭgareanŭ [76])) showed that the helicity is the asymptotic linking number:

**Theorem 6.26 (Helicity Theorem).** $\mathcal{H}(V) = \ell k(V)$

As a corollary, the helicity is an invariant of $V$ under the action of volume-preserving diffeomorphisms of $\mathbb{S}^3$, since linking numbers are unchanged by such. This is of great significance in fluid dynamics, since the velocity field of an ideal fluid evolves in time according to the Euler equations (Example 6.17), and the energy of the fluid (the integral of the norm of the velocity field) is bounded below by helicity – $\mathcal{H}$ is a topological measure of a fluid's inability to relax. The proof of the Helicity Theorem in [16] uses currents on $\mathbb{S}^3 \times \mathbb{S}^3$ to capture linking behavior. ⊚

**Example 6.27 (Normal and conormal cycles)** 

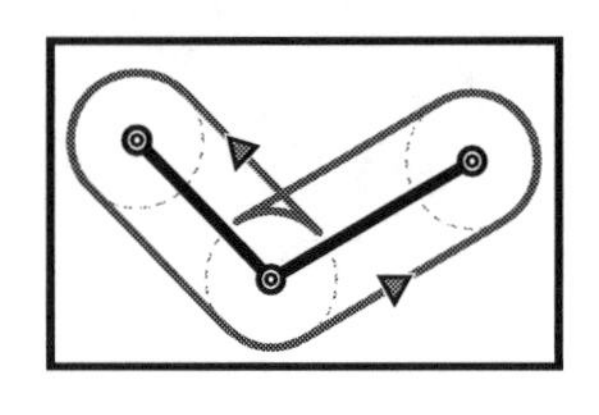

Many of the constructs of Chapter 3 concerning Euler characteristic and intrinsic volumes have a representation in terms of currents [234]. The **normal cycle** of a tame set $A \subset \mathbb{R}^n$ is a special $(n-1)$-current $\mathbf{N}^A$ on the unit cotangent bundle $UT^*\mathbb{R}^n \cong \mathbb{S}^{n-1} \times \mathbb{R}^n$. For $A \subset \mathbb{R}^n$ of positive codimension, the normal cycle is best visualized as having support on the set of points a *'unit'* distance from $A$.

The **conormal cycle** of a tame set $A \subset \mathbb{R}^n$ is a particular $n$-current $\mathbf{C}^A \in \Omega_n(T^*\mathbb{R}^n)$ on the cotangent bundle. It is, for lack of a better explanation in this text, the *cone* over the normal cycle. Each of the intrinsic volumes $\mu_k$ of §3.10, including Euler characteristic $\chi = \mu_0$, can be defined as the integral of a canonical form ($\alpha_k \in \Omega_c^{n-1}(UT^*\mathbb{R}^n)$ or $\omega_k \in \Omega_c^n(T^*\mathbb{R}^n)$) over the appropriate cycle:

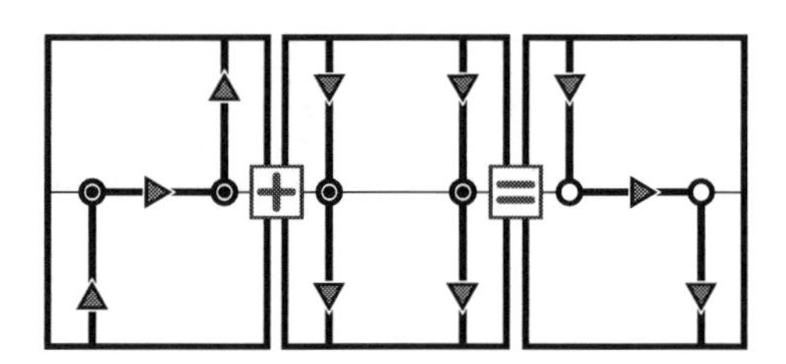

$$\mu_k(A) = \int_{\mathbf{N}^A} \alpha_k = \int_{\mathbf{C}^A} \omega_k.$$

Additivity of the intrinsic volumes is expressed in terms of additivity of currents: *e.g.*, $\mathbf{N}^{A\cup B} = \mathbf{N}^A + \mathbf{N}^B - \mathbf{N}^{A\cap B}$ and likewise for $\mathbf{C}$. This means, *e.g.*, that one can see the difference between the Euler characteristic of a compact disc $\chi(\mathbb{D}^n) = 1$ and of its interior $\chi(\mathbb{D}^n - \partial\mathbb{D}^n) = (-1)^n$ as being a *reflection*. When subtracting the conormal cycle $\mathbf{C}^{\partial\mathbb{D}^n}$ from $\mathbf{C}^{\mathbb{D}^n}$, the support in $\mathbb{R}^n$ is the same, but the orientation in each axis is reversed. For $n$ odd, this results in an orientation-reversal, reflected in the sign change. ⊚

## 6.12 Laplacians and Hodge Theory

With the addition of a geometric structure, there is another manifestation of Poincaré duality in cohomology for manifolds via partial differential equations. Recall that for a vector space $V$ of dimension $n$, the algebra of forms, $\mathbf{\Lambda}(V)$, displays a combinatorial duality: $\dim \mathbf{\Lambda}^p(V) = \binom{n}{p} = \dim \mathbf{\Lambda}^{n-p}(V)$. Fix a geometry on $V$ in the form of an inner product $\langle\cdot,\cdot\rangle$ and choose an orthonormal basis $\{x_i\}_1^n$. Fix also an orientation on $V$ in the form of an equivalence class of orderings $[x_i]_1^n$ of basis elements up to even permutations. Define the **Hodge star** $\star\colon \mathbf{\Lambda}^p(V) \to \mathbf{\Lambda}^{n-p}(V)$ on basis elements as follows:

$$\star dx_{i_1} \wedge dx_{i_2} \wedge \cdots \wedge dx_{i_p} := dx_{i_{p+1}} \wedge \cdots \wedge dx_{i_{n-1}} \wedge dx_{i_n},$$

where the ordering $[x_{i_j}]_{j=1}^n$ respects orientation. Extend to all of $\mathbf{\Lambda}(V)$ via linearity. The Hodge star depends only on the inner product and the orientation, not on the basis itself. It satisfies a signed duality $\star\star = (-1)^{p(n-p)}\mathrm{Id}$.

If an oriented manifold $M$ is Riemannian (see Example 6.17) then the Hodge star extends to $\star\colon \Omega^p \to \Omega^{n-p}$. For example, in Euclidean $\mathbb{R}^3$ with the standard basis as per Example 6.15, $\star\alpha_{\vec{F}} = \beta_{\vec{F}}$. Every oriented Riemannian manifold has a well-defined **volume form** $\mu \in \Omega^n$ which, in local orthonormal coordinates, is $dx_1 \wedge \cdots \wedge dx_n$ and which is given by $\mu = \star\mathbb{1}_M$. The Hodge star yields an inner product on each $\Omega^p$ via integration:

$$\langle\alpha,\beta\rangle := \int_M \alpha \wedge \star\beta.$$

With this geometry in place, one may define a codifferential $\delta\colon \Omega^p \to \Omega^{p-1}$ given by the adjoint: $\langle\alpha, d\beta\rangle = \langle\delta\alpha,\beta\rangle$; more explicitly, $\delta := (-1)^{p(n-p)}\star d\star$. The **Laplacian** is the operator $\Delta\colon \Omega^\bullet \to \Omega^\bullet$ given by:

$$\Delta := (d+\delta)^2 = d\delta + \delta d.$$

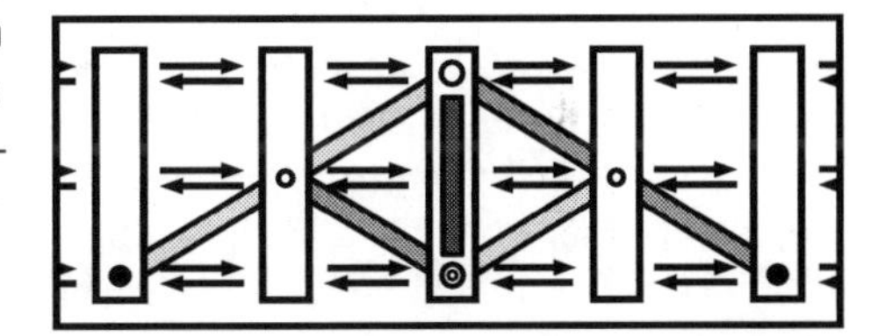

Note that the Laplacian is degree zero, and for $p = 0$ is the familiar second-order differential operator. The Laplacian blends analytic, geometric, and topological features. The **harmonic** forms are defined as ${}_\Delta H^\bullet(M) := \ker \Delta$, the kernel of the Laplacian.

**Theorem 6.28 (Hodge Theorem).** *For $M$ a compact oriented Riemannian manifold, $\Omega^p(M)$ has an orthogonal decomposition:*

$$\Omega^p = d\Omega^{p-1} \oplus {}_\Delta H^p \oplus \delta\Omega^{p+1}.$$

**Corollary 6.29.** *For $M$ a compact oriented Riemannian manifold, ${}_\Delta H^\bullet(M) \cong {}_{dR}H^\bullet(M)$.*

The bother of working with geometry has the following payoff: the Hodge star $\star$ is an incarnation of Poincaré duality. Let $\alpha \in {}_\Delta H^p$ be a harmonic form. Theorem 6.28 implies that $d\alpha = \delta\alpha = 0$. This implies that $\star\alpha$ is also harmonic, since

$$\Delta(\star\alpha) = (-1)^{p(n-p)}(d\star d\star + \star d\star d)\star\alpha = (d\star d + \star d\delta)\alpha = 0.$$

Thus, one may realize Poincaré duality as the isomorphism $\star\colon {}_{\Delta}H^k(M) \to {}_{\Delta}H^{n-k}(M)$; *cf.* Theorem 6.22.

One of the benefits of using differential-topological constructs is the ability to import and export ideas between smooth and discrete frameworks. There is a simple simplicial analogue of the Hodge theorem which has the advantage of requiring no forms, differentiability, or manifold structures, but merely an implicit geometry. Consider a cell complex $X$ with cellular cochain complex $\mathcal{C} = (C^\bullet, d)$ in $\mathbb{R}$ coefficients. Choose an inner product $\langle \cdot, \cdot \rangle$ so that indicator functions over the cells of $X$ are orthogonal. The implicit choice is in the **weight** of each simplex $\langle \mathbb{1}_\sigma, \mathbb{1}_\sigma \rangle = \omega_\sigma \in \mathbb{R}$. With this inner product structure, use the adjoint $\delta$ of $d$ to define the **discrete Laplacian**: $\Delta = d\delta + \delta d$. As in the smooth theory, the harmonic cochains are ${}_{\Delta}H = \ker \Delta$. The following discrete Hodge theorem becomes a simple exercise.

**Theorem 6.30 (Discrete Hodge Theorem).** *For $X$ a finite cell complex with choice of weights, $C^p = dC^{p-1} \oplus {}_{\Delta}H^p \oplus \delta C^{p+1}$.*

**Example 6.31 (Graph Laplacian)** ◎

Discrete Laplacians have seen the greatest use in graph theory under the guise of the **graph Laplacian**. For $X$ an undirected graph and $f\colon V(X) \to \mathbb{R}$ a function on the nodes, the graph Laplacian of $f$ is defined as

$$(\Delta_X f)(v) = \sum_{w\colon (v,w) \in E(X)} f(w) - f(v).$$

The reader may verify that this agrees with the Laplacian on $C^0(X)$ under the obvious inner-product structure on chains. It is clear that a harmonic 0-chain is one in which each node's value is the average of its neighbors'. Graph Laplacians have found extensive use in algorithms [281], image processing [71] and much more. The principal (smallest nonzero) eigenvalue of the graph Laplacian controls the behavior of random walks on a graph and points to interesting generalizations in random simplicial complexes [171]. ◎

**Example 6.32 (Distributed homology computation)** ◎

The Laplacian is a local operator and, as such, is well-suited to distributed computation. The work of Tahbaz-Salehi and Jadbabaie [267] details the use of the simplicial Laplacian to distributed computation of the homological coverage criterion of §5.6. By Theorems 5.10 and 6.30, verified coverage in a sensor network in $\mathbb{R}^2$ follows from showing that $\ker \Delta = 0$ on discrete 1-forms of the flag complex F of the network. It is easy to see that the **heat equation**, $\frac{d}{dt}\alpha = \Delta\alpha$, has 0 as an asymptotically stable solution if and only if $\ker \Delta = 0$. Thus, by solving a heat equation with random initial conditions, one can safely (to the degree one trusts in random initial conditions) conclude coverage if the solution converges to zero. That this equation can be solved locally and in a distributed manner [235] should come as no surprise to the reader who has spent time with the heat equation, though the convergence to the solution can be slow. This can be improved by instead using a wave equation approach [265]. ◎

## 6.13 Circular coordinates in data sets

In §5.14 the problem of determining the topology of a point cloud was addressed by means of persistent homology. A cohomological approach becomes the appropriate tool for addressing a related problem of coordinatizing a point cloud.

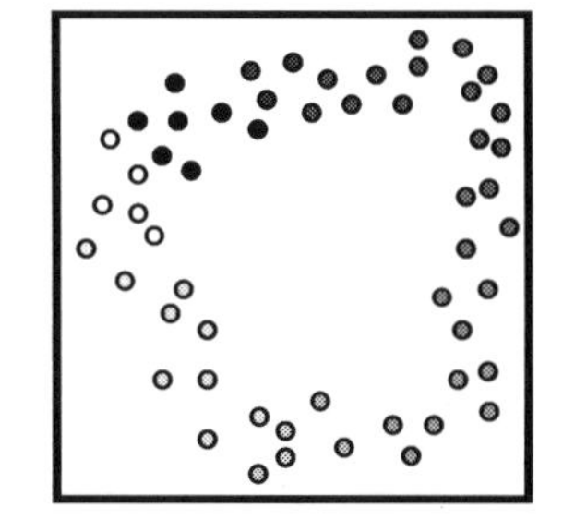

Assume for the present that $\mathcal{Q} \subset \mathbb{R}^n$ is a point-cloud whose topology is known or suspected to be sufficiently *circular* so as to merit outfitting circular coordinates $\Theta\colon \mathcal{Q} \to \mathbb{S}^1$ in a manner that respects the underlying topology of the space $X \subset \mathbb{R}^n$ (homotopic to $\mathbb{S}^1$) that $\mathcal{Q}$ is presumed to sample. Many of the existing algorithms for assigning circular coordinates to a point cloud [263, 288] have implicit convexity assumptions.

The solution of de Silva, Morozov, and Vejdemo-Johansson [89] is a slick application of algebraic topology that highlights the particular benefits of cohomology and the role of coefficients. The outline of their work is as follows.

1. One begins with the following result from homotopy theory: for any space $X$, the group of (basepoint-preserving) homotopy classes $[X, \mathbb{S}^1]$ of maps $X \to \mathbb{S}^1$ is isomorphic to $H^1(X;\mathbb{Z})$ (see §8.6). The coordinatization function $\Theta\colon X \to \mathbb{S}^1$ therefore is naturally approachable via cohomology.
2. To find a cohomology class for $X$ based on a sampling of nodes $\mathcal{Q}$, compute the persistent cohomology of a sequence of Vietoris-Rips complexes, as in §5.13-5.14. For the Structure Theorem (Theorem 5.21), field coefficients are required; for numerical reasons (to avoid roundoff errors), coefficients in a *finite* field $\mathbb{F}_p$ are preferred.
3. A (persistent) class $[\alpha_p] \in H^1(X;\mathbb{F}_p)$ can be converted to a integral class $[\alpha] \in H^1(X;\mathbb{Z})$ by means of the following process. The short exact sequence of coefficients $0 \longrightarrow \mathbb{Z} \xrightarrow{\cdot p} \mathbb{Z} \longrightarrow \mathbb{F}_p \longrightarrow 0$ yields a short exact sequence of cochain complexes on $X$ and, via Lemma 5.5, a long exact sequence on cohomology:

$$\longrightarrow H^1(X;\mathbb{Z}) \longrightarrow H^1(X;\mathbb{F}_p) \xrightarrow{\delta} H^2(X;\mathbb{Z}) \xrightarrow{\cdot p} H^2(X;\mathbb{Z}) \longrightarrow$$

   The kernel of $\cdot p\colon H^2(X;\mathbb{Z}) \to H^2(X;\mathbb{Z})$ consists of $p$-torsional cohomology classes: for $p > 2$ these would seem to be rare occurrences in *organic* spaces $X$ living behind data sets. By exactness, $\ker(\cdot p) = \operatorname{im}\delta$; assuming this is zero, it implies that $H^1(X;\mathbb{Z}) \to H^1(X;\mathbb{F}_p)$ is surjective, and (persistent) classes in $\mathbb{F}_p$ coefficients therefore lift to integral classes.
4. The resulting integer cocycle $\alpha$ is perhaps a poor $\mathbb{S}^1$-coordinatization – all the circular motion may be concentrated over a small subset of $X$. To relax $\alpha$ to a smooth circular coordinate system, lift to $\mathbb{R}$ coefficients and find a cohomologous harmonic cocycle $\overline{\alpha} \in {}_{\Delta}H^1(X)$. Thanks to the local-averaging properties of the Laplacian, this 1-cocycle integrates to a well-regulated coordinate function $\Theta\colon X \to \mathbb{S}^1$.

For details on computational aspects and implementation, see [89]. This work illustrates well the utility of cohomology, while highlighting the delicate interplay between real, integral, and cyclic coefficients.

# Notes

1. This chapter is woefully incomplete: a short, motivational text cannot do justice to cohomology theory. The interested reader should resolve to learn the theory properly. Hatcher [176] is, as ever, the best place to begin. For the de Rham theory, Bott and Tu [46] is the classic lucid source, and Fulton [135] is more elementary still.
2. The conflation of objects with duals is ubiquitous and insidious. Examples include confusing gradient 1-forms with vector fields and defining simplicial chains as functions from simplices to coefficients.
3. The idea of the impossible tribar as a 1-cocycle was suggested by Penrose [243]. One can imagine more interesting Escherian illusions based on $H^2$.
4. One of the many cohomology theories not covered in this text is related to configuration spaces. Given $X$ a topological space and $\mathbf{R}$ a ring, let $C^k$ be the set of functions (not necessarily maps!) $f\colon X^{k+1} \to \mathbf{R}$. This gives a complex $\mathcal{C}$ with differential $d$ taking $f(x_0,\ldots,x_k)$ to the sum $\sum_i(-1)^i f(x_0,\ldots,x_{i-1},x_{i+1},\ldots,x_k)$. The set of all functions which vanish in a neighborhood of the grand diagonal $(x,\ldots,x)$ forms a subcomplex $\mathcal{C}^0$. The **Alexander-Spanier cohomology** of $X$ is $H^\bullet(\mathcal{C}/\mathcal{C}^0)$. It is, for reasonable spaces, isomorphic to the singular $H^\bullet(X)$ [280]. This theory seems suspiciously relevant to applications in configuration spaces.
5. Helly's Theorem is important in a wide array of combinatorics and optimization problems [9]. That is has a purely topological proof is a testament to the power of topological methods. The homological proof was known to Helly and many topological generalizations exist. It is remarkable how many experts nevertheless consign Helly's Theorem to convex geometry.
6. There is, as one might suspect, a deeper form of duality, of which Poincaré, Alexander, and Lefschetz are emanations. **Verdier duality** for sheaves is perhaps the best encapsulation of the scope and power of duality theorems in co/homology. Chapter 9 will provide some of the requisite background for that theory.
7. The cup product is more important than may at first appear. It is good to visualize it using differential forms. Even better is its homological adjoint. Theorem 6.22 (and an illustration or two) hints at a relationship between $\smile$ and homology: the **cap product** $\frown\colon H_p(X)\times H^q(X)\to H_{p-q}(X)$ is defined on a chain $\sigma$ and a cochain $\alpha$ via

$$(\sigma \frown \alpha) := \alpha\left(\sigma|[v_0,v_1,\ldots,v_q]\right)\cdot\sigma|[v_q,v_{q+1},\ldots,v_p].$$

   In field coefficients, $\frown$ and $\smile$ are related via $\beta(S\frown\alpha) = (\alpha\smile\beta)(S)$ for a cycle $S$ and cocycles $\alpha,\beta$.
8. The reader may wonder, given the utility of cohomology with compact supports, where is the corresponding homology with compact supports? This exists and goes under the name of **Borel-Moore homology**.
9. A very clean proof of the de Rham Theorem (6.21) uses a double complex [46] (see the notes to Chapter 5).
10. Hodge theory is merely a hint at how partial differential equations (in this case, Laplace's equation) on geometric manifolds can lead to topological invariants. Many subtler and deeper invariants come from other PDEs using auxiliary structures and have implications in string theory, algebraic geometry, knot theory, and more.

11. Simplicial/cellular Hodge theory is having significant impact in numerical analysis [12, 13, 32, 94] via **discrete exterior calculus**. The discretization of space and time can destroy auxiliary structures or symmetries within the underlying differential equations: *e.g.*, the symplectic structure implicit in celestial mechanics. By working with simplicial $p$-forms and discretizing the conservation laws and symmetries themselves, one is led to more accurate numerical solutions: see Example 10.23.

## Chapter 7

# Morse Theory

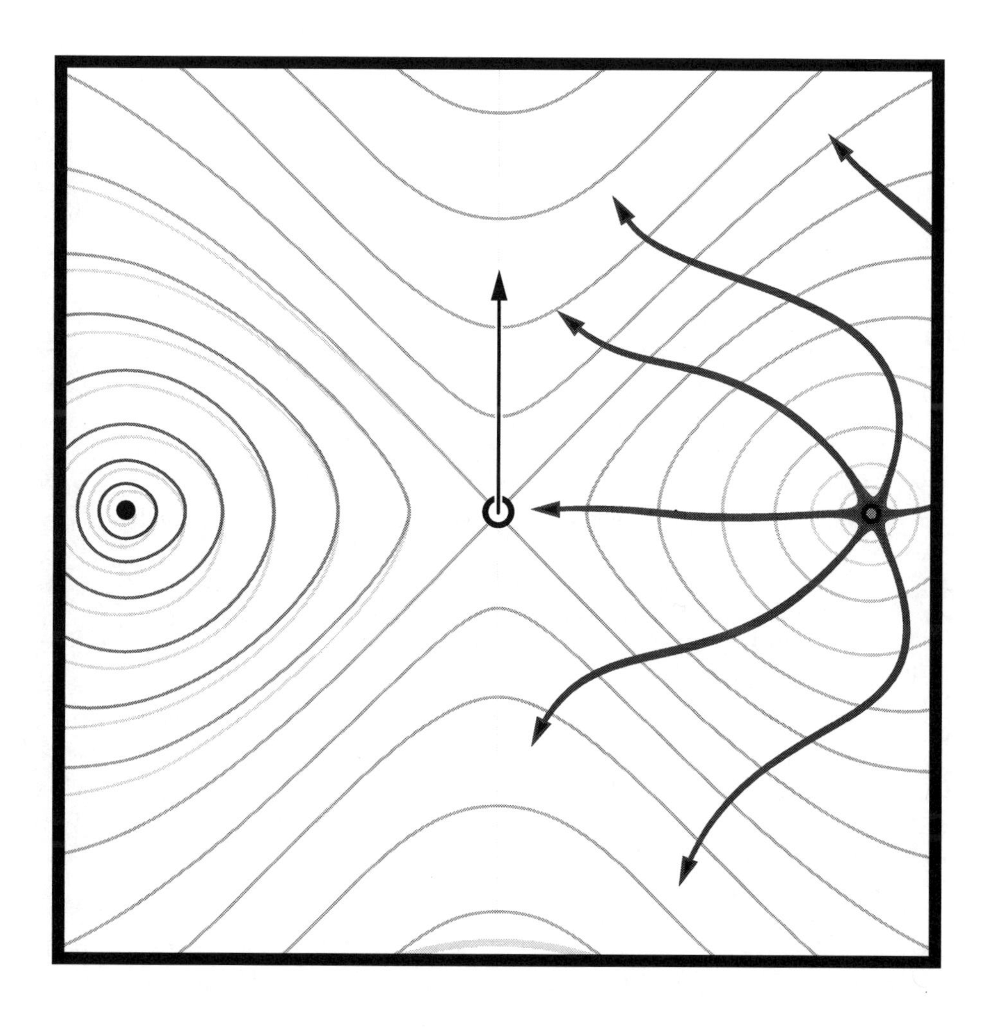

The two great themes of smooth manifolds and assembled complexes compete politely in topology. One locus of synthesis between the two lies in the eponymous theory of Morse. This chapter integrates all of the previous chapters into a suite of perspectives, tools, and applications connecting local behavior to global.

## 7.1 Critical points

In any homology theory, one counts certain objects with respect to an appropriate cancelation, usually with some ancillary structure imposed to keep counts finite. Morse theory uses a *height function* to facilitate homological counting.

Fix $M$ a compact Riemannian manifold without boundary. Morse theory operates via a real-valued function and the dynamics of its gradient flow. Fix $h\colon M \to \mathbb{R}$ a smooth function and consider the gradient field $-\nabla h$ on $M$. The dynamics of this vector field are simple: solutions either are fixed points (critical points of $h$) or flow *downhill* from one fixed point to another. Let $\mathrm{Cr}(h)$ denote the set of critical points, and assume for the sake of simplicity that all such critical points are **nondegenerate** – the second derivative (or *Hessian*) is nondegenerate (has nonzero determinant) at these points. Equivalently, the gradient field $-\nabla h$, thought of as a section of the tangent bundle $T_*M$, is transverse to the zero section (*cf.* §1.6), whence it follows that nondegeneracy is generic. These nondegenerate critical points are the basis elements of Morse theory.

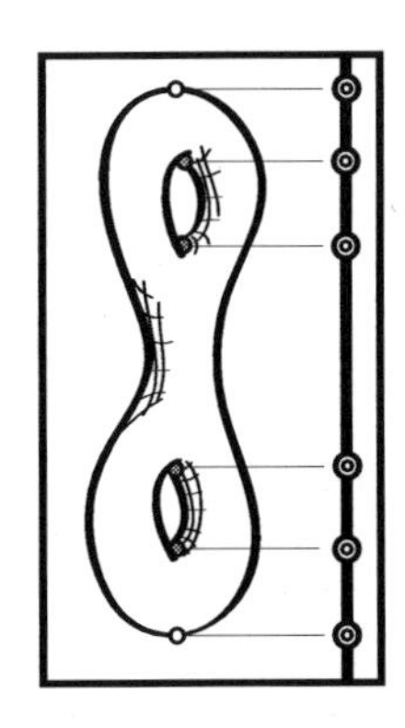

The critical points have a natural grading – the **Morse index**, $\mu(p)$, of $p \in \mathrm{Cr}(h)$ – the number of negative eigenvalues of the Hessian of second derivatives of $h$ at $p$. This has the more topological interpretation as the dimension of the unstable manifold of the vector field $-\nabla h$ at $p$ (recall Example 1.6): $\mu(p) = \dim W^u(p)$. The Morse index measures how unstable a critical point is: minima have the lowest Morse index; maxima the highest. Balancing a three-legged stool on $k$ legs leads to an index $\mu = 3 - k$ equilibrium.

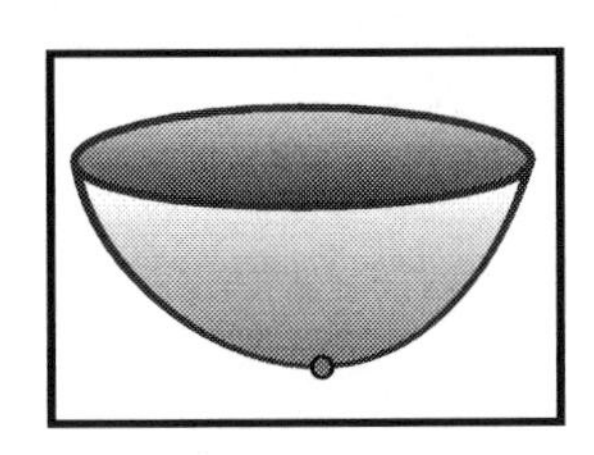

Classical Morse theory begins by observing the (lower) excursion sets $M_t := \{h \leq t\}$ of a Morse function $h$ on a compact manifold $M$. The story, in brief, is as follows. As $t \in \mathbb{R}$ increases, the family $M_t$ gives a filtration of spaces that begins with the empty set and ends with $M$. In the beginning, a disc appears *ex nihilo* and evolves, pinching and branching as critical points are passed, ultimately capping at the maximum. The critical observation: $M_t$ changes homeomorphism type *only* at critical values of $h$.

The local picture tells all. Morse theory asserts the following:

**Lemma 7.1.** *Consider a small compact ball $B$ about a critical point $p \in \mathrm{Cr}(h)$ of Morse index $\mu$. Denote by $E$ the* lower set *$E = B \cap \{h \leq h(p) - \epsilon\}$ for $\epsilon > 0$ small, and consider $U$, the closure of $B - E$, the complementary* upper set.

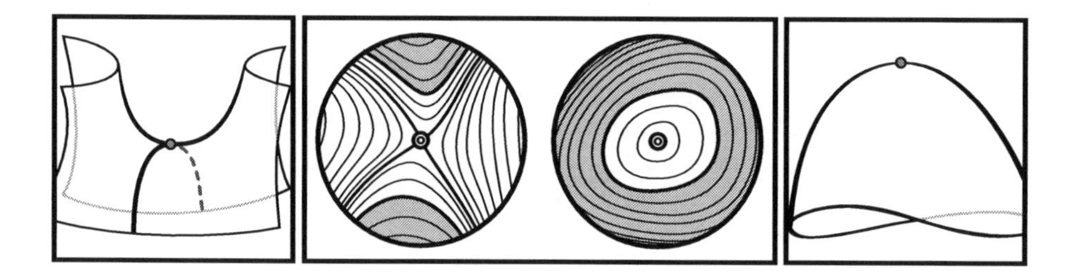

1. $U$ *is homeomorphic to* $\mathbb{D}^n$ *for* $n = \dim M$;
2. $E$ *is homeomorphic to* $\mathbb{D}^{n-\mu+1} \times \mathbb{S}^{\mu-1}$; *and*
3. $E \cap U$ *is homeomorphic to* $\mathbb{D}^{n-\mu} \times \mathbb{S}^{\mu-1}$.

One says that there is a **surgery** in the neighborhood of $p$ that attaches a product disc $U \cong \mathbb{D}^{n-\mu} \times \mathbb{D}^{\mu}$ glued along $E \cap U \cong \mathbb{D}^{n-\mu} \times \partial\mathbb{D}^{\mu}$.

## 7.2 Excursion sets and persistence

The filtration of $M$ by lower excursion sets $M_t$ of a Morse function $h\colon M \to \mathbb{R}$ fits perfectly into the picture of persistence sketched in §5.13-5.15. In this setting (sometimes called **sublevel set persistence** [104]), one considers the persistence barcode of the filtration of $M$ by subsets $M_t$, where the parameter $t$, is suitably discretized.

In keeping with the idea of Morse theory, the sublevel sets change their topological type only at critical points; hence barcodes for sublevel set persistence are tethered to the critical values. The meanings of the barcodes are as in §5.13-5.15, in that a long bar connotes significance; however, in sublevel set persistence, one is not trying to find an optimal cut-off $t$ to approximate $M$; rather, one wants to know which topological features of a manifold are important when it is stretched out along a table ruled by $h$. Clearly a small bar in a barcode for excursion sets indicates something like a *wrinkle* in the Morse function: a transient hole. Long bars in the sublevel set barcode indicate a large-scale feature as seen by $h$. In the many applications of persistence, the lingering problem of noise is pertinent, since a wiggling of $h$ leads to many small spurious bars. There is a large collection of stability-type results for persistence: see [70] for the first of these, which concerns sublevel set persistence and asserts an (*interleaving*) distance on barcodes with continuity guarantees under addition of noise to $h$, *cf.* §10.6.

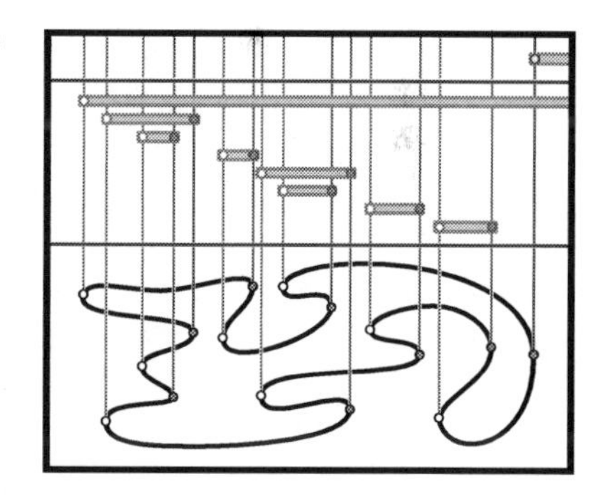

More salient to the themes of this chapter is the phenomenon of cancellation. The birth and death implicit in the barcodes for lower excursion sets reveal one of the great perspectives of Morse theory: critical points and the topological features they generate are naturally *paired* in a cancellative manner. Lemma 7.1 implies the following:

**Corollary 7.2.** *For a Morse function $h$ on a compact manifold $M$, births and deaths in the sublevel set homology barcode of grading $k$ implicate only critical points of Morse index $k$ and $k+1$ respectively.*

This cancellation of features foreshadows a self-contained homology theory for Morse functions.

## 7.3 Morse homology

Classical Morse theory concerns equivalence up to homeomorphism, based on the uniform behavior of nondegenerate critical points. By relaxing to a more homological view, the theory will connect better with the rest of the text and will naturally suggest extensions to non-Morse functions.

The constructs of the previous sections are perfect for homology. One has a natural set of objects ($\mathrm{Cr}(h)$) and a grading ($\mu$). One obtains the **Morse complex**, $\mathcal{C}^h = (MC_\bullet, \partial)$, with $MC_k$ the vector space with basis $\{p \in \mathrm{Cr}(h)\,;\, \mu(p) = k\}$. The boundary maps encode the global flow of the gradient field: $\partial_k$ counts (modulo 2 in the case of $\mathbb{F}_2$ coefficients) the number of **connecting orbits** – flowlines from a critical point with $\mu = k$ to a critical point with $\mu = k-1$. One hopes (or assumes) that this number is well-defined.

The difficult business is to demonstrate that $\partial^2 = 0$: this involves careful analysis of the connecting orbits, as in, e.g., [21, 274]. The use of $\mathbb{F}_2$ coefficients is highly recommended: dis-orientation is a plus. The ensuing **Morse homology**, $MH_\bullet(h)$, captures information about $M$.

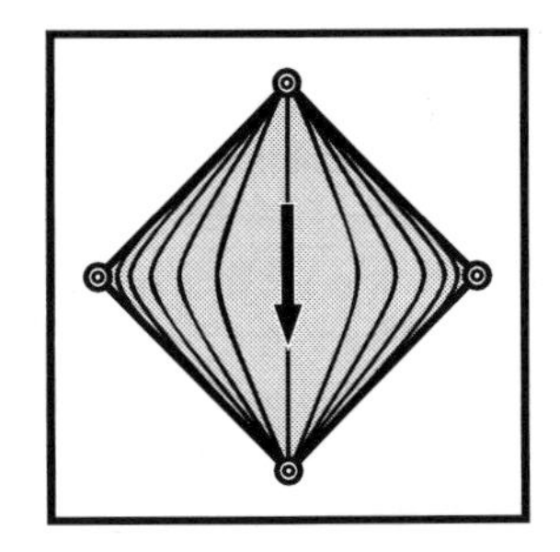

**Theorem 7.3 (Morse Homology Theorem).** *For $M$ compact and $h\colon M \to \mathbb{R}$ Morse, $MH_\bullet(h;\mathbb{F}_2) \cong H_\bullet(M;\mathbb{F}_2)$, independent of $h$.*

The conceptually simplest proof involves an isomorphism to the cellular (CW) homology of $M$, where the cell structure is that given by the $W^u(p)$ for $p \in \mathrm{Cr}(h)$. The **Stable Manifold Theorem** from dynamical systems asserts that these unstable manifolds are all cells homeomorphic to $\mathbb{R}^{\mu(p)}$, and a transversality argument gives acceptable attaching maps. The proofs are clearest for **Morse-Smale functions**, for which all stable and unstable manifolds of critical points are transverse. Morse-Smale functions, like Morse functions, are generic, and for such, there is an isomorphism at the level of chain complexes: *cf.* Example 2.2.

**Example 7.4 (Morse homology)** ⊙

For the particularly simple height function $h$ on $\mathbb{S}^2$ with two maxima, one minimum, and (of necessity) one saddle, the Morse complex in $\mathbb{F}_2$ coefficients is of the form,

$$0 \longrightarrow \mathbb{F}_2^2 \xrightarrow{[1\,|\,1]} \mathbb{F}_2 \xrightarrow{0} \mathbb{F}_2 \longrightarrow 0\ ;$$

thus, $MH_2(h) \cong \mathbb{F}_2$, $MH_1(h) \cong 0$, and $MH_0(h) \cong \mathbb{F}_2$; one notes that in this example $MH_\bullet(h) \cong H_\bullet(\mathbb{S}^2; \mathbb{F}_2)$. ⊚

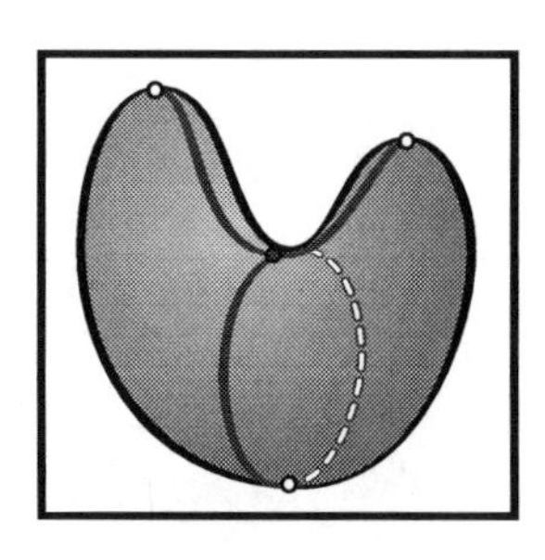

One efficient means of encoding all the critical point data of a Morse function is by means of the **Morse polynomial** of $h$, defined as $M_h(t) := \sum_{p \in \mathrm{Cr}(h)} t^{\mu(p)}$. This polynomial in the abstract variable $t$ has as its coefficients $c_i \in \mathbb{N}$ the number of critical points of $h$ with Morse index $i$. Theorem 7.3 implies a relationship between the Morse polynomial and the Poincaré polynomial, $P(t) = \sum_i \dim H_i(M)t^i$, which, recall, encodes the singular homology of the manifold $M$. At the very least, the $i^{\text{th}}$ coefficient of $M_h(t)$ must be greater than or equal to that of $P(t)$. A stronger inequality uses the polynomial algebra explicitly.

**Corollary 7.5 (Strong Morse Inequalities).** *For $h\colon M \to \mathbb{R}$ Morse,*

$$M_h(t) = P(t) + (1+t)Q(t), \tag{7.1}$$

*where $Q \in \mathbb{N}[t]$ is a polynomial with all coefficients in $\mathbb{N}$.*

**Corollary 7.6 (Euler characteristic).** *For $h\colon M \to \mathbb{R}$ Morse on a compact manifold $M$,*

$$\chi(M) = \sum_{p \in \mathrm{Cr}(h)} (-1)^{\mu(p)}.$$

**Proof.** Use Corollary 7.5, Lemma 5.17, and $t = -1$. ⊙

Morse theory offers a painless demonstration of one manifestation of homological Poincaré duality:

**Corollary 7.7 (Poincaré Duality).** *The $\mathbb{F}_2$-homology of a compact $n$-manifold $M$ is symmetric in its grading: $H_p(M; \mathbb{F}_2) \cong H_{n-p}(M; \mathbb{F}_2)$ for all $0 \le p \le n$.*

**Proof.** For $h$ a Morse function, the function $-h$ is also Morse. Changing from $h$ to $-h$ reverses the direction of the gradient flow, preserving the critical points and connecting orbits, but exchanging stable and unstable manifolds. At the chain complex level,

$$\begin{array}{ccccc} \xrightarrow{\partial} & MC_p(h) & \xrightarrow{\partial} & MC_{p-1}(h) & \xrightarrow{\partial} \\ & \downarrow \cong & & \downarrow \cong & \\ \xleftarrow[\partial]{} & MC_{n-p}(-h) & \xleftarrow[\partial]{} & MC_{n-p+1}(-h) & \xleftarrow[\partial]{} \end{array}$$

and thus $H_p(M) \cong MH_p(-h) \cong MH_{n-p}(h) \cong H_{n-p}(M)$. $\odot$

## 7.4 Definable Euler integration

Corollary 7.6 hints at the role of Morse theory in integration with respect to Euler characteristic. The integral operator $\int d\chi \colon \mathrm{CF}(X) \to \mathbb{Z}$ of Chapter 3 does not readily extend to continuous real-valued integrands: there is an extension of the integral to real-valued integrands by Rota, then Chen, that vanishes on all continuous integrands [262, 65]. Recent work [25] has revealed a novel Euler calculus for $\mathbb{R}$-valued integrands, with interesting Morse-theoretic connections. Fix an o-minimal structure, as in §3.5, and denote by $\mathrm{Def}(X)$ the **definable functionals** $h \colon X \to \mathbb{R}$ – those (compactly supported) functions whose graphs in $X \times \mathbb{R}$ are definable sets. Recall, these are not necessarily continuous functions, as they include the constructible functions $\mathrm{CF}(X)$. Given $h \in \mathrm{Def}(X)$, define the integral of $h$ as a limit of discretizations:

$$\int_X h\,\lfloor d\chi\rfloor := \lim_{n\to\infty}\frac{1}{n}\int_X \lfloor nh\rfloor\,d\chi \quad : \quad \int_X h\,\lceil d\chi\rceil := \lim_{n\to\infty}\frac{1}{n}\int_X \lceil nh\rceil\,d\chi\,. \tag{7.2}$$

These limits exist and are well-defined, though *not equal* in general. The **triangulation theorem** for $\mathrm{Def}(X)$ [293] states that to any $h \in \mathrm{Def}(X)$, there is a definable triangulation (a definable bijection to a disjoint union of open affine simplices in some Euclidean space) on which $h$ is affine on each open simplex. From this, one may reduce all questions about the integrals over $\mathrm{Def}(X)$ to questions of affine integrands over individual open simplices, using the additivity of the integral. Using this reduction technique, one proves the following analogue of Equation (3.10):

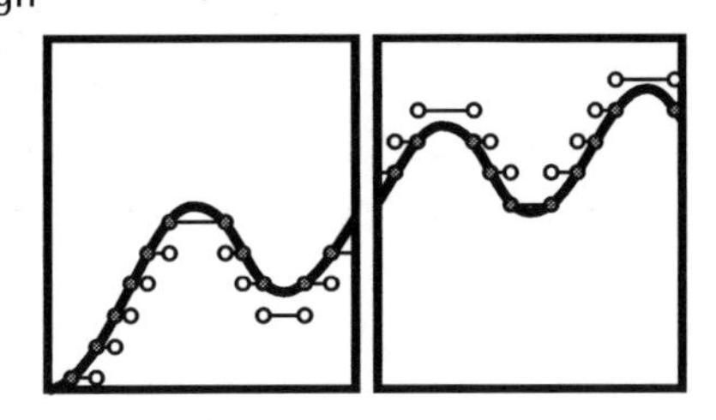

**Proposition 7.8 ([25]).** *For* $h \in \mathrm{Def}(X)$,

$$\int_X h\,\lfloor d\chi\rfloor = \int_{s=0}^{\infty} \chi\{h \geq s\} - \chi\{h < -s\}\,ds = -\int_X -h\,\lceil d\chi\rceil. \tag{7.3}$$

In its favor, the integral is coordinate-free, in the sense that $\int_X h \circ \phi\lfloor d\chi\rfloor = \int_X h\lfloor d\chi\rfloor$ for $\phi$ a homeomorphism of $X$. Less pleasant is that these integral operators are *not linear*, nor even homogeneous with respect to negative coefficients, by (7.3). The compelling feature of the valuations $\lfloor d\chi\rfloor$ and $\lceil d\chi\rceil$ is their relation to Morse theory. The following theorem has the effect of *concentrating* the "measure" $\lfloor d\chi\rfloor$ on the critical points of the integrand.

**Proposition 7.9 ([25]).** *If $h$ is a Morse function on a compact $n$-manifold $M$, then:*

$$\int_M h\,\lfloor d\chi\rfloor = \sum_{p\in\mathrm{Cr}(h)} (-1)^{n-\mu(p)}h(p) \quad : \quad \int_M h\,\lceil d\chi\rceil = \sum_{p\in\mathrm{Cr}(h)} (-1)^{\mu(p)}h(p).$$

This result does not require having a Morse function: see Theorem 7.12. The theory of Euler integration on CF is the highly degenerate setting where the entire domain is critical.

Signal processing – with applications ranging from radar imagery to sensor networks – is fueled by integral transforms. The Euler integral of Chapter 3 yields some interesting novel examples of integral transforms with the limiting feature of being an integer-valued theory, and thus perhaps not directly applicable to, say, image processing. The extension of the Euler integral to $\mathbb{R}$-valued functions allows for a wider array of applications thanks to some novel integral transforms [79, 154].

**Example 7.10 (Euler-Fourier transforms)** ⊚

There is an integral transform $\mathcal{F}\colon \mathrm{CF}(\mathbb{R}^n) \times (\mathbb{R}^n)^\vee \to \mathrm{Def}(\mathbb{R}^n)$ that is best thought of as an Eulerian version of a Fourier transform. For $h \in \mathrm{CF}(\mathbb{R}^n)$ and $\xi \in (\mathbb{R}^n)^\vee$, define the **Euler-Fourier transform** of $h$ in the direction $\xi$ via

$$\mathcal{F}h(\xi) := \int_{-\infty}^{\infty} \int_{\xi^{-1}(s)} h \, d\chi \, ds.$$

It is clear that for $A$ a compact convex subset of $\mathbb{R}^n$ and $\|\xi\| = 1$, $(\mathcal{F}\mathbb{1}_A)(\xi)$ equals the projected length of $A$ along the $\xi$-axis. This points to the following theorem of [154]. Let $A$ be a tame compact $n$-dimensional submanifold of $\mathbb{R}^n$ with boundary $\partial A$, decomposed into positive $\partial^+ A$ and negative $\partial^- A$ components (depending on whether the oriented normal to $\partial A$ has positive or negative dot product with $\xi$). Then, the Euler-Fourier transform of $A$ can be reduced to an integral over $\partial A$:

$$\mathcal{F}\mathbb{1}_A(\xi) = \int_{\partial^+ A} \xi \lfloor d\chi \rfloor - \int_{\partial^- A} \xi \lceil d\chi \rceil. \tag{7.4}$$

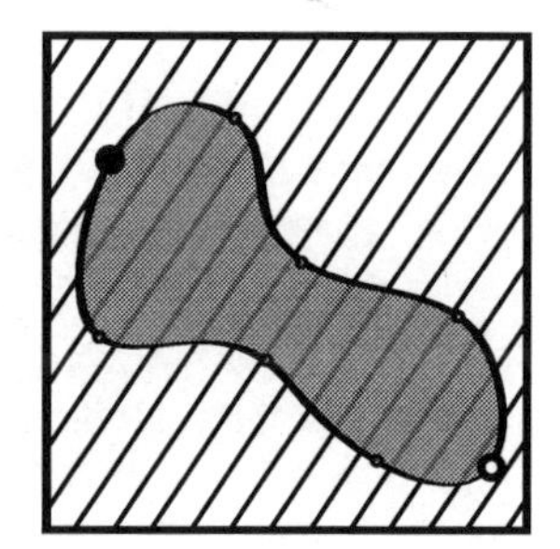

Note the resemblance to Stokes' theorem: an integral over the interior equals an integral of an "anti-derivative" over the boundary. One must be careful with orientations and signs, as with Stokes' theorem. Since $\partial A$ is a manifold and $\xi\colon \partial A \to \mathbb{R}$ is generically a Morse function, Proposition 7.9 implies that $\mathcal{F}h(\xi)$ is an alternating sum of critical values, graded by Morse index, providing a vast generalization of the $\xi$-projected width observation for $A$ convex. A polar version of this integral transform (called an **Euler-Bessel transform**) has a similar index formula and is useful in shape detection from enumerative sensor data [154]. ⊚

## 7.5 Stratified Morse theory

The initial emphasis of Morse theory on nondegenerate Morse functions and local coordinate representations leaves students with the impression that manifolds and nondegeneracy are a *sine qua non*. Though it is convenient to assume a Morse function, nature often interferes, necessitating a degenerate Morse theory. Morse Theory can be adapted to settings where the object of interest is not a manifold, but rather a stratified space, built from manifold pieces, assembled in a sufficiently tame manner. This leads to some complications, the consequence of increased generality.

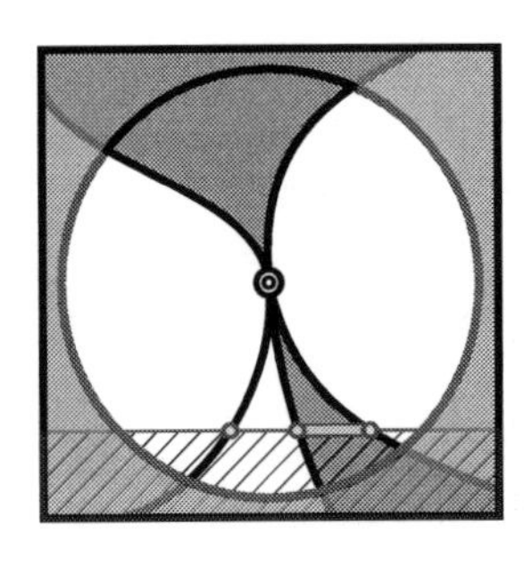

The theory of Goresky and MacPherson [163] recreates Morse theory for stratified spaces (see §1.8). This large and technical body of work cannot be summarized quickly and accurately: the following is elementary, at the expense of carefully-stated theorems. Instead of the usual (Whitney-type) stratified spaces, consider (*cf.* [272]) a fixed o-minimal structure. Let $Y \subset \mathbb{R}^n$ be tame and $h\colon \mathbb{R}^n \to \mathbb{R}$ a definable function – the tame analogue of a Morse function on $Y$. Stratified Morse theory delineates when the excursion sets $Y_t = \{h \leq t\}$ change their topological type. This occurs by defining a Morse index and classifying attachments in the manner of Lemma 7.1.

In classical Morse theory, the index of a critical point is a natural number, and attaching is by surgery of a disc along a sphere of dimension $\mu$. In this stratified setting, one defines the **local Morse data** of $Y$ at a point $p \in Y$ under $h$ to be the *homeomorphism type* of the space

$$\mathrm{LMD}(p) = \mathrm{LMD}(Y, p, h) := \lim_{\epsilon' < \epsilon \to 0^+} \overline{B_\epsilon(p) \cap Y} \cap \{h > h(p) - \epsilon'\}.$$

The limit exists thanks to tameness [293] and is sometimes expressed in terms of a pair $(B, E)$ of compact spaces: $B = \overline{B_\epsilon(p) \cap Y}$ and $E = B \cap \{h \leq h(p) - \epsilon'\}$ (sufficiently small), and $\mathrm{LMD} = B - E$. What is the appropriate Morse index in this setting? For a numerical value, it would be appropriate to call the **Euler-Morse index** of $h$ the constructible function $\mathfrak{I}_h \in \mathrm{CF}(Y)$ given by $\mathfrak{I}_h(x) := \chi(\mathrm{LMD}(Y, x, h)) = \chi(B) - \chi(E)$. Note that one must be careful with open versus closed cells, as in Chapter 3. For a classical Morse function $h$, $\mathfrak{I}_h$ is zero except at the critical points, at which the index takes on a value of $(-1)^\mu$, for $\mu$ the Morse index. A richer index would be the relative homology $H_\bullet(B, E)$, which, for a nondegenerate critical point of a Morse function on a manifold, would be concentrated in grading $\mu$. For more degenerate critical points or critical sets, this index can capture some local topological behavior.

Stratified Morse theory was developed for applications in algebraic geometry that lie outside the bounds of this text, and only the very first step – the local Morse data – has been touched upon. Technical aspects of stratified spaces are numerous and crucial to the theory. The significant steps lie in a *tangential* versus *normal* decomposition of the local Morse data, with instructions as to how local changes in excursion sets are controlled by this data. All of this culminates in yet another homology theory – **intersection homology** – which requires the constructible sheaves of Chapter 9 to fully appreciate (but see [34]). Despite its rarefied origins, stratified Morse theory has found applications in several contexts, including problems in grasping and manipulation in robotics [249], in which potential functions on the relevant configuration spaces can lead to interesting critical sets, on which one cannot simply "compute derivatives" to determine stability.

**Example 7.11 (Euler integration)** ◎

A blending of the Poincaré-Hopf Theorem 3.5 and Corollary 7.6 yields a link to Euler integration in that $\int_Y \mathfrak{I}_h \, d\chi = \chi(Y)$. This, Proposition 7.9, and more follow from a connection between stratified Morse theory and Euler integration over Def($X$).

**Theorem 7.12 ([25]).** *For $h \in$ Def($X$) continuous,*

$$\int_X h \lfloor d\chi \rfloor = \int_X h \mathfrak{I}_{-h} \, d\chi \quad : \quad \int_X h \lceil d\chi \rceil = \int_X h \mathfrak{I}_h \, d\chi$$

Thus, the appearance of not one but two valuations, $\lfloor d\chi \rfloor$ and $\lceil d\chi \rceil$, on Def($X$), is not an anomaly, but rather another manifestation of Morse-theoretic Poincaré duality: $h \leftrightarrow -h$. This result, like so much of Morse theory done properly, does not require either a manifold or a nondegeneracy condition beyond tameness. ◎

## 7.6 Conley index

Stratified Morse theory allows one to relax to non-manifolds and degenerate gradients: with the proper approach, it is also possible to ignore the Morse function altogether. One of the best approaches for doing generalized Morse theory is due to Conley [72], and has enjoyed great success in applications to mathematical biology [44, 181], rigorous numerics [292, 81, 228], experimental time-series inference [229], and more. What follows is a simple version of Conley's theory in the continuous-time setting.

Consider the case of a gradient field $-\nabla h$ of a Morse function $h: M \to \mathbb{R}$ on a manifold $M$. Choose a fixed point $p \in \mathrm{Cr}(h)$ and consider a small ball $B$ about $p$. The boundary $\partial B$ is partitioned into an *exit set* on which the gradient field $-\nabla h$ points *out* of $B$; an *entrance set* on which the field points *in*; and the remaining points of tangency to $\partial B$. If $p$ is a critical point of Morse index $k$, then the exit set is homotopic to a sphere of dimension $\mu(p) - 1$: Lemma 7.1. This example prompts a more general index. *The Conley index is not an integer, but a homotopy type of spaces.*

One simple approach is as follows. Let $X$ be a complete locally compact metric space with a continuous flow $\varphi_t : X \to X$ (which may or may not come from a smooth vector field). The Conley index is associated to a suitable compact subset, $B \subset X$. The **invariant set** $S = \mathrm{Inv}(B; \varphi)$ of the flow on a set $B$ is the set of all points $x \in B$ such that $\varphi_t(x) \in B$ for all $t$. One says a compact $B$ is an **isolating block** if $\mathrm{Inv}(B; \varphi)$ lies strictly in the interior of $B$, and, for all $x \in \partial B$, the flowline through $x$ exits $B$ either in arbitrarily small forwards or backwards time or both. *No internal "tangencies" are permitted.* The isolating block condition is a loose type of transversality: small perturbations to $B$ remain isolating blocks.

The Conley index of an invariant set $S$ with isolating block $B$ collates the topology of $B$ relative to how the flow exits $\partial B$. The **exit set** of $B$ is $E := \{x \in \partial B : \varphi_\epsilon(x) \notin B, \forall\, 0 < \epsilon \ll 1\}$. The **Conley index** of $S$ is the pointed homotopy type $\mathrm{Con}(S) := \mathrm{h}[B/E, \{E\}]$. The index *is* the quotient space $B/E$ (up to homotopy) with $E$ remembered as an abstract basepoint. The **homological Conley index** of $S$ is the relative singular homology $\mathrm{Con}_\bullet(B) := H_\bullet(B, E)$. The index is well-defined in that any two isolating blocks with the same invariant set have equivalent Conley indices.

**Example 7.13 (Morse index)** ⊚

The Conley index of a nondegenerate critical point of a Morse function with Morse index $\mu$ is the (homotopy type of the) sphere $\mathbb{S}^\mu$ (with basepoint). The basepoint, which initially seems extraneous, is vital when considering the case $\mu = 0$. There, the fixed point is a sink, and the exit set $E$ is the empty set, which, when remembered as an abstract basepoint, gives a Conley index of $\mathbb{S}^0$, the disjoint union of two points. Notice that (1) the (local) topology of the unstable manifold of the invariant sets figures prominently in the Conley index; and (2) the Conley index depends only on the type of critical point $B$ surrounds, not on $B$ itself. ⊚

Isolating blocks are, unfortunately, not as abundant or flexible as one would like. The solution is to allow for a larger exit set than simply what lies on the boundary of $B$. Consider an invariant set $S$ of the flow $\varphi$. One defines an **index pair** $(B, E)$ of $S$ to be compact subsets $E \subset B$ of $X$ satisfying:

1. **Isolation:** $S = \mathrm{Inv}(\overline{B - E}; \varphi)$ lies in the interior of $B - E$;
2. **Invariance:** any flowline starting in $E$ and staying in $B$ lies within $E$; and
3. **Exit:** any flowline exiting $B$ does so via $E$.

This $B$ analogous to an isolating block, and $E$ its exit set. The resulting Conley index is well-defined in its homotopical $\mathrm{Con}(S) := \mathrm{h}[B/E, \{E\}]$ and homological $\mathrm{Con}_\bullet(S) := H_\bullet(B, E)$ instantiations. Wonderful to tell, the Conley index generalizes the Morse index greatly:

1. Fixed points need not be nondegenerate, discrete, or stratified.
2. Vector fields need not be smooth or gradient.
3. Domains need not be manifolds or stratified/definable spaces.

Warning: some flows (*e.g.*, integrable Hamiltonian) admit few index pairs, and some minimal amount of local compactness is convenient. Computation of the Conley index is aided by the following:

**Theorem 7.14 (Conley Index Theorem (see [266])).** *The Conley index has the following properties:*

1. **[Invariance]** *The Conley index depends only on $S$, not on the choice of index pair $(B, E)$ for $S$.*
2. **[Continuation]** *If $(B_\lambda, E_\lambda)$ is a (Hausdorff-) continuous family of index pairs for a continuous family of flows, then $\mathrm{Con}_\lambda = \mathrm{Con}(B_\lambda, E_\lambda)$ is constant.*

3. **[Additivity]** *If* $(B, E)$ *and* $(B', E')$ *are disjoint index pairs for* $S$ *and* $S'$, *then*

$$\mathrm{Con}(S \sqcup S') = \mathrm{Con}(S) \vee \mathrm{Con}(S').$$

**Example 7.15 (Forcing theory)** ⊙

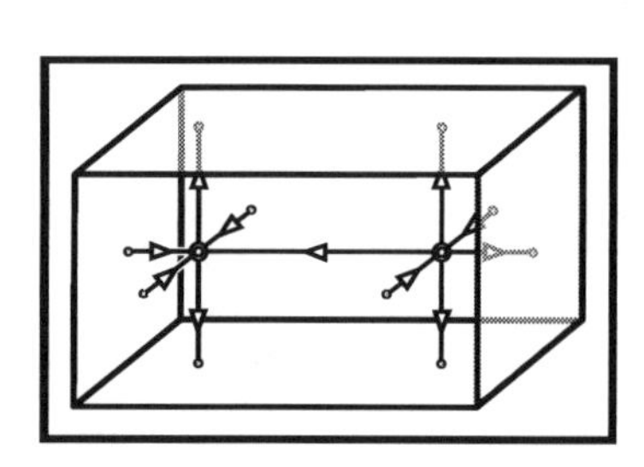

The signature application of index theory in dynamics is forcing the existence of invariant sets. The Poincaré-Hopf Theorem (Theorem 3.5), the Lefschetz Theorem (Theorem 5.19), and the Morse inequalities (Corollary 7.5) all can be used to force the existence of fixed points. The Conley index version is more general still: if $\mathrm{h}[B/E, \{E\}] \neq 0$ (*i.e.*, the pointed homotopy type is not that of a point), then $\mathrm{Inv}(B-E; \varphi) \neq \varnothing$. This invariant set may or may not be a fixed point – heteroclinic orbits, periodic orbits and even chaotic invariant sets are detectable [54, 141].

Consider the simple example of the 3-d system $\dot{x} = x^2 - x$, $\dot{y} = -y$, and $\dot{z} = z$. This gradient flow has two fixed points, which can be characterized by their Euler-Poincaré indices ($-1$ and $+1$), their Morse indices (1 and 2), or their Conley indices ($\mathbb{S}^1$ and $\mathbb{S}^2$), each computed using local information from small neighborhoods of the fixed points. If, instead of small neighborhoods of the fixed points, one chooses a rectangular prism $B$ surrounding the pair of fixed points, then, clearly, the fixed-point index on $B$ is $\mathcal{I}(B) = 1 - 1 = 0$; likewise, the Conley index of this isolating block is zero as well, since $E \subset \partial B$ is contractible. However, $\mathrm{Con}(B) = 0 \neq \mathbb{S}^1 \vee \mathbb{S}^2$ as would follow from Theorem 7.14 if there were no other invariant sets in $B$. One therefore concludes that there is another invariant set in $B$ besides the fixed points: it is in fact the heteroclinic orbit connecting the two fixed points.

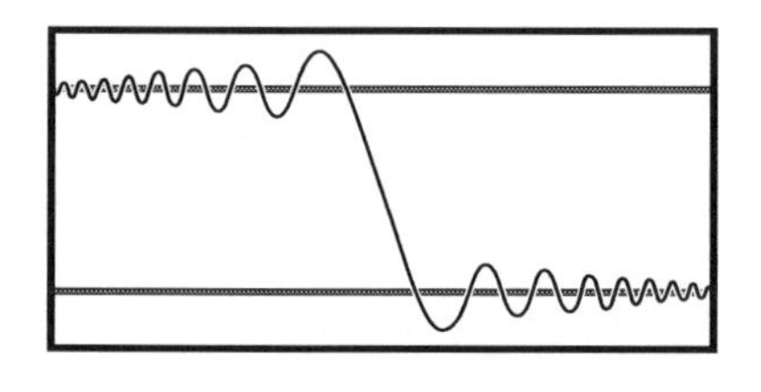

This simple example of an explicit gradient field is trivial, but it captures the spirit of deeper applications. One of the earliest uses of the Conley index was to show the existence of a **traveling wave** solution $u(x, t)$ to a class of reaction-diffusion partial differential equations of the form $u_t = u_{xx} + f(x, u, u_x)$ arising from mathematical biology (population dynamics [73], nerve impulses [279], and more). A pair of constant stationary solutions $u_0(x, t) = u_0$, $u_1(x, t) = u_1$ (*i.e.*, fixed points of the PDE flow) admits a traveling wave if there is a solution $u(x, t)$ that approaches $u_1$ as $t \to \infty$ and $u_0$ as $t \to -\infty$ (*i.e.*, a heteroclinic orbit of the PDE flow). Index pairs around each fixed point yield a nonzero Conley index, but a suitable index pair for the union of the two has trivial Conley index, as computed by 'deforming' the PDE and using continuation. An additivity argument *proves* that a travelling wave exists in cases where direct numerical simulation is inconclusive. Further extensions of these ideas have been very useful in finding stationary, time-periodic, and travelling-wave solutions to very general classes of PDEs [84, 156, 182, 292]. ⊙

**Example 7.16 (Attractor-repeller pairs)** ⊚

Rigorous arguments for the existence of a connecting orbit require a bit more machinery [129, 203, 222, 266]. The simplest example of such is as follows.

Consider an isolated invariant set $S \subset X$ for the flow $\varphi$. An **attractor-repeller pair** is a pair of disjoint compact sets $(A, R)$ in $S$ such that, for every $x \in S-(A \sqcup R)$, the flowline $\varphi_t(x) \to A$ as $t \to \infty$ and $\varphi_t(x) \to R$ as $t \to -\infty$. Thus, $S$ decomposes as $A$, $R$, and $S-(A \cup R)$ the connecting orbits. Given any attractor-repeller pair, there exists an **index triple** $N_0 \supset N_1 \supset N_2$ of compact subsets of $X$, where $(N_0, N_2)$ is an index pair for $S$, $(N_1, N_2)$ is an index pair for $A$, and $(N_0, N_1)$ is an index pair for $R$. The homological long exact sequence of the triple $(N_0, N_1, N_2)$ (from the end of §5.3) becomes an exact sequence relating Conley indices:

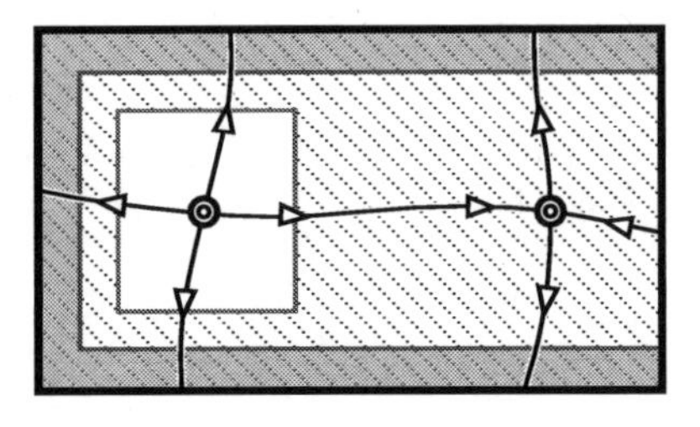

$$\cdots \xrightarrow{\delta} \mathrm{Con}_k(A) \xrightarrow{H(i)} \mathrm{Con}_k(S) \xrightarrow{H(j)} \mathrm{Con}_k(R) \xrightarrow{\delta} \cdots,$$

This can be used for forcing connecting orbits from $R$ to $A$.

**Lemma 7.17.** *If the connecting homomorphism $\delta$ on an index triple is nonvanishing, then there exists a connecting orbit from $R$ to $A$ in $S$.*

**Proof.** Assume that $S-(A \sqcup R) = \varnothing$. Then $S = A \sqcup R$ and, as per Theorem 7.14, the Conley index decomposes as a wedge sum. Passing to homology, one obtains $\mathrm{Con}_\bullet(S) = \mathrm{Con}_\bullet(A) \oplus \mathrm{Con}_\bullet(R)$. Exactness then implies that $\delta = 0$. ⊙⊚

**Example 7.18 (Floer homology)** ⊚

One of the great triumphs of the Conley's approach to Morse theory is in an infinite-dimensional version due to Floer [122] (see also [21, 224, 274]). It has been known since the original works in Morse theory that the methods are applicable to gradient flows on certain infinite-dimensional settings[1] when a Morse index can be defined. Unfortunately, those functionals for which Morse index is finite are rare. More common is the case of functionals whose critical points have Hessians with an infinite number of positive and negative eigenvalues. The breakthrough of Floer was to mimic the Conley approach in settings where the critical points whose linearizations yield Fredholm operators: the positive and negative eigenspaces are both infinite-dimensional, but the difference of their dimensions – the **Fredholm index** – is finite. By combining these insights with the pseudoholomorphic curve technology of Gromov [166] and an appropriate (nontrivial) array of analytic results, Floer and those who followed construct a chain complex over $\mathbb{F}_2$ freely generated by critical points of the functional and graded by the Fredholm index. The resulting **Floer homology** $\mathrm{Fl}_\bullet$ possesses similar properties as the Conley homology, including continuation. Most remarkable is the fact that,

[1] For example, Banach manifolds – spaces locally modeled on a fixed Banach space.

since the Fredholm index takes values in $\mathbb{Z}$ as opposed to $\mathbb{N}$, Floer homology is graded over the full integers: it is common to have $\mathsf{Fl}_k \neq 0$ for negative values of $k$. ⊙

**Example 7.19 (Arnol'd Conjecture)** ⊙

The first achievement of Floer's theory was a resolution of the **Arnol'd Conjecture**. Fix a compact $2n$-manifold $M$ with a **symplectic form**, a closed nondegenerate[2] 2-form $\omega \in \Omega^2(M)$. A symplectic manifold allows for Hamiltonian dynamics as follows. Given a function $H: M \to \mathbb{R}$, let $V_H$ be the (unique) vector field satisfying $\omega(V_H, \cdot) = -dH$. This **Hamiltonian field** $V_H$ is a twisted analogue of the gradient $-dH$, and it follows from the Morse inequalities that $V_H$ has at least $\sum_k \dim H_k(M; \mathbb{R})$ fixed points.

Choose a smooth $\mathbb{S}^1$-parameter family of functions $H_t$, $t \in \mathbb{S}^1$, and let $V_t = X_{H_t}$ be the corresponding $t$-dependent vector field. Arnold's conjecture – that the number of 1-periodic orbits of the family $V_t$ is bounded below by $\sum_k \dim H_k(M)$ – is a deceptively simple-sounding analogue of the Strong Morse Inequalities. It was first proved using Floer-theoretic arguments as follows. Periodic orbits of $V_t$ satisfy a variational principle with respect to *action*. This action yields a real-valued function on the space of loops in $M$, extrema of which correspond to 1-periodic orbits. The action functional is unbounded and the loop space of $M$ is infinite dimensional, thus necessitating Floer-theoretic arguments to find critical points of action. An equivalence between the Floer homology of the action functional and the homology of $M$ – a variant of Theorem 7.3 – provides the key to the Arnol'd Conjecture. Details are beyond the scope of this text (and are largely analytic), but the theme of counting certain invariant sets by means of a specialized homology theory and relating it to classical homology is fully in the spirit of Morse theory. ⊙

## 7.7 Lefschetz index, redux

As hinted at in §5.10, there is a way to compute the Lefschetz index of a self-map $f: X \to X$ that is localized at the fixed point set. This turns out to have a deep connection to both stratified Morse theory and the Conley index. The number and types of Lefschetz theorems are difficult to keep track of. Let the reader keep in mind that the utility of fixed-point theorems in economics, game theory, differential equations, and dynamical systems justifies the sometimes prickly technical machinery that arises.

**Example 7.20 (Degree-theoretic Lefschetz)** ⊙

The fixed point index $\mathfrak{I}$ of a vector field (from §3.4) is intimately related to the Euler characteristic, thanks to Poincaré-Hopf (Theorem 3.5). It is also easily computed as a degree, as per Example 4.23. This perspective lifts to the Lefschetz theorem as well [98]. Consider first the case where $U \subset \mathbb{R}^n$ is open and $f: U \to \mathbb{R}^n$ has compact fixed

[2]Nondegenerate means that $\omega^n = \omega \wedge \cdots \wedge \omega \neq 0$ is a volume form.

point set $F = \text{Fix}(f)$. Then the fixed point index, $\mathfrak{I}_f(F)$, is defined to be the degree of $\text{Id} - f$:

$$\mathfrak{I}_f(F) := \deg\left(\left(U, U-F\right) \xrightarrow{\text{Id}-f} \left(\mathbb{R}^n, \mathbb{R}^n-0\right)\right). \tag{7.5}$$

The domain and codomain local homologies (in $\mathbb{Z}$-coefficients) have rank one in dimension $n$; thus the degree is well-defined and has the usual properties, including homotopy invariance and additivity. Thus, the index can be computed as the sum $\mathfrak{I}_f = \sum_F \mathfrak{I}_f(F)$, where the sum is over all $F$ disjoint connected components of $\text{Fix}(f)$. When $f: X \to X$ is not defined on open neighborhoods in $\mathbb{R}^n$, but only on $X$ a neighborhood retract (suitably small open neighborhoods of $X$ in $\mathbb{R}^n$ retract to $X$), then, given a neighborhood $U$ of a fixed point component $F$, $\mathfrak{I}_f(F)$ can be computed as $\deg(\text{Id} - f \circ r)$, where $r : U \to X$ is the retraction. In either setting, each term is determined by the local behavior near $F$. The deep result is that this sum of *local* fixed point indices is equal to the *global* Lefschetz index:

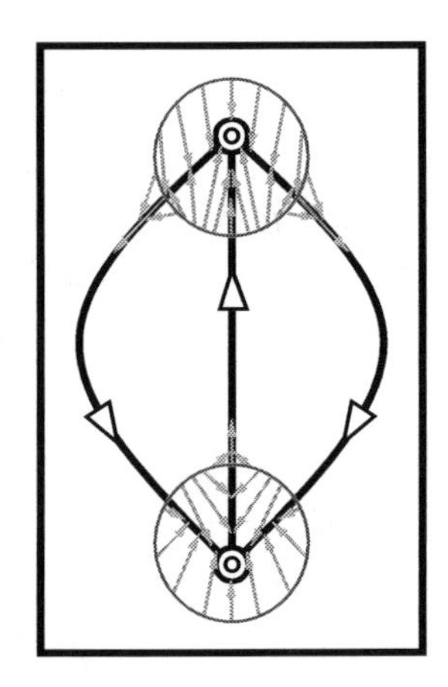

**Theorem 7.21 (Lefschetz-Hopf Theorem).** *For $X \subset \mathbb{R}^n$ a neighborhood retract and $f: X \to K \subset X$ a map to a compact subset $K$, the Lefschetz index (on $H_\bullet(X;\mathbb{R})$) equals the fixed point index:*

$$\mathfrak{I}_f = \tau_f = \sum_k (-1)^k \text{trace}\left(H(f): H_k X \to H_k X\right).$$

The proof is beyond what can be reasonably done briefly: see, *e.g.*, [49, 98]. It is very instructive to draw some pictures of neighborhoods of fixed point sets and compute the degrees *by hand*. It is the best way to see the connection to Morse theory, since some fixed point components are *attracting*, some are *repelling*, and some have *mixed* behavior. ⊚

**Example 7.22 (Stratified Morse-theoretic Lefschetz)** ⊚

Equation (7.5) is at best of limited use given the requirement that $X$ have an explicit embedding in $\mathbb{R}^n$: at the very least, it is poor form to work extrinsically. An intrinsic approach works with a Morse-type assumption on $f$, as pointed out in [164]. What follows is a slight reformulation. Assume that $f: X \to X$ is definable. Then $\text{Fix}(f)$ is automatically compact and decomposed into connected components.

Following the index pair construction of §7.6, one could define for each connected component $F$ of $\text{Fix}(f)$ an **index pair** $(B, E)$ of compact subsets $E \subset B$ of $X$ satisfying:

1. **Isolation:** $F = \text{Fix}(f|B - E)$ lies in the interior of $B - E$;
2. **Invariance:** $f(E) \cap (B - E) = \varnothing$; and

3. **Exit:** if $x \in B$ and $f(x) \notin B$, then $x \in E$.

For any such index pair, there is an induced map $f : (B/E, \{E\}) \to (B/E, \{E\})$ since points in $B - E$ either remain in $B - E$ or are sent to $E$. The choice of an index pair is by no means unique; nor is the pointed homotopy type of $(B/E, \{E\})$, in contradistinction to the Conley index. However, remarkably, the analogue of a localized relative Lefschetz index *is* well-defined:

$$\tau_f(F) = \sum_k (-1)^k \text{trace}\left(H(f)\colon H_k(B,E) \to H_k(B,E)\right). \tag{7.6}$$

In this setting, the Lefschetz-Hopf theorem becomes a theorem about the action of $f$ on the local homologies of the fixed point components, relative to the appropriate exit sets:

$$\mathcal{I}_f(F) = \tau_f(F) \quad : \quad \mathcal{I}_f = \sum_F \tau_f(F)$$

There is yet a better version of the theorem that uses the tools of stratified Morse theory. For $X$ compact and $f\colon X \to X$ definable, the fixed point set not only splits into a finite number of compact components; each of these is further stratified into disjoint open simplices on which the local behavior of $f$ is well-defined and 'constant' in the sense that for each stratum $F_\alpha$ of Fix$(f)$, there is a well-defined local Lefschetz index $\tilde{\tau}_f(F_\alpha) \in \mathbb{Z}$. The definition of this index is similar in spirit to Equation (7.6), but for a localized index pair. By defining the local Lefschetz index to be zero off of Fix$(f)$, one can interpret $\tilde{\tau}_f$ as a constructible function on $X$. It was shown by Goresky and MacPherson [164] that

$$\tau_f = \int_X \tilde{\tau}_f \, d\chi.$$

This beautiful result is sadly under-appreciated, in part because the construction of $\tilde{\tau}$ (general enough to apply to multivalued mappings $F\colon X \rightrightarrows X$) requires techniques from the theory of sheaves that will only be hinted at in Chapters 9 and 10. ⊙

## 7.8 Discrete Morse theory

An idea as deep as Morse theory has emanations throughout all of Mathematics. This chapter has focused primarily on the smooth or continuous theory; however, there is a discrete version of Morse theory that has of late yielded powerful results in combinatorics [197, 198], braid groups [118], computational homology [230] and certain problems in computer science. The original papers of Forman [123, 126] are complemented by the recent book of Kozlov [198] and a growing literature.

Consider for concreteness a simplicial or cell complex $X$. The critical ingredient for Morse theory is *not* the Morse function but rather its gradient flow. A **discrete vector field** is a pairing $V$ which partitions the cells of $X$ (graded by dimension) into pairs $V_\alpha = (\sigma_\alpha \lhd \tau_\alpha)$ where $\sigma_\alpha$ is a codimension-1 face of $\tau_\alpha$. All leftover cells of $X$ not paired by $V$ are the **critical cells** of $V$, Cr$(V)$. A **discrete flowline** is a sequence

$(V_i)$ of distinct paired cells with codimension-1 faces, arranged so that

$$\overbrace{\sigma_1 \lhd \tau_1}^{V_1} \unrhd \overbrace{\sigma_2 \lhd \tau_2}^{V_2} \unrhd \cdots \unrhd \overbrace{\sigma_N \lhd \tau_N}^{V_N} .$$

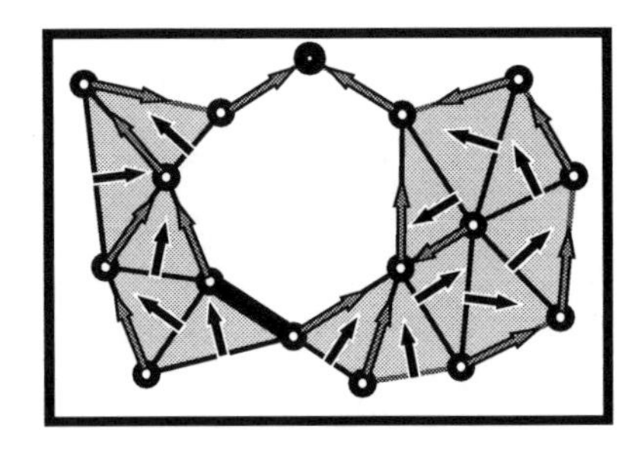

A flowline is **periodic** if $\tau_N \unrhd \sigma_1$ for $N > 1$. A **discrete gradient field** is a discrete vector field devoid of periodic flowlines.

It is best to lift everything to algebraic actions on the chain complex $\mathcal{C} = (C_\bullet^{\text{cell}}, \partial)$ associated to the cell complex $X$. By linearity, the vector field $V$ induces a chain map $V\colon C_k \to C_{k+1}$ induced by the pairs $\sigma \lhd \tau$ – one visualizes an arrow from the face $\sigma$ to the cell $\tau$. As with classical Morse homology, $\mathbb{F}_2$ coefficients is simplest; when oriented, one specifies $V\colon \sigma \mapsto [\sigma\colon \tau]\tau$ using incidence numbers. The **discrete flow** of $V$ is generated by the degree-zero chain map $\Phi$ given by

$$\Phi := \text{Id} + \partial V + V\partial,$$

with iterations of $\Phi$ describing how cells descend along the gradient field $V$. Unlike continuous time flows, the discrete flow has a limit: $\Phi^\infty = \lim_{n\to\infty} \Phi^n$ is constant for $n$ sufficiently large. To every discrete gradient field is associated a discrete Morse complex, $\mathcal{C}^V = (MC_\bullet, \tilde{\partial})$ with $MC_k$ the vector space (or module) with basis the critical cells $\{\sigma \in \text{Cr}(V)\, ;\, \dim(\sigma) = k\}$. Note that dimension plays the role of Morse index.

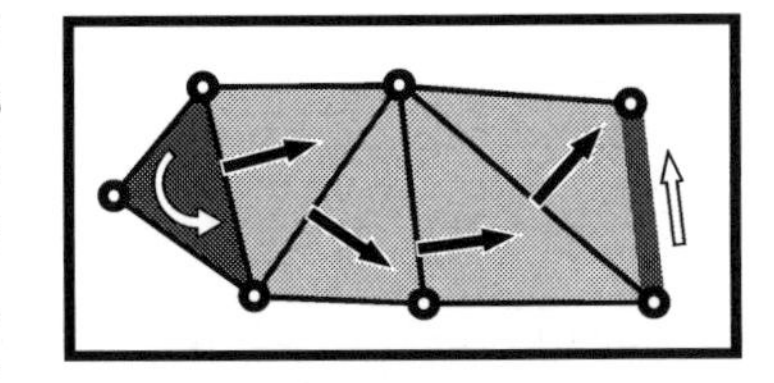

The boundary maps $\tilde{\partial}_k$ count (modulo 2 in the case of $\mathbb{F}_2$ coefficients; with a complicated induced orientation else) the number of discrete flowlines from a critical simplex of dimension $k$ to a critical simplex of dimension $k-1$. Specifically, given $\tau$ a critical $k$-simplex and $\sigma$ a critical $(k-1)$-simplex, the contribution of $\tilde{\partial}_k(\tau)$ to $\sigma$ is the number of gradient paths from a face of $\tau$ to a coface of $\sigma$. In the case that $\sigma \lhd \tau$, then this number is 1, ensuring that the trivial $V$ for which all cells are critical yields $\mathcal{C}^V$ the usual cellular chain complex. It is not too hard to show that $\tilde{\partial}^2 = 0$ and that, therefore, the homology $MH_\bullet(V) = H_\bullet(\mathcal{C}^V)$ is well-defined. As usual, the difficulty lies in getting orientations right for $\mathbb{Z}$ coefficients.

**Theorem 7.23 ([123]).** *For any discrete gradient field $V$, $MH_\bullet(V) \cong H_\bullet^{\text{cell}}(X)$.*

It follows from the proof that the strong Morse inequalities, Equation (7.1), hold with Morse polynomial $M_V(t) = \sum_{\sigma\in\text{Cr}(V)} t^{\dim\sigma}$. Discrete Morse theory, like the Conley index theory, shows that the classical constraints – manifolds, smooth dynamics, nondegenerate critical points – are not necessary. Applications of discrete Morse theory are numerous and expansive, including to combinatorics [198], mesh simplification [208], image processing [252], configuration spaces of graphs [117, 118], and, most strikingly, efficient computation of homology of cell complexes [230].

**Example 7.24 (20 questions)** ⊙

One of the early applications by Forman was to a problem in decision theory related to evasiveness [124]. Let $\Delta$ be an abstract $n$-simplex on the vertex set $\{v_i\}_0^n$ and let $K \subset \Delta$ be a known subcomplex. There is a *hidden* (*i.e.*, unknown) simplex $\sigma$ of $\Delta$, and the goal of the *20 questions* game is to determine whether $\sigma \subset K$ by asking questions of the form *"Is $v_i$ in $\sigma$?"* Clearly, one can win the game with $n+1$ questions by interrogating each vertex. One says that $K$ is **nonevasive** if there is a strategy for determining whether $\sigma \subset K$ in strictly less than $n+1$ questions, independent of $\sigma$; else, $K$ is **evasive**. For example, if $K = \partial\Delta$, then determining if $\sigma \subset K$ is clearly evasive, since one must check that all $v_i \in \sigma$. However, there is only one **evader** – only one simplex $\tau$ of $\Delta$ has the property that all $n+1$ questions must be asked in order to determine if $\tau \subset K$.

The insight of discrete Morse theory is that a guessing algorithm for determining if $\sigma \subset K$ induces a discrete gradient field on $\Delta$, with the twist that one of the vertices is paired with the formal basepoint $\varnothing$, leading to reduced homology. This yields Morse-theoretic proofs of the following [124]:

1. If $K$ is nonevasive, $K$ collapses to a point.
2. If $\tilde{H}_\bullet(K) \neq 0$, then $K$ is evasive.
3. The number of evaders is $\geq 2\sum_i \dim \tilde{H}_i(K)$.

One clever application of this result is to independence tests for random variables. Let $\mathcal{X} = \{X_i\}$ be a collection of random variables and recall from Example 2.1 the independence complex $\mathcal{I}_{\mathcal{X}} \subset \Delta^n$ of $\mathcal{X}$. Given an unknown subcollection $\sigma \subset \mathcal{X}$ of the random variables, how many trials of the form *"Is $X_i$ a member of $\sigma$"* are required to determine if the collection is statistically independent? According to the results cited above, statistical independence is evasive if $\mathcal{I}_{\mathcal{X}}$ is not acyclic: any nontrivial homology class in $\tilde{H}_\bullet(\mathcal{I}_{\mathcal{X}})$ is an obstruction to evasiveness of statistical independence. How many such evasive collections of random variables are there? It is at least twice the total dimension of $\tilde{H}_\bullet(\mathcal{I}_{\mathcal{X}})$. ⊙

## 7.9 LS category

Given a space, how complicated is it? One means of characterizing topological complexity is co/homology. Another approach might involve critical points and other Morse-theoretic constructs. There is a more primal measure of topological complexity dating back to the work of Lusternik and Schnirelmann in the 1930s that goes under the (suboptimal) name of **category**. This classical measure of complexity for spaces, living in the shadow of Morse theory, informs various contemporary problems ranging from statistics to motion-planning in robotics.

Given $X$, the **LS category** of $X$, $\mathrm{LScat}(X)$, is the minimal number $\#\alpha$ of elements in an open cover $\{U_\alpha\}$ of $X$ by sets which are nullhomotopic in $X$, meaning that

$U_\alpha$ is contractible in $X$, not that $U_\alpha$ is necessarily a contractible (or even connected) set. If a finite cover does not suffice, one sets LScat $= \infty$. The **geometric category** of $X$, gcat($X$), is the minimal such number $\#\alpha$ where each $U_\alpha$ is a contractible set. Invariance of LScat and gcat under, respectively, homotopy type and homeomorphism type, follows from standard results: see [74] for a comprehensive introduction.

**Example 7.25 (LS category)** ◎
A sphere has $\text{LScat}(\mathbb{S}^n) = \text{gcat}(\mathbb{S}^n) = 2$ for all $n$. A compact surface $S_g$ of genus $g > 0$ has $\text{LScat}(S_g) = \text{gcat}(S_g) = 3$. It is a nontrivial exercise to find a space on which LScat and gcat differ. One simple example is the space $X = \mathbb{S}^2 \vee \mathbb{S}^1$ obtained by gluing together $\mathbb{S}^2$ and $\mathbb{S}^1$ at a single point: $\text{gcat}(X) = 3$, but $\text{LScat}(X) = 2$. ◎

The original motivation for investigating LS category was (in modern parlance) degenerate Morse theory. The Morse inequalities (Corollary 7.5) give a lower bound on the number of critical points of a smooth non-degenerate functional on a manifold $M$. In the case where the functional is not necessarily non-degenerate or $M$ not necessarily a manifold, the LS category gives the correct lower bound.

**Theorem 7.26 ([74]).** *Any $C^2$ function $h\colon M \to \mathbb{R}$ on a compact manifold $M$ must have at least* LScat($M$) *critical points.*

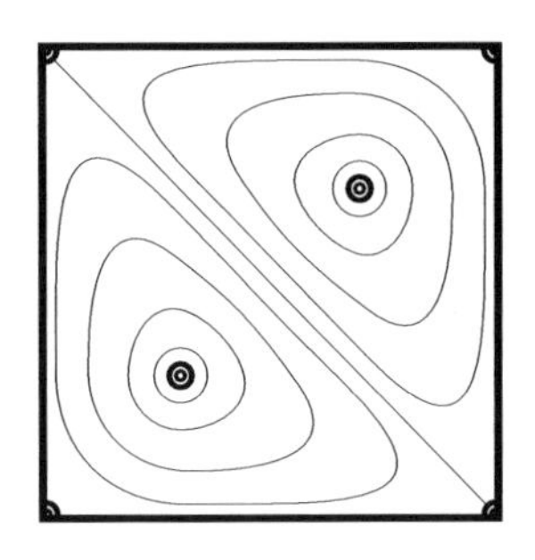

For example, any smooth functional on the 2-torus $\mathbb{T}^2$ must have at least three critical points. For a Morse function, the smallest number of critical points is

$$\sum_k \dim H_k(\mathbb{T}^2; \mathbb{R}) = 1 + 2 + 1 = 4.$$

Given a space $X$, it is typically difficult to compute the category of $X$, either geometric or LS. For example, the **Ganea conjecture**, open from 1971 until its disproval in 1998, was the deceptively simple statement that $\text{LScat}(X \times \mathbb{S}^n) = \text{LScat}(X) + 1$ for $n > 0$ and $X$ a smooth closed manifold. Given such subtleties, one adopts a strategy of estimation, which, fortunately, has some reasonable steps.

**Theorem 7.27 ([74]).** *The LS category of a path-connected CW complex $X$ is bounded by*

$$\text{cup}(X) \leq \text{LScat}(X) - 1 \leq \dim X.$$

The **cup length**, cup, is the smallest $N$ such that there are $N$ cohomology classes $\alpha_i \in H^\bullet(X)$ with nonzero grading and nonzero cup product $\alpha_1 \smile \cdots \smile \alpha_N \neq 0$. Cup length may depend on the coefficient ring used; the bound above does not. The theorem holds for more general (locally-contractible paracompact) spaces, at the cost of using *covering dimension* in the upper bound. These elementary bounds are the beginning of a rich theory of complexity for topological spaces. It complements (classical) Morse theory in its insensitivity to nondegeneracy.

## 7.10 Unimodal decomposition in statistics

LS category inspires definitions of topological complexity in several settings. Distributions on $\mathbb{R}^n$ form one excellent example relevant to statistics and mode-counting. Let $\mathfrak{D} = \mathfrak{D}(\mathbb{R}^n)$ denote the set of all compactly supported continuous functions $f: \mathbb{R}^n \to [0, \infty)$ and consider the statistical problem of **mode counting**. Given $f \in \mathfrak{D}$, assume it is the result of a sum of basis Gaussian distributions, or **modes**, of unknown mean, variance, and height. How many modes are there? This is an ill-defined question, but the minimal number of such modes is a reasonable measure of the distribution's complexity.

The problem becomes more topological in the coordinate-free setting where the distribution $f$ is not known in terms of a fixed coordinate system, as might occur if the function values are sampled over a network of non-localized sensors. In this context, the following coordinate-free notion of a mode is relevant: $u \in \mathfrak{D}$ is **unimodal** if the non-empty upper excursion sets $u^c = u^{-1}([c, \infty))$ are contractible.

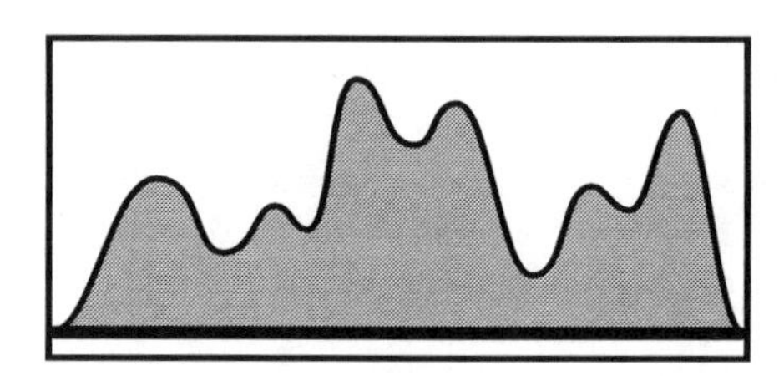

All Gaussians and other typical basis modes are, as the name connotes, unimodal. Following [26], define the **unimodal category** of a distribution $f \in \mathfrak{D}$ to be the minimal number ucat of unimodal distributions $u_\alpha$, for $\alpha = 1, \ldots,$ ucat such that $f$ is a combination of unimodals:

$$f(x) = \sum_\alpha u_\alpha(x).$$

Unimodal category is invariant under changes of coordinates, as follows. For $u \in \mathfrak{D}$ unimodal and $\varphi: \mathbb{R}^n \to \mathbb{R}^n$ a homeomorphism, $(u \circ \varphi)^c = \varphi(u^c)$, which, being the homeomorphic image of a contractible set, is contractible. Thus, $\text{ucat}(f)$ is a topological invariant of $f$. In the same way that one lifts the Euler characteristic from subsets of a space to distributions over subsets (integer or real valued) via the Euler integral, one lifts gcat from subsets of a space to distributions thereon. The unimodal category of the constant distribution $\mathbb{1}_U$ is, simply, $\text{gcat}(U)$.

In general, the computation of unimodal category is, as with LScat or gcat, difficult. There is a simple algorithm [26] for the case of a univariate distribution: a greedy sweep of the distribution from left to right (or, via topological invariance, right to left) suffices. The correctness of this algorithm is based on the following result:

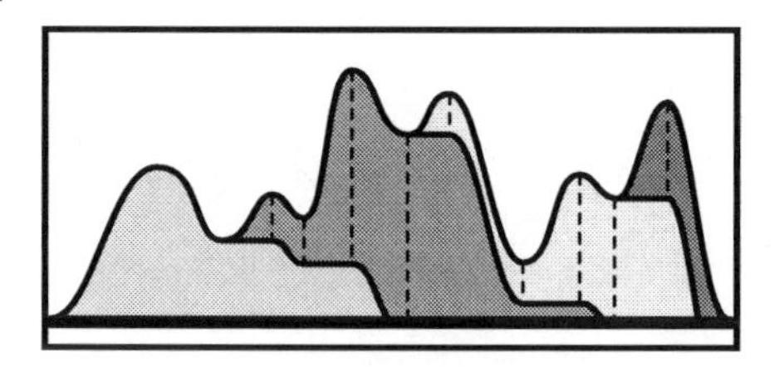

**Proposition 7.28 ([26]).** *For any $f \in \mathfrak{D}(\mathbb{R})$, $\text{ucat}(f)$ is equal to the maximal number of closed intervals $I_k$ covering the support of $f$ such that*

$$\int_{I_k} f\lfloor d\chi \rfloor < 0$$

*for all k.*

From Theorem 7.12, this criterion can be translated to critical value data. For distributions over $\mathbb{R}^n$, an algorithmic solution is unknown and appears difficult.

# Notes

1. One of the lessons of discrete Morse theory, Conley index theory, and nearly all modern variants of Morse theory is that it is not the *function* but the *dynamics* that matters. The initial emphasis on the Morse Lemma in classical texts was, in this author's opinion, an unfortunate distraction.
2. The subject of topological signal processing is embryonic. The Euler-Fourier transform of Example 7.4, the book of Robinson [256] and a few papers [27, 79, 154, 257, 253] are the starting points of deriving qualitative features of environments via low-fidelity signals. Already, Morse theory seems to play a prominent role (but this may be the author's bias).
3. The intrinsic volumes $\mu_k$ of §3.10 are likewise liftable to measures $d\mu_k$ on $\mathrm{CF}(\mathbb{E}^n)$ and then to $\lfloor d\mu_k \rfloor$ and $\lceil d\mu_k \rceil$ on $\mathrm{Def}(\mathbb{E}^n)$ via procedures analogous to those given here for $\lfloor d\chi \rfloor$ and $\lceil d\chi \rceil$. This requires extensive use of currents [28].
4. The definitions involved in the Conley index require more care than is given in the brief overview of §7.6, particularly in defining index pairs, as there are multiple formulations in the literature, with subtle differences in applicability. In the definition of an attractor-repeller pair, it is more proper to use the omega-limit set of the flow, see [167, 258]. The definition as given here is suitable for intuition only, and is but the beginning of a more refined Morse decomposition of the flow.
5. Conley index has been defined for maps (discrete-time dynamics) as well as flows. An **index pair** for $f$ is a pair $(B, E)$ of compact sets satisfying the three properties as listed in Example 7.22, with the exception that the isolation property requires $\mathrm{Inv}(\overline{B-E}; f)$ to be in the interior of $B-E$; invariance and exit properties are the same. The homotopy type of $(B/E, \{E\})$ is not unique. However, one can obtain a well-defined class by looking at the action on this homotopy type up to a certain equivalence [128, 286].
6. The work of Vandervorst *et al.* [155, 156] has adapted Conley and Floer indices to *braids* (see Example 1.8 and §8.3), using, in some cases, the flow of a parabolic PDE on $\mathbb{S}^1$ to set up a stratified Morse theory on the spaces of braids. This leads to some novel examples of forcing, where a single stationary solution to a PDE can force chaotic dynamics, complete with an infinite collection of forced stationary braided solutions.
7. Floer theory is at the moment multifarious, bubbling into many branches of topology, (symplectic, contact, and knot-theoretic). Though the perspective of this chapter is dynamical, much of the current work in Floer theory is symplectic in nature. Among topologists, Floer homology tends to be spoken of as a black box, unfortunately. It remains to incarnate Floer theory into a computational toolset for more directly applied problems. Several authors are progressing to this end [10, 90, 188, 261], but much work remains.
8. Stronger results on discrete Morse theory than presented here are proved in, *e.g.*, [197, 123]. In particular, homotopy-theoretic results about CW complexes are given. The analogue of Forman's discrete Morse theory for differential forms and cohomology is presented in [125]. Forman derived Morse inequalities for arbitrary (non-gradient) discrete fields by counting periodic flowlines properly. A discrete-Morse-theoretic analogue for the Conley index theory appears in the recent monograph of Nicaolescu [240]

(which is also an excellent source for Conley index theory).

9. Farley and Sabalka [117, 118] use discrete Morse theory to explicitly compute the cohomology ring of the (discretized) configuration space $\mathcal{UC}^n(T)$ of unlabeled points on a tree (a cycle-free graph). Their insight was to find a well-suited gradient flow which illuminates a small number of critical cells, the identification of which reveals not merely the homology, but the full cohomology product structure.
10. Mischaikow and Nanda have recently implemented a discrete Morse theory algorithm for computing homology and persistent homology quickly [230].
11. LS category is an abbreviation of Lusternik and Schnirelmann. The abbreviation is convenient vis-a-vis parsimony and frequent variations in spelling past the first letters. The tragic story of these two mathematicians present at the discovery of this invariant is told in [74]. The author apologizes for not using the normalization convention of [74]: the LS category there is one less than that here. The (excellent) motivations for normalizing in this way do not enter the picture in the elementary applications presented here.
12. One can generalize ucat to the **unimodal $p$-category** of $f$ – the minimal number $\mathbf{ucat}^p$ of unimodal distributions $u_\alpha, \alpha = 1, \ldots, \mathbf{ucat}^p$ such that $f$ is pointwise an $\ell^p$ combination of the unimodals:

$$f(x) = \left(\sum_\alpha (u_\alpha(x))^p\right)^{\frac{1}{p}} \quad \text{or} \quad f(x) = \max_\alpha \{u_\alpha(x)\} \text{ when } p = \infty.$$

The unimodal category $\mathbf{ucat} = \mathbf{ucat}^1$ adopts a simple additive model of interference between modes; $\mathbf{ucat}^2$ measures something akin to an *energy* of a distribution; and $\mathbf{ucat}^\infty$ is a natural 'tropical' measure for problems in which mode interference is negligible and the *strongest mode wins*.

Chapter 8

# Homotopy

Deformation is the root operation in topology. Homotopy is the primal deformation, leading to homotopy equivalence and then homotopy theory. As compared to co/homology theory, homotopy theory is intuitive, winsome, and largely immune to computational methods. The intuition of homotopy combines with the practicality of co/homology to forge a more complete picture of algebraic topology.

## 8.1 Group fundamentals

Homological methods are almost entirely comprehensible (and computable) via linear algebra, usually over the reals or over $\mathbb{F}_2$. Homology with $\mathbb{Z}$ coefficients is a bit more subtle, but even the novice is so familiar with this ring that no detailed explanations are required for either intuition or computation. The general setting for homology is best managed using **R**-modules, and this structure has been both alluded to and exploited. In homotopy theory, it is no longer possible to avoid the use of general groups, though, in most every case, it will suffice to work with finitely presented groups described grammatically in terms of generators and relations: see Appendix A.2 for the appropriate keywords.

Let $X$ be a space and $x_0 \in X$ be a designated **basepoint**. A *loop based at* $x_0$ is defined to be a map $\alpha\colon [0,1] \to X$ with $\alpha(0) = \alpha(1) = x_0$. The **fundamental group** $\pi_1(X, x_0)$ is defined on the set of homotopy classes of loops at $x_0$. That is, two loops $\alpha$ and $\beta$ are equivalent if there is a homotopy of loops at $x_0$, $F_t\colon [0,1] \to X$, deforming $F_0 = \alpha$ to $F_1 = \beta$. Note that the basepoint is kept fixed throughout the homotopy, and it is this that permits a group operation given by concatenation of loops as follows:

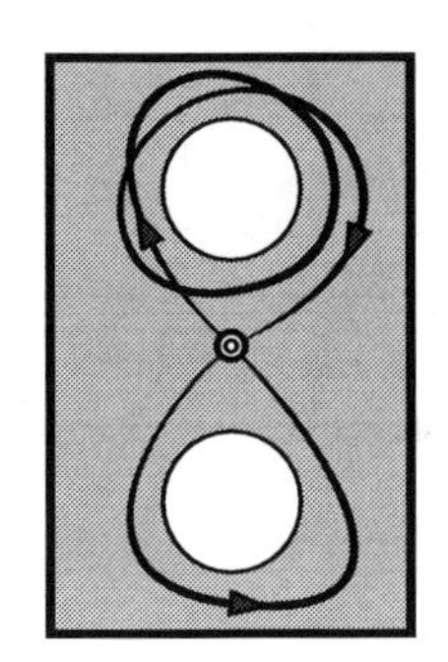

$$\alpha{\scriptstyle\bullet}\beta\colon [0,1] \to X \quad : \quad t \mapsto \begin{cases} \alpha(2t) & : \ 0 \le t \le \frac{1}{2} \\ \beta(2t-1) & : \ \frac{1}{2} \le t \le 1 \end{cases} .$$

It is to be checked that this extends to a well-defined associative operation of homotopy classes of loops: $[\alpha]{\scriptstyle\bullet}[\beta] := [\alpha{\scriptstyle\bullet}\beta]$. The **trivial loop** is the constant map $e\colon [0,1] \to \{x_0\}$. A loop is **contractible** if it is homotopic to the trivial loop. The inverse of a loop $\alpha$ is $\alpha^{-1}(t) := \alpha(1-t)$ and provides a true inverse on homotopy classes: $[\alpha][\alpha^{-1}] = [e] = [\alpha^{-1}][\alpha]$.

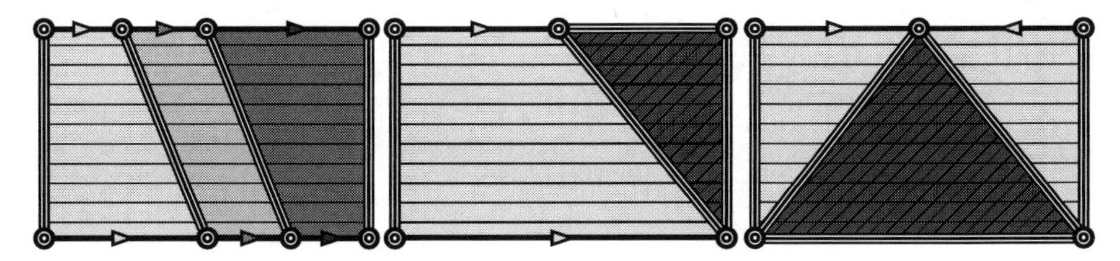

The basepoint is largely irrelevant, in that for $X$ a path-connected space, changing the basepoint from $x_0$ to $x_1$ by means of a path $\gamma\colon [0,1] \to X$ leads to an isomorphism $\pi_1(X, x_0) \cong \pi_1(X, x_1)$ via $[\alpha] \mapsto [\gamma \cdot \alpha \cdot \gamma^{-1}]$. The notation $\pi_1(X)$ or $\pi_1$ will therefore be used when $X$ is path-connected.

While $H_1$ comports with a linear-algebraic sensibility, $\pi_1$ insists upon the full algebraic regalia of a group: in general $\pi_1$ is *not* abelian. This is a frustrating and wonderful fact. Wonderful, in that $\pi_1$ yields information not captured by homology. Frustrating, in that *any* (finitely presented) group can arise as $\pi_1$ of a space – even such simple spaces as finite 2-dimensional cell complexes or compact smooth 4-manifolds. Determining such facts as whether a loop is contractible, or whether two given loops are homotopic, leads to provably uncomputable problems over finitely presented groups. Any computational homotopy questions must be limited to spaces from a suitably inoffensive class. This does not make the theory inapplicable, but it does dampen one's hopes. On the other hand, $\pi_1$ is made for working with homotopy theory. Compare the following to the task of proving why simplicial homology or Euler characteristic is a homotopy invariant.

**Lemma 8.1.** *Fundamental group $\pi_1$ is a homotopy invariant of spaces.*

**Proof.** Let $f: X \to Y$ be a map. There is an **induced homomorphism**,

$$\pi(f): \pi_1(X, x_0) \to \pi_1(Y, f(x_0)),$$

that sends the homotopy class of a loop $\alpha: [0, 1] \to X$ to the homotopy class of the loop $f \circ \alpha$ in $Y$. One observes that the induced homomorphism is, as in the case of co/homology, **functorial**, respecting identities and composition. For $f_t : X \to Y$ a homotopy of maps, the loops $f_t \circ \alpha$ are all homotopic. Thus, a homotopy-equivalence induces an isomorphism on $\pi_1$. ⊙

**Example 8.2 (Examples of $\pi_1$)** ◎

Certain simple spaces have abelian $\pi_1$: for example, $\pi_1(\mathbb{S}^1) \cong \mathbb{Z}$, since any loop $\alpha$ to $\mathbb{S}^1$ has a well-defined degree that fixes its homotopy and homology class (*cf.* the Hopf Theorem of §4.12). The annulus and Möbius strip are homotopic to a circle and thus have the same $\pi_1$. The 2-sphere $\mathbb{S}^2$ has $\pi_1(\mathbb{S}^2) \cong 1$, since any loop can be homotoped to one which is not onto $\mathbb{S}^2$ and thus factors through a punctured sphere $\mathbb{S}^2 - \star$, which is contractible. In like manner it is shown that spheres $\mathbb{S}^n$ for $n > 1$ are all **simply connected**, meaning that they are connected and have $\pi_1 \cong 1$. One predicts an accord between $\pi_1$ and $H_1$: the torus has $\pi_1(\mathbb{T}^2) \cong \mathbb{Z}^2$, and the real projective plane has $\pi_1(\mathbb{P}^2) \cong \mathbb{Z}_2$. However, nonabelian fundamental groups abound. The plane $\mathbb{R}^2$ with $N$ points removed has $\pi_1$ a free group on $N$ generators. Furthermore, every compact closed surface of genus $g > 1$ has a nonabelian fundamental group. These **surface groups** are beautiful, but hyperbolically complex. For example, a compact oriented genus $g$ surface $S_g$ has fundamental group presented as:

$$\pi_1(S_g) \cong \langle x_1, y_1, x_2, y_2, \ldots, x_g, y_g \;:\; x_1 y_1 x_1^{-1} y_1^{-1} x_2 y_2 x_2^{-1} y_2^{-1} \cdots x_g y_g x_g^{-1} y_g^{-1} = 1 \rangle \quad \odot$$

**Lemma 8.3.** *For $X$ and $Y$ pointed path-connected spaces:*

1. $\pi_1(X \times Y) \cong \pi_1(X) \times \pi_1(Y)$, *the Cartesian product of groups*
2. $\pi_1(X \vee Y) \cong \pi_1(X) * \pi_1(Y)$, *the free product of groups*

Any finite connected graph is homotopic (by collapsing out a spanning tree) to $\bigvee_1^N \mathbb{S}^1$, a wedge of $N$ circles, which has as $\pi_1$ the free product on $N$ elements: the group of all words on $N$ symbols and their inverses, with no relations. Notice how this differs from $H_1$, which is the abelianization to $\mathbb{Z}^N$ that forgets the order in which one traverses loops. In contrast, the $N$-torus $\mathbb{T}^N$ has enough "room" to reverse the order of loops and has $\pi_1(\mathbb{T}^N) \cong \mathbb{Z}^N$. First homology $H_1$ cannot distinguish graphs from tori; $\pi_1$ can.

Computing $\pi_1$ in general is "easy" in that there is an almost-mechanical procedure for assembling $\pi_1$ from pieces, much like the Mayer-Vietoris sequence does for homology. Instead of using exact sequences, a more explicit statement using presentations is preferable.

**Theorem 8.4 (Van Kampen Theorem).** *Let $U \overset{\iota_U}{\hookleftarrow} U \cap V \overset{\iota_V}{\hookrightarrow} V$ be open and path-connected with finitely-presented fundamental groups:*

$$\pi_1(U) \cong \langle u_i \;:\; r_j = 1 \rangle_{i,j}$$
$$\pi_1(V) \cong \langle v_k \;:\; s_\ell = 1 \rangle_{k,\ell}$$
$$\pi_1(U \cap V) \cong \langle w_m \;:\; t_n = 1 \rangle_{m,n}$$

*Then the union $U \cup V$ has fundamental group with presentation:*

$$\pi_1(U \cup V) \cong \langle u_i, v_k \;:\; r_j = 1,\; s_\ell = 1,\; \pi(\iota_U)(w_m) = \pi(\iota_V)(w_m) \rangle_{i,j,k,\ell,m}. \tag{8.1}$$

In other words, one takes the union of the generators and relations of $U$ and $V$ and declares new relations identifying generators $\pi_1(U \cap V)$ as mapped via $\iota_U \colon U \cap V \hookrightarrow U$ with those mapped via $\iota_V \colon U \cap V \hookrightarrow V$. The need for $U$ and $V$ open can be relaxed if, *e.g.*, they are subcomplexes of a cell structure. The construction permits induction, and stronger versions can be stated [221]. The difficulties of comparing presentations should not be underestimated: this theorem, though constructive, is not a panacea.

## 8.2 Covering spaces

Fundamental groups and induced homomorphisms display their power in classification theorems, the best example of which is for covering spaces. For the remainder of this section, all spaces will be assumed path-connected. A **cover** of a space $X$ is a (covering) space $\tilde{X}$ and a map $p\colon \tilde{X} \to X$ which is a **local homeomorphism**. This means that to each $x \in X$, there is a small neighborhood $U \subset X$ of $x$ with the property that $p^{-1}(U)$ is a disjoint union of homeomorphic copies of $U$, projected by $p$. The **fibers** $p^{-1}(x)$ are all discrete and have the same cardinality. Covers of $X$ have the same local topology but (often) different global topology.

**Example 8.5 (Circles)** ⓞ

The canonical example of a covering is the map $\mathbb{R} \to \mathbb{S}^1$ given by $t \mapsto e^{2\pi i t} \in \mathbb{S}^1 \subset \mathbb{C}$. However, there are other covers of $\mathbb{S}^1$ – the maps $e^{2\pi i t} \mapsto e^{2\pi n i t}$ for any $n \neq 0 \in \mathbb{Z}$ give an $|n|$-fold covering $\mathbb{S}^1 \to \mathbb{S}^1$. ⓞ

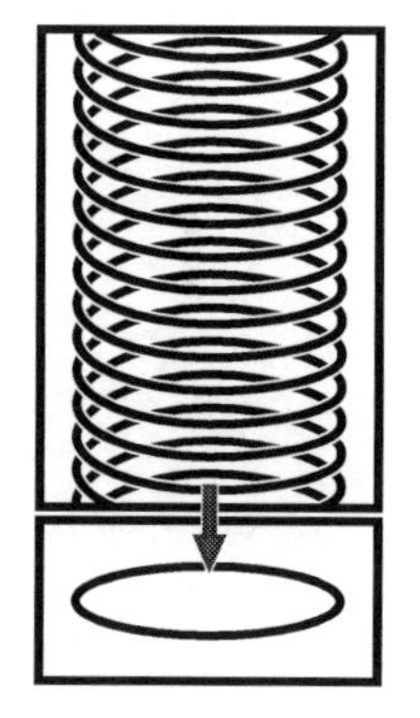

Two covers $p_1\colon \tilde{X}_1 \to X$ and $p_1\colon \tilde{X}_2 \to X$ are said to be *equivalent* if there is a homeomorphism $f\colon \tilde{X}_1 \to \tilde{X}_2$ such that $p_1 = p_2 \circ f$. For example, $e^{2\pi i t} \mapsto e^{\pm 2\pi n i t}$ are equivalent covers from $\mathbb{S}^1$ to $\mathbb{S}^1$ via the antipodal map. A **universal cover** is a cover $p : \tilde{X} \to X$ with $\tilde{X}$ simply connected. For all reasonable (connected and semi-locally simply connected) spaces $X$, a universal cover exists and is unique up to equivalence. Hence, if $X$ is simply connected, it is its own universal cover, as is the sphere $\mathbb{S}^n$ for $n > 1$ – it has no nontrivial covers. On the other hand, $\mathbb{S}^n$ for $n > 1$ *is* the universal cover of lots of interesting quotient spaces, such as 3-dimensional **lens spaces** with finite $\pi_1$. For example, $\mathbb{S}^n \to \mathbb{P}^n$ is a double cover (the fiber has cardinality 2).

The problem of distinguishing between a space and a cover is salient in robot navigation and mapping. Assume a robot that moves about in an unknown environment and can use primitive vision/sensing to patch together explored local neighborhoods into a rough map of the environment based on (random or deterministic) exploration. While local patching is possible, global recurrence is more problematic (*cf.* being lost in the woods – *"Have we been here before?"*). It is a persistent problem to determine whether or not the robot has accurately mapped the region or one of its covering spaces.

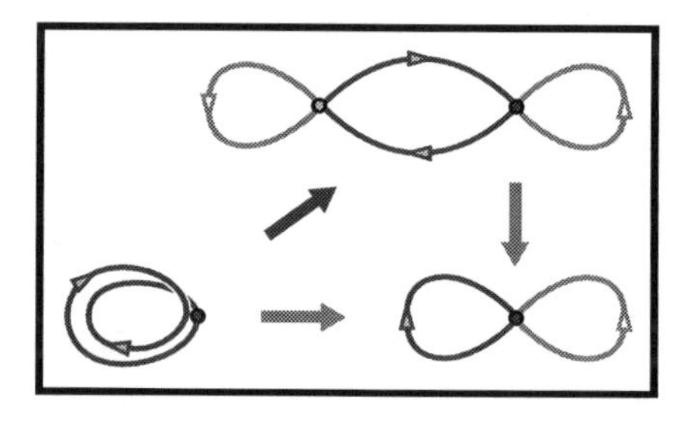

This partially motivates the question of classifying and distinguishing different covers of a fixed space $X$: this is a neat-and-tidy theory that mirrors the fundamental group perfectly. The reader is encouraged to try to classify all the different covers of $\mathbb{S}^1 \vee \mathbb{S}^1$. This is not an easy exercise, either in the setting of finite or infinite covers. However, the general theory *is* elementary (as elementary as is possible within the whirl of homotopy theory). This depends crucially on a single concept: a *lift*. A **lift** of a map $f\colon Y \to X$ to a cover $p\colon \tilde{X} \to X$ is a map $\tilde{f}\colon Y \to \tilde{X}$ such that $f = p \circ \tilde{f}$. The reader should augment this definition with some examples from the simplest types of covers: covers of $\mathbb{S}^1$, $\mathbb{T}^2$, and $\mathbb{S}^1 \vee \mathbb{S}^1$.

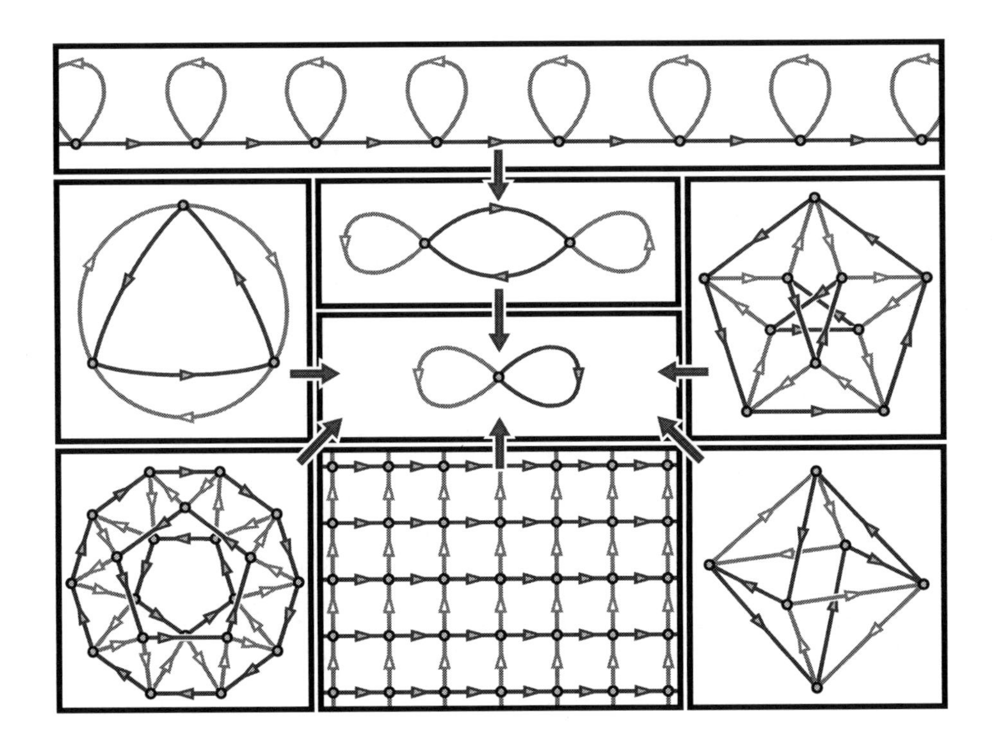

The following encapsulates the main results of covering space theory, with an emphasis on lifts:

**Theorem 8.6 (Covering Space Theory).** *Let $X$ and $Y$ be path-connected, locally path-connected, and locally simply connected spaces*[1], *and let $p\colon (\tilde{X}, \tilde{x}_0) \to (X, x_0)$ be a cover.*

1. **[Lifting criterion]**: *A map $f\colon (Y, y_0) \to (X, x_0)$ lifts if and only if $\pi(f) < \pi_1(p)$: that is, nontrivial loops in the image of $f$ are also nontrivial in the image of $p$.*
2. **[Homotopy lifting]**: *Any homotopy $f_t\colon (Y, y_0) \to (X, x_0)$ with an initial lift $\tilde{f}_0\colon (Y, y_0) \to (\tilde{X}, \tilde{x}_0)$ lifts uniquely to a homotopy $\tilde{f}_t\colon (Y, y_0) \to (\tilde{X}, \tilde{x}_0)$.*
3. **[Classification]**: *Covers of $(X, x_0)$ up to covering space equivalence are in bijective correspondence with subgroups of $\pi_1(X, x_0)$.*

These results are sharp and tightly connected. Together, they provide a complete understanding of covering spaces (topological objects) in terms of subgroups of $\pi_1$ (algebraic objects). Of course, the correspondence is bidirectional, and as often as algebra enlightens topology, topology returns the favor: the best proof that every subgroup of a free group is free is a simple application of covering space theory [176, 218].

**Example 8.7 (Euler angles)** ⊚

[1]These can be relaxed with care [176].

As per §4.11, the orientation of an object (*e.g.*, an airplane or a *wiimote*) in $\mathbb{R}^3$ is as an element of $SO_3$ and can be written as an orthogonal matrix with determinant $+1$ or as a point in $\mathbb{P}^3$. It is more common in applications, however, to use angles to describe the object's orientation – in aviation, *e.g.*, one uses *roll*, *pitch*, and *yaw*. These angles – the *Euler angles* or any other choice of three cyclic variables – implicitly define a map $\mathbb{T}^3 \to \mathbb{P}^3$. Since the covers of $\mathbb{P}^3$ are classified by subgroups of $\pi_1(\mathbb{P}^3) \cong \mathbb{Z}_2$, there is only the trivial ($\mathbb{P}^3$) and universal ($\mathbb{S}^3$) cover. As $\mathbb{T}^3$ cannot be a cover, there is *no* good coordinate system that is everywhere a local homeomorphism to $\mathbb{P}^3$: the coordinates cannot have full rank at all image points. Here, $\pi_1$ acts both as a means of classification and obstruction. Note, however, that $\mathbb{S}^3$ is a perfectly good cover, meaning that one can faithfully use **quaternions** (the group structure on $\mathbb{S}^3$) without experiencing the same degeneracies. ⊙

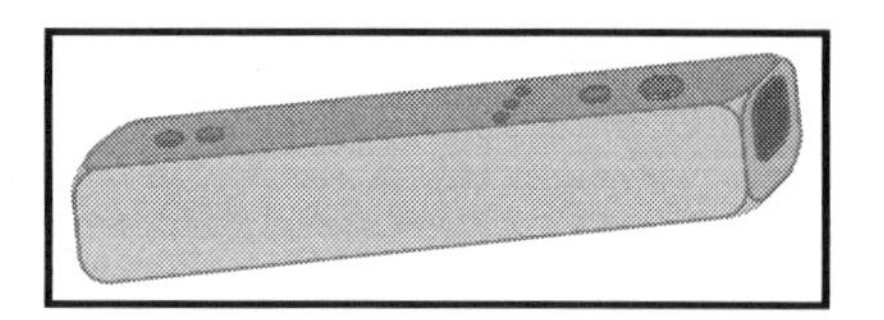

## 8.3 Knot theory

The fundamental group is well-suited to the theory of knots and links, a beautiful subject for visual topology [260]. Recall from Example 4.24 that a **knot** is an embedding of $\mathbb{S}^1$ into $\mathbb{S}^3$. Two knots are said to be equivalent (or of the same *knot type*) if there is an **ambient isotopy** – a homotopy of homeomorphisms – of $\mathbb{S}^3$ carrying one knot to the other. This fits with the intuition of deforming the strands without cutting or pulling a knot so tight as to cause it to vanish. The **unknot** is a knot equivalent to a standard $\mathbb{S}^1 \subset \mathbb{R}^2 \subset \mathbb{R}^3$.

The topological type of the complement $\mathbb{S}^3 - K$ of a knot is, clearly, an invariant, since an ambient isotopy drags the complement of one homeomorphically to that of the other. Thus, any algebraic-topological invariant provides a potential means of discriminating knot types. One is at first tempted to use homology; however, this is insufficient to the task. Every knot complement in $\mathbb{S}^3$ has the homology type of the circle, since, by Alexander duality, $\tilde{H}_k(\mathbb{S}^3 - K) \cong \tilde{H}^{2-k}(\mathbb{S}^1)$. The fundamental group is a much stronger, though not a complete, invariant.

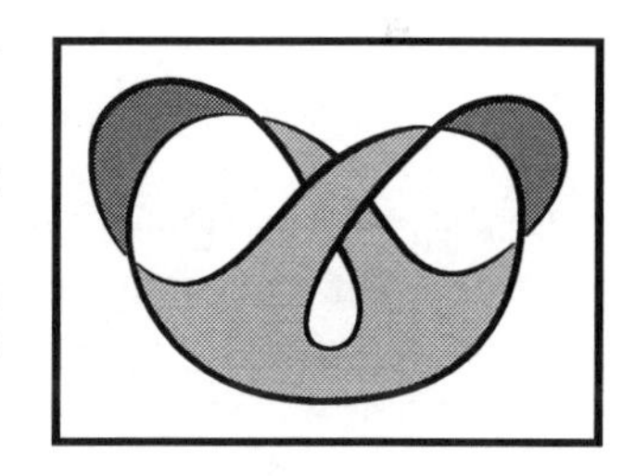

**Example 8.8 (Genus)** ⊙

The question of knot equivalence has led to a dizzying array of invariants, drawn on tools ranging from combinatorial trickery, covering spaces, Euler characteristic, geometry, Morse theory, Floer theory, and more. Among the simplest of invariants is the **genus** of a knot. A **Seifert surface** of a knot $K \subset \mathbb{S}^3$ is a punctured orientable surface $S \subset \mathbb{S}^3 - K$ embedded in the complement that *spans* the knot ($K = \partial S$). Such an $S$ is homeomorphic to a punctured orientable surface of genus $g$. The *minimal* such $g$ is defined to be the genus of the knot. This is by definition an invariant, since an ambient isotopy of $\mathbb{S}^3$ deforms a spanning surface along with the knot. Genus is by no means a complete invariant, since many distinct knot types have equal genus. However, genus is *unknot detecting* in the sense the genus of $K$ is zero if and only if $K$ is the unknot. ⊚

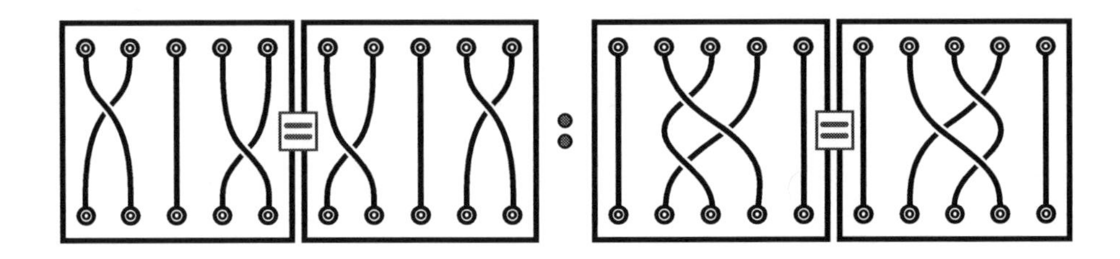

**Example 8.9 (Braids)** ⊚

Recall from Example 1.8 that one can describe periodic motions of robots (labeled or unlabeled) using loops in a configuration space ($\mathcal{C}^n(\mathbb{R}^2)$ or $\mathcal{UC}^n(\mathbb{R}^2)$ respectively). The description given is Chapter 1 was necessarily *ad hoc.* A better language is now available. The **braid group** on $n$ strands is defined to be $B_n = \pi_1(\mathcal{UC}^n(\mathbb{R}^2))$; the **pure braid group** on $n$ strands is $P_n = \pi_1(\mathcal{C}^n(\mathbb{R}^2))$. These are both, naturally, groups. The identity element is the constant loop; that is, *nobody moves.* Composition in the braid group is concatenation of braids: *first this, then that.* The inverse of a braid reverses motion. The braid group $B_n$ has a clean presentation whose generators $\sigma_i$ consist of crossing the $i^{\text{th}}$ strand over the $(i+1)^{\text{st}}$:

$$B_n \cong \left\langle \sigma_1, \ldots, \sigma_{n-1} \ : \ \begin{array}{c} \sigma_i\sigma_j = \sigma_j\sigma_i \ : \ |i-j|>1 \\ \sigma_i\sigma_{i+1}\sigma_i = \sigma_{i+1}\sigma_i\sigma_{i+1} \end{array} \right\rangle$$

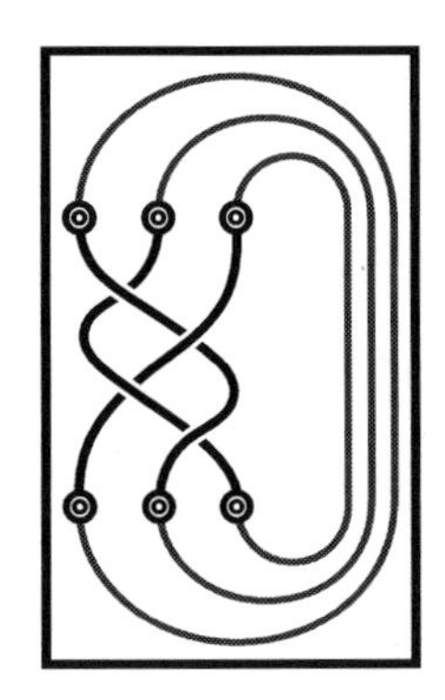

Braids provide not only an efficient language to describe robot motions but also algebraic descriptions of knots and links. A **closed** $n$**-braid** is the link obtained from a braid in $B_n$ by connecting the points on the *bottom* of the braid to those on the *top* via $n$ strands in the simplest possible manner. Equivalent braids give rise to isotopic closures. A theorem of Alexander [7] confirms the suspicion that every link can be represented as the closure of some braid. The smallest $n$ for which a braid in $B_n$ can be closed to form a given knot is a topological invariant called the **braid index** of the knot.

The $\pi_1$-based definition of braid groups extends to braids on any domain. One can consider, *e.g.*, braids on surfaces other than $\mathbb{R}^2$. A great deal of interesting structure is to be found in braid groups of graphs [118, 145]. ⊚

**Example 8.10 (DNA and enzyme actions)** ⊚

Protein chains, power cords, and DNA strands can coil into conformations one is tempted to call *Gordian*; however, a typical such chain is not a loop, and thus cannot be knotted. That has not prevented knot theorists from investigating knotting and linking phenomena in DNA, which *sometimes* comes in circular substrate molecules – loops. With a combination of electron micrography and gel electrophoresis, it is possible to sort out collections of cyclic DNA strands by knot type. This makes it possible to use knotted DNA as a test bed for determining the action of certain enzymes that aid in recombination. Since knotting, linking, and writhing of the chain prevents a simple *parallel* replication from separating from the parent chain, there must be some agents that aid in disassembly and reassembly of the chain. These are enzymes (*recombinase, topoisomerase, etc.*) whose actions are localized, vital, and largely hidden. By applying selective enzymes and analyzing changes to global knot type, the functionality of these enzymes can be inferred, quantified, and characterized rigorously [285]. ⊚

**Example 8.11 (Flowlines)** ⊚

Three-dimensional flows exhibit all kinds of knotting. Some of these flows are physical: smoke-rings and other types of vortices (as in, *e.g.*, superconducting fluids) provide beautiful dynamic examples of embedded loops in fluids, some of which are knotted/linked. Numerous authors [14, 16, 131, 232] have contributed to understanding lower bounds on the energy of a perfect fluid flow by means of knotting and linking of the flowlines (*cf.* helicity in Example 6.25), with parallel investigations in magentohydrodynamics: the text [16] is a good resource for this body of ideas.

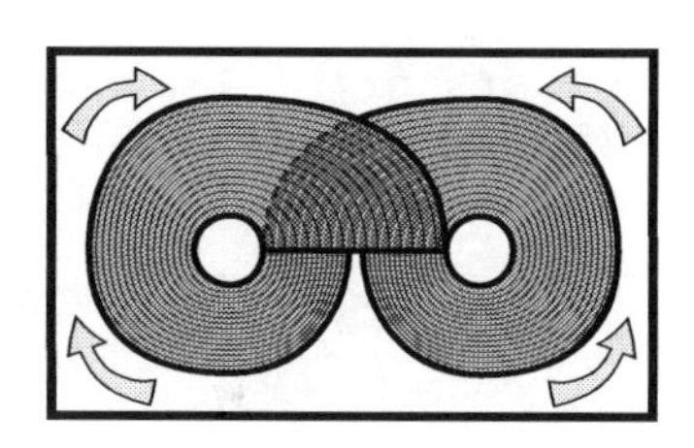

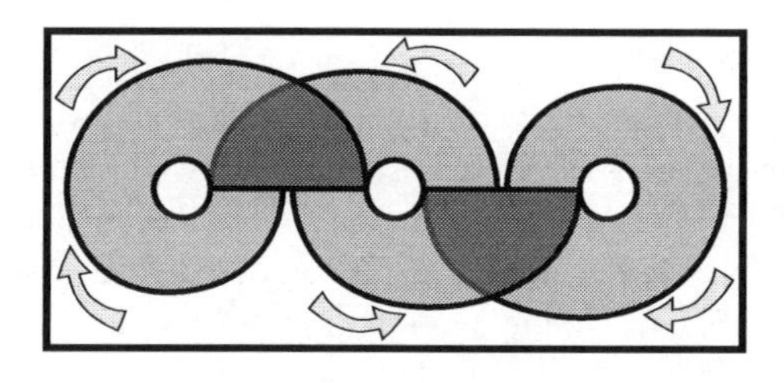

In general, any vector field on $\mathbb{S}^3$ or $\mathbb{R}^3$ may have periodic orbits, each of which is (by uniqueness of solutions to ODEs) an embedded loop – a knot. Together, these form a link of periodic orbits, which may or may not be a finite link. The basic question *"Which link types are possible?"*, even in the context of a sufficiently smooth or tame nonsingular vector field, is delicate. The empty link is possible, thanks to the solution to the **Seifert conjecture** [202]. Any finite link is possible (a simple exercise). For flows exhibiting chaotic dynamics (and thus infinitely many periodic orbits), Birman and Williams [39, 40] showed how to collapse sufficiently hyperbolic invariant sets onto a **template** – an embedded branched surface $\mathcal{T} \subset \mathbb{S}^3$ with a semiflow[2] – in a manner that preserves all knot and link data of periodic orbits. Their seminal work on the geometric **Lorenz attractor**

[2] A semiflow is an action of $\mathbb{R}^+$ instead of $\mathbb{R}$. One can flow forward in time uniquely, but not backward.

showed that although infinitely many knots types exist as periodic orbits in this flow, only certain types of knots and links can arise [39]. In contrast, there exist **universal templates** which contain *all* knots and (finite) links as periodic orbits of the semiflow [143]. These can arise in a number of interesting physical settings – explicit ODEs on $\mathbb{R}^3$ possessing *all* knots and links as solutions [147].

⊚

## 8.4 Higher homotopy groups

The notation $\pi_1(X)$ for fundamental group foreshadows the higher homotopy groups. The notation does not predict the resulting surprises.

Fix a space $X$ and a basepoint $x_0 \in X$. The **homotopy group** $\pi_n$ of $X$ at $x_0$ measures the number of ways to map a sphere $\mathbb{S}^n$ with a fixed basepoint $s_0$ into $X$ up to (basepoint-preserving) homotopy. That is, $\pi_n$ consists of homotopy classes of maps $f\colon (\mathbb{S}^n, s_0) \to (X, x_0)$. Note that this reduces to the loop-based definition in the case $n = 1$. For $n = 0$, $\pi_0(X, x_0)$ is a set whose cardinality measures the number of path-connected components of $X$. For all other $n > 0$, $\pi_n(X, x_0)$ has the structure of a group under the following multiplication operation. Given $f, g\colon (\mathbb{S}^n, s_0) \to (X, x_0)$, define the product $f{\bullet}g\colon (\mathbb{S}^n, s_0) \to (X, x_0)$ by sending an equator of $\mathbb{S}^n$ (containing $s_0$) to $x_0$, and mapping via $f$ on the *upper* hemisphere and $g$ on the *lower*. It is clear that the identity element is represented by the map $\mathbb{S}^n \mapsto x_0$. It is an easy exercise to show that, for connected spaces $X$, the homotopy groups are isomorphic for different choices of basepoint $x_0$: as in the case of $\pi_1$, basepoints are often suppressed unless explicitly needed.

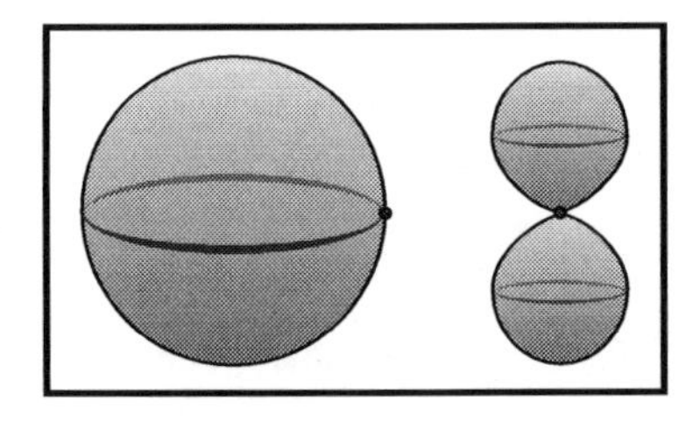

The definition of homotopy groups is more elementary than that of co/homology. Indeed, as many algebraic topology courses begin with homotopy groups and only later turn to co/homology, the experienced reader may be frustrated at this late-in-time treatment of so fundamental a species. There is good reason to beware homotopy groups as an admixture of the divine and the devilish. A good example of a homotopy group computation is that of a sphere. One begins simply enough: $\pi_k(\mathbb{S}^n)$ is trivial for $0 \leq k < n$, and $\pi_n(\mathbb{S}^n) \cong \mathbb{Z}$. This compares favorably with the homology of spheres: $\pi_k(\mathbb{S}_n) \cong H_k(\mathbb{S}^n; \mathbb{Z})$ for all $1 \leq k \leq n$ (see Theorem 8.14 to come). But what of the higher homotopy groups? These *seem* even simpler than the sometimes problematic $\pi_1$. Any topologist who can't prove the following with a picture, isn't:

**Proposition 8.12.** *For $k > 1$, $\pi_k$ is abelian.*

The computation of $\pi_3(\mathbb{S}^2)$ is the first hint at the mysterious nature of higher homotopy groups. Surprisingly unlike homology, $\pi_k(X)$ does not necessarily vanish for $k > \dim X$. One is misled by the case of $\mathbb{S}^1$, whose universal cover is contractible.

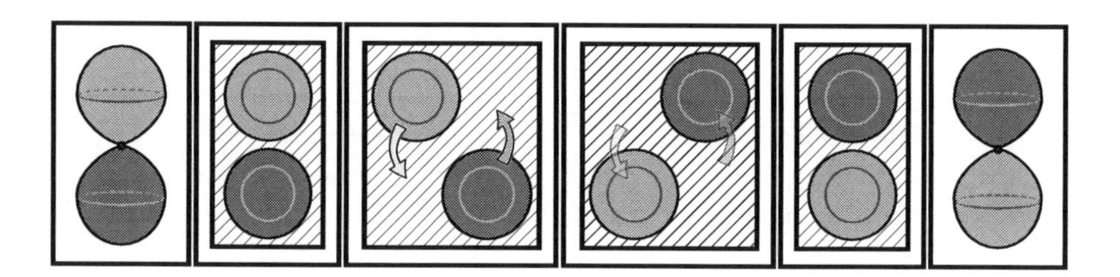

In §8.10, it will be shown that $\pi_3(\mathbb{S}^2) \cong \mathbb{Z}$. Higher homotopy groups do not offer much in the way of bubbly optimism for computational topology: perhaps the largest unsolved problem in algebraic topology is the computation of $\pi_k(\mathbb{S}^n)$ for large values of $k > n$. For example, $\pi_k(\mathbb{S}^2)$ as a function of $k > 0$ is:

$$0\,,\ \mathbb{Z}\,,\ \mathbb{Z}\,,\ \mathbb{Z}_2\,,\ \mathbb{Z}_2\,,\ \mathbb{Z}_{12}\,,\ \mathbb{Z}_2\,,\ \mathbb{Z}_2\,,\ \mathbb{Z}_3\,,\ \mathbb{Z}_{15}\,,\ \mathbb{Z}_2\,,\ \mathbb{Z}_2^2\,,\ \mathbb{Z}_2 \times \mathbb{Z}_{12}\,,\ \mathbb{Z}_{84} \times \mathbb{Z}_2^2\,,\ \mathbb{Z}_2^2\,,\ \ldots$$

There is good reason to believe this problem will not be readily solved, not because of a lack of pattern in existing data on $\pi_\bullet(\mathbb{S}^n)$, but, as with much of Mathematics, because of a wild abundance of pattern.

Some computations *are* possible. Like $\pi_1$, higher homotopy groups are functorial: a map $f\colon X \to Y$ induces homomorphisms on $\pi_n$ for all $n > 0$, and homotopic maps yield the same homomorphism (modulo basepoint considerations). Some maps preserve (higher) homotopy groups without necessarily being a homotopy equivalence of spaces. The following is a direct consequence of the lifting criterion of Theorem 8.6:

**Corollary 8.13.** *Covers induce isomorphisms on $\pi_n$ for all $n > 1$.*

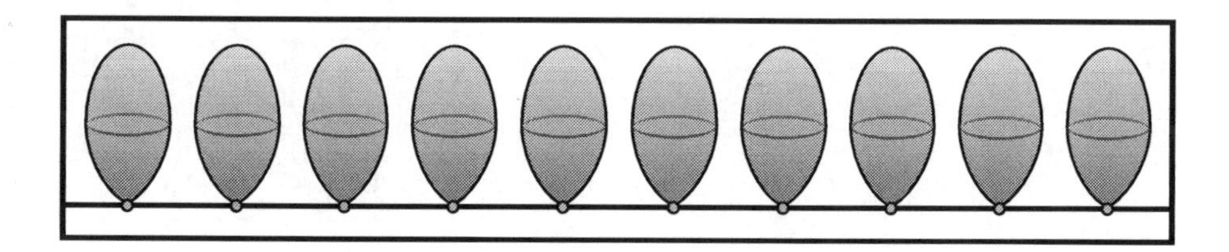

This leads to some interesting computations.

1. All graphs have $\pi_n \cong 0$ for $n > 1$.
2. All surfaces except $\mathbb{S}^2$ and $\mathbb{P}^2$ have $\pi_n \cong 0$ for $n > 1$.
3. $\mathbb{S}^2$ and $\mathbb{P}^2$ have all $\pi_n$ isomorphic except for $n = 1$.
4. $\pi_2(\mathbb{S}^2 \vee \mathbb{S}^1) \cong \mathbb{Z}^\infty$, since the universal cover has an infinite number of nonhomotopic copies of $\mathbb{S}^2$.

## 8.5 Biaxial nematic liquid crystals

Recall from Example 4.25, nematic liquid crystals in $\mathbb{R}^2$ and $\mathbb{R}^3$ are composed of molecules whose idealized form is that of an axisymmetric rod. The corresponding singularities in the crystals are completely described by degree theory using homology (in $\mathbb{Z}$ coefficients for the 2-d case and $\mathbb{F}_2$ for 3-d).

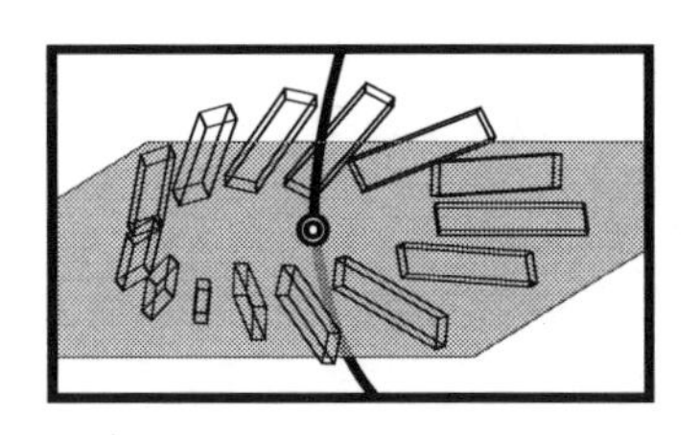

Let the reader note that in cases where the degree is computed from the director field $\xi$ by means of a loop, then the homological degree of $\xi\colon \mathbb{S}^1 \to \mathbb{P}^1$ or $\xi\colon \mathbb{S}^1 \to \mathbb{P}^2$ classifies the singularity type: switching to the fundamental group $\pi_1$ returns no new information. However, in the case of a point-defect in $\mathbb{R}^3$, the relevant map of a surrounding sphere gives $\xi\colon \mathbb{S}^2 \to \mathbb{P}^2$. In homology, this degree is $\mathbb{F}_2$-valued, but in homotopy, one has $\pi(\xi)\colon \pi_2(\mathbb{S}^2) \to \pi_2(\mathbb{P}^2)$ which, being a map from $\mathbb{Z} \to \mathbb{Z}$, reveals a finer invariant.

The advantages of homotopy groups become more pronounced in more general liquid crystal structures. One important (though only more recently investigated) class of liquid crystals are the 3-d **biaxial nematics**, whose molecules are not axisymmetric, but rather have the form of a rectangular prism [6]. Recall that for the axisymmetric (nematic) case in $\mathbb{R}^3$, the director field takes values in the quotient of the rotation group $SO_3$ by the group of symmetries of an axisymmetric rod: $\mathbb{S}^1$. This quotient is clearly $\mathbb{P}^2$. However, in the biaxial setting, the director field takes values in the quotient of $SO_3$ by $D_2$, the symmetry group of a rectangle in the plane.

With a bit of work, this space can be shown to be homeomorphic to the quotient of $\mathbb{S}^3$ by $Q_8$, the **unit quaternions**. This is the (unique) non-abelian group $Q_8 = \{\pm 1, \pm i, \pm j, \pm k\}$ of order eight. This group, familiar from physics and 3-d vector calculus, expresses noncommutativity in the relations $ij = k = -ji$ (and permutations thereof). Because the action of $Q_8$ on $\mathbb{S}^3$ is regular, the quotient $\mathbb{S}^3/Q_8$ is a cover; since $\mathbb{S}^3$ is simply connected, the quotient has fundamental group $\pi_1 \cong Q_8$. It is intriguing that this director field has noncommutative fundamental group: it provides a much richer dictionary for curves of defect singularities than homology can [6]. Note also that because $\pi_2(\mathbb{S}^3/Q_8) \cong \pi_2(\mathbb{S}^3) = 0$, there are no point-like singularities up to homotopy.

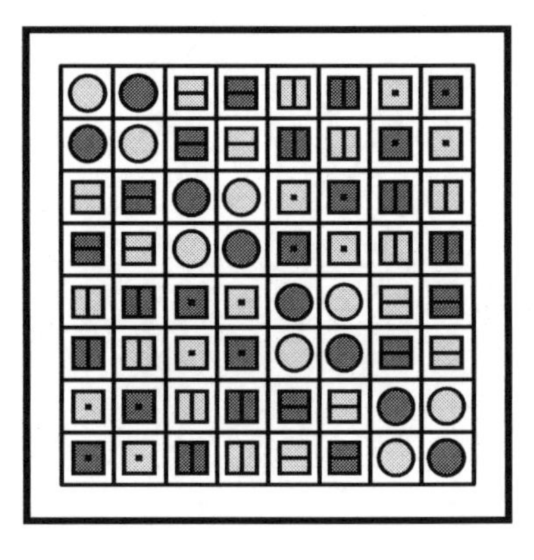

There are many other interesting materials whose internal structure reveals a director field expressible as a quotient of the group $SO_3$ or $SO_2$ by some subgroup. These include not only various types of liquid crystals, but also metallic glasses, ferromagnets, and superfluid helium [29], all of which can exhibit disclinations and defects of various types. The best way to categorize these is via induced maps on homotopy groups.

## 8.6 Homology and homotopy

The relationships between $\pi_\bullet$, $H_\bullet$, and $H^\bullet$, are too many and too deep to encapsulate. One begins with the elementary observation that, while $\pi_1(\mathbb{S}^1 \vee \mathbb{S}^1)$ is a free group $\mathbb{Z} * \mathbb{Z}$, $H_1(\mathbb{S}^1 \vee \mathbb{S}^1)$ is the free-abelian group $\mathbb{Z} \oplus \mathbb{Z}$. The following is one of the few simple results that binds homotopy and homology groups together. It is the natural extension of the pattern seen with spheres.

**Theorem 8.14 (Hurewicz Theorem).** *There are homomorphisms*

$$\mathrm{Hur}_n : \pi_n(X) \to H_n(X; \mathbb{Z}),$$

*which, for $n = 1$, is abelianization. If $n > 1$ and $\pi_k(X)$ is trivial for $0 \le k \le n-1$, then* $\mathrm{Hur}_n$ *is an isomorphism and* $\mathrm{Hur}_{n+1}$ *is surjective.*

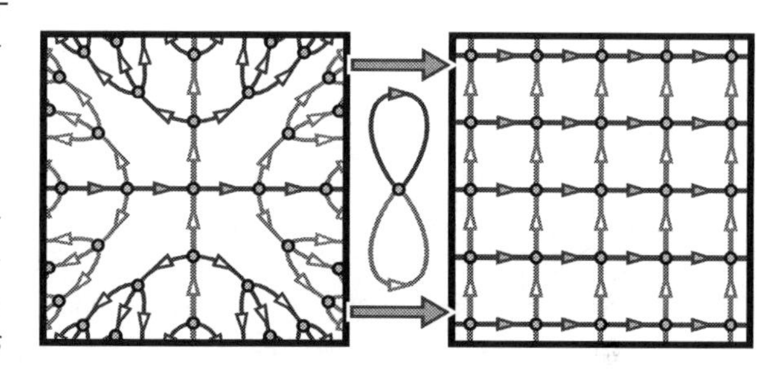

When a mapping between spaces is involved, the Hurewicz theorem pairs well with an extremely powerful result for proving homotopy equivalence of spaces.

**Theorem 8.15 (Whitehead Theorem).** *If $f : X \to Y$ is a map of cell complexes with $\pi(f) : \pi_n(X) \to \pi_n(Y)$ an isomorphism for all $n$, then $f$ is a homotopy equivalence.*

Since homotopy equivalences are difficult to construct by hand, it is helpful to have implicit tools. One must be careful not to misread the result as saying that spaces with isomorphic homotopy groups are homotopic: it is the mapping and the induced homomorphisms that carry the theorem.

**Example 8.16 (Eilenberg-MacLane spaces)** ⊚

In homotopy theory, there are numerous approaches for decomposing spaces. One type of building block is an **Eilenberg-MacLane space**. Denoted, $K(\mathbf{G}, n)$, this is a (connected) space, unique up to homotopy type, whose homotopy groups are trivial, with the lone exception that $\pi_n(K(\mathbf{G}, n)) \cong \mathbf{G}$. For example, $\mathbb{S}^1$ is a $K(\mathbb{Z}, 1)$ since the circle has contractible universal cover and all higher homotopy groups vanish. It is not so easy to find Eilenberg-MacLane spaces within the class of finite cell complexes: existence results, though constructive, yield infinite-dimensional spaces as examples. The easiest-to-find finite-dimensional Eilenberg-MacLane spaces are of type $K(\mathbf{G}, 1)$. These include:

1. All knot complements in $\mathbb{S}^3$;
2. Configuration spaces $\mathcal{C}^n(\mathbb{R}^2)$ of points in the plane;
3. All state complexes (§2.11); hence, configuration spaces of graphs.

These spaces serve as the bridge to a surprising relationship between cohomology and homotopy. It is a theorem that the cohomology of a cellular space $X$ is expressible in terms of homotopy classes of maps of $X$ into Eilenberg-MacLane spaces; specifically, $H^n(X;\mathbf{G}) \cong [X, K(\mathbf{G}, n)]$, where $[X, Y]$ denotes a group of basepoint-preserving homotopy classes of maps $X \to Y$ (where the group structure is not entirely obvious: see [176, §4.3]). This is the first hint of the depth of the relationship between co/homology groups and homotopy groups. One simple example of this is that $\pi_1(X) \cong [\mathbb{S}^1, X]$ while $H^1(X;\mathbb{Z}) \cong [X, \mathbb{S}^1]$, thus revealing a type of duality linking $\pi_1$ and $H^1$; *cf.* §6.13. ⊚

## 8.7 Topological social choice

All of the applications to Economics in this text have thus far relied upon homological tools. Homotopy theory has something to contribute. Economists have long considered the problems associated with social choice and preferences. The following is a topological version of a classical social choice problem. Consider a set of preferences that is topologized as a space, $X$; examples include preferred prices, budget allocation ratios, or relative rankings of politicians. Given a population of $n$ agents, each with a fixed preference, the state of that population's preferences is an $n$-tuple of points $\xi \in X^n$. The conversion of individual (local) preferences into a single (global) choice is via a **social choice** map $\Xi\colon X^n \to X$. To reflect reasonable conditions, such a map is required to satisfy the following properties:

1. **Continuity:** $\Xi$ is continuous, so that small shifts in local preferences have small impact on the aggregate preference;
2. **Unanimity:** $\Xi$ is the identity on the grand diagonal in $X^n$, so that a unanimous vote is accepted; and
3. **Anonymity:** $\Xi$ is invariant under the action of a permutation on the factors of $X^n$.

The question of existence of a choice map sounds suspiciously like that of existence of an equilibrium in price- or game-theory. Here, instead of universal existence, there is a near-universal non-existence. The following theorem provides the basis for a nonexistence result.

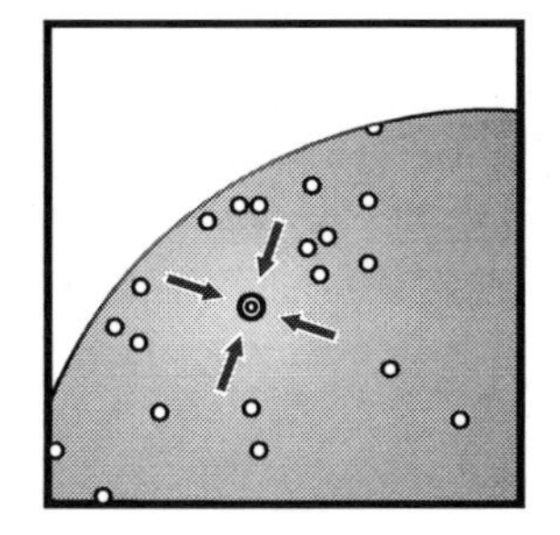

**Theorem 8.17 ([102]).** *If $X$ admits a social choice map for some $n > 1$, then for each $k > 0$, $\pi_k(X)$ is abelian and uniquely divisible by $n$.*

**Corollary 8.18.** *If $X$ is homotopic to a cell complex with finitely-generated $H_\bullet(X;\mathbb{Z})$ and has a social choice map for some $n > 1$, then $X$ must be contractible.*

**Proof.** *(assuming Theorem 8.17)* As $\pi_1$ is abelian, it is isomorphic to $H_1$ via the Hurewicz theorem. Finiteness and divisibility imply both are zero. By Hurewicz again, $\pi_2 = H_2 = 0$. Induct to show that $\pi_k = 0$ for all higher $k$. Since $X$ is (homotopic to) a cell complex and the trivial map $X \to \star$ induces isomorphisms on homotopy groups, the Whitehead theorem implies that $X$ is contractible. $\odot$

The reader familiar with Arrow's Impossibility Theorem for voting will note the similarities: the Arrow theorem is in the case where $X$ is a finite set of rankings and the anonymity is not enforced (allowing for dictatorial outcomes) [22]. It is interesting to note that there are non-contractible spaces which do admit social choice maps. They are necessarily infinite-dimensional and algebraically subtle: Weinberger [301] shows that the infinite-dimensional real projective space $X = \mathbb{P}^\infty$ admits a social choice map for any $n$ odd, but never for $n > 0$ even. It would also be interesting to connect this work with certain difficulties associated with managing swarms of mobile robots by generating a **consensus** in subspaces of the individual robot configuration spaces (*e.g.*, bearing or pose) [287].

## 8.8 Bundles

In linear algebra, a surjective linear transformation of vector spaces is characterized by the kernel and the image. Surjective maps $f\colon X \to Y$ between spaces are potentially wilder. The nicest type of nonlinear surjection has a homogeneity not unlike the linear case: the fiber [kernel] and the base [image] tell all, locally.

A **(fiber) bundle** is a space $E$ together with a projection map $p\colon E \to B$ to a **base space** having **fibers** $p^{-1}(b)$ all homeomorphic to some fixed $F$, so that, on sufficiently small open sets $U \subset B$, $p^{-1}(U) \cong U \times F$. One thinks of the **total space** $E$ of the fiber bundle as a family of $F$ parameterized by $B$. In the case where $E = F \times B$, one says the bundle is **trivial**. Trivial bundles are all alike: every nontrivial bundle is nontrivial in its own way, bound up in the topology of the base and fiber. Simple examples include:

1. Covering spaces $p\colon \tilde{X} \to X$ are fiber bundles in which $E = \tilde{X}$ is the cover, $B = X$ the base, and the fiber $F$ is discrete.
2. A Klein bottle is a nontrivial bundle over $\mathbb{S}^1$ with fiber $\mathbb{S}^1$. The difference between this and the trivial bundle $\mathbb{T}^2$ lies in a flip of the fiber.
3. The configuration space $\mathcal{C}^n(M)$ of $n$ points on a connected manifold $M$ is a fiber bundle over base $M$ with fiber $\mathcal{C}^{n-1}(M-\star)$ for $\star$ a point.
4. The 3-sphere $\mathbb{S}^3$ is a nontrivial bundle over $\mathbb{S}^2$ with fiber $\mathbb{S}^1$. This elegant structure is called the **Hopf fibration** and is important in integrable Hamiltonian dynamics [45] and more. Each pair of fibers in $\mathbb{S}^3$ has linking number 1.
5. The **unit tangent bundle**, $UT_*M$, of a manifold $M$ is the collection of unit tangent spheres in $T_*M$, expressed as a bundle over $M$ with fiber $F \cong \mathbb{S}^{\dim M-1}$. This bundle yields the Hopf fibration for $M = \mathbb{S}^2$.

A **vector bundle** is a bundle $p\colon E \to B$ whose fiber $F$ is a vector space such that addition and scalar multiplication on the fibers extend to continuous maps on all of

$E$. Examples of vector bundles include tangent and cotangent bundles of a manifold, but others also exist.

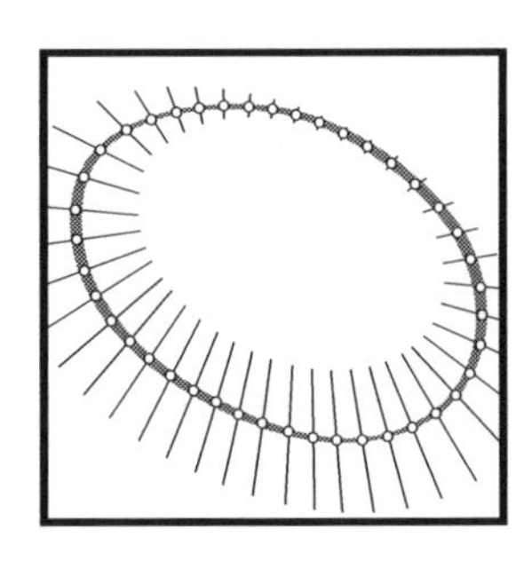

For example, there are, up to a natural equivalence, exactly two vector bundles over $\mathbb{S}^1$ with 1-dimensional fibers: one (the trivial $\mathbb{R}$-bundle) is equivalent to $T_*\mathbb{S}^1 \cong \mathbb{S}^1 \times \mathbb{R}^1$; the other (non-trivial) one is the **twisted bundle** with $E$ homeomorphic to a Möbius band without boundary. Note that both these examples are homotopic to the base $\mathbb{S}^1$. Many of the constructs of this text take on richer meanings in the context of the cohomology of vector bundles [226]. For example, the Euler characteristic of a oriented connected compact $n$-manifold $M$ lifts to an **Euler class** – a particular cohomology class $e \in H^n(M;\mathbb{Z}) \cong \mathbb{Z}$. This is a generalization of Euler characteristic in that the pairing of $e$ with the fundamental class, the generator of $[M] \in H_n(M;\mathbb{Z})$, yields $e([M]) = \chi(M)$.

**Example 8.19 (Fibered knots and magnetic fields)** ⊚

A knot $K \subset \mathbb{S}^3$ is said to be **fibered** if the complement is a bundle over $\mathbb{S}^1$ with fibers of $p\colon \mathbb{S}^3 - K \to \mathbb{S}^1$ the soap-film-like Seifert surfaces of Example 8.8. The unknot is fibered (with fibers homeomorphic to a disc). Trefoil knots, as well as the classic figure-8 knot, are also fibered, though in general fibered knots are not common.

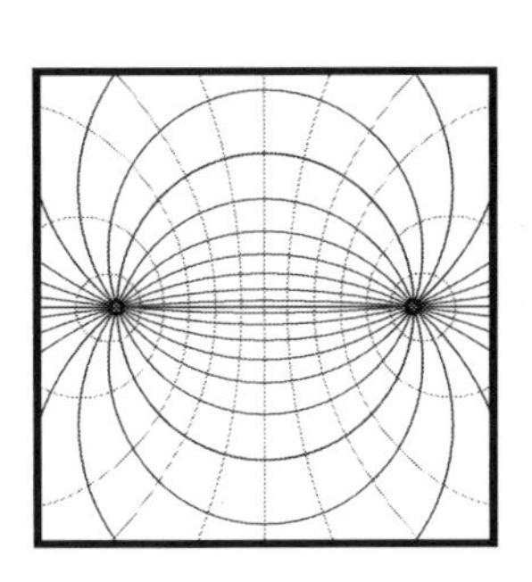

Fibered knots may be related to electromagnetics. A wire bent into a knot $K$ in $\mathbb{S}^3$ emits, upon passing a current through it, an induced magnetic field on the complement. The magnetic vector field *coils* about the wire. For a fibered knot $K$, the magnetic field is transverse to the fiber bundle near $K$ – the magnetic flowlines cross the fibers. One guesses [40] that the magnetic field is everywhere transverse to the fibers; this seems a reasonable conjecture for *relaxed* embeddings of $K$. *If* this is true, then any (fiber-transverse) magnetic field induced by a current through a figure-8 knot [143] (and many other fibered knots [148]) possesses closed field lines spanning all possible knot and link types. ⊚

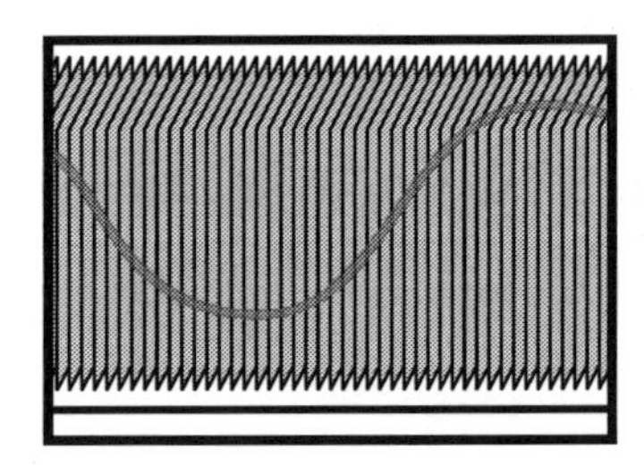

Nontrivial bundles are globally so: by definition, every bundle is locally trivial. What is the measure of nontriviality of a bundle? Euler characteristic is not a good invariant of the bundle structure, since $\chi(E) = \chi(B)\chi(F)$ for *all* bundles $E$ over $B$ with fiber $F$, whether trivial or not. Homology and cohomology of $E$ can sometimes distinguish between bundles (*cf.* $\mathbb{T}^2$ and $K^2$), but not always; *e.g.*, a vector bundle is homotopic to its base. The question of triviality is related to that of the existence of sections. Recall from §1.4 that a vector field on a manifold is a section of the tangent bundle. For a general bundle $p\colon E \to B$, a **section** is a map $s\colon B \to E$ satisfying $p \circ s = \mathrm{Id}_B$. Not all bundles have sections: for example, the existence of a non-vanishing vector field on a manifold $M$ is equivalent to the existence of a section of the unit tangent bundle $UT_*M$, and this does not exist if $\chi(M) \neq 0$. Trivial bundles always have sections. The degree to which sections fail to exist provides a measure of complexity relevant to applications.

## 8.9 Topological complexity of path planning

Recall from §1.5 the importance of configuration spaces in motion-planning problems for robotics: a configuration space converts the problem of physical motion planning to topological path planning. The insight of Farber [112] is to consider motion planning as a problem parameterized by the start and end configurations, and to crystalize this parametrization in terms of fiber bundles and invariants thereof. Given a path-connected configuration space $X$, consider the **path space** $\mathcal{P}(X)$, the space of all maps $\gamma\colon [0,1] \to X$ with the usual (compact-open) topology. There is a projection mapping $p\colon \mathcal{P}(X) \to X^2$ taking a path $\gamma$ to its ordered endpoints $(\gamma(0), \gamma(1))$. With respect to this projection, the path space $\mathcal{P}(X)$ is a fiber bundle (assuming $X$ is locally homogeneous, as in the case of a manifold – else it is a *fibration*: see §8.10). A **path planner** is a section of the path bundle: a continuous map $s\colon X^2 \to \mathcal{P}(X)$ satisfying $p \circ s = \mathrm{Id}$. The following result echoes the obstruction to the inverse kinematic map in §4.11.

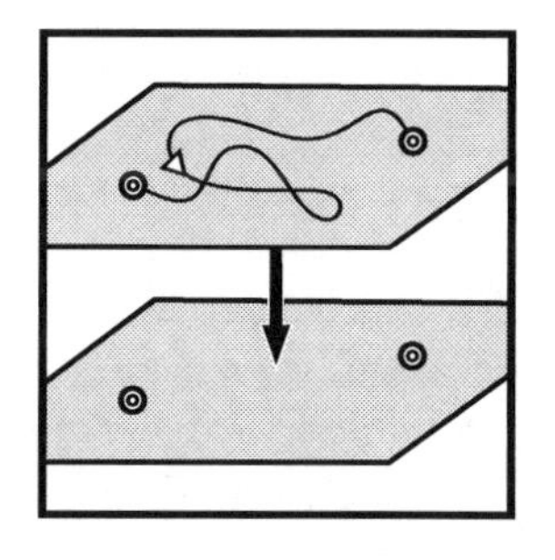

**Lemma 8.20 (Farber [112]).** *The bundle $p\colon \mathcal{P}(X) \to X^2$ possesses a section if and only if $X$ is contractible.*

**Proof.** If $s\colon X^2 \to \mathcal{P}(X)$ is a section, then to each pair $(x_0, x_1) \in X^2$ is associated a path $s(x_0, x_1)\colon [0,1] \to X$ connecting $x_0$ to $x_1$, continuous in these parameters. Fix $x_1 \in X$ a basepoint, and consider $s(\cdot, x_1)\colon X \times [0,1] \to X$, which is a homotopy from $\mathrm{Id}\colon X \to X$ to the constant map $X \to \{x_1\}$. The argument is reversible: a deformation to a fixed basepoint can be unwound to yield a path planner (with all paths passing through the basepoint). ⊙

This initially discouraging result says that there are, in general, no stable motion-planning algorithms; fixing a motion-plan and varying the endpoints can lead to a discontinuous change in the plan, much in the same way that a portable GPS trip planner exhibits instabilities with respect to small changes in point-of-origin. However, similar negative results for the kinematics of a robot arm (§1.5) have not prevented the ubiquitous use of coupled rotation joints: lack of a continuous section simply means that the problem is more complicated and may exhibit instabilities. To what degree?

Bundles prompt a parametric version of the LS category of §7.9, yielding a notion of complexity relevant to the path-planning problem. The **sectional category**, secat, of a fiber bundle is the minimal number of open sets $U_\alpha$ covering $B$ such that $p^{-1}(U_\alpha) \cong U_\alpha \times F$. This records the minimal number of trivial bundles needed to cover $E$. As with LScat, secat is difficult to compute, but can be bound by algebraic-topological invariants in the manner of Theorem 7.27. In the context of path-planning, sectional category is the degree of instability of the problem. After the initial question of obstruction – *can* or *cannot do* – there remains the issue of complexity. If one were to solve the stable path-planning problem piece-wise, how many pieces would be required? Farber defines the (reduced) **topological complexity** of the motion planning problem on $X$ as:

$$\mathrm{TC}(X) := \mathrm{secat}\,(p\colon \mathcal{P}(X) \to X \times X) - 1. \tag{8.2}$$

**Example 8.21 (Topological complexity)** ⊚

The following examples of TC computations are surveyed in [113]:

1. **Spheres:** $\mathrm{TC}(\mathbb{S}^n)$ equals 1 for $n$ odd; 2 for $n$ even.
2. **Surfaces:** For a closed orientable surface $S_g$ of genus $g$, $\mathrm{TC}(S_g) = 2$ for $g \leq 1$; $\mathrm{TC}(S_g) = 4$ for $g > 1$.
3. **Rotations:** $\mathrm{TC}(\mathrm{SO}_3) = 3$.
4. **Graphs:** For $X$ a graph, $\mathrm{TC}(X)$ equals 0 (if $X$ is a tree), 1 (if $X \simeq \mathbb{S}^1$), or 2 (else).
5. **Projective space:** $\mathrm{TC}(\mathbb{P}^n)$ is known only for $n < 24$. In general, it is *very* hard to compute [116]. ⊚

Theorem 7.27 can be used to give bounds on TC: upper bounds are regulated by dimension, and lower bounds are regulated by cohomology: for $X$ a reasonably tame space, the topological complexity satisfies $\mathrm{cup}(X \times X) - 1 \leq \mathrm{TC}(X) \leq 2\dim X$.

**Example 8.22 (Configuration spaces)** ⊚

Those examples of TC that come closest to being relevant to robotics are for configuration spaces of points $\mathcal{C}^n(Y)$ or $\mathcal{UC}^n(Y)$. In particular, for a fixed space $Y$ the computation of $\lim_{n\to\infty} \mathrm{TC}(\mathcal{C}^n(Y))$ gives a measure of asymptotic difficulty of collision avoidance. This is in general difficult to compute, as can be guessed from previous examples. It has been shown [115] that for labeled configuration spaces,

$$\mathrm{TC}(\mathcal{C}^n(Y)) = \begin{cases} 2n-2 & : \quad Y = \mathbb{R}^{2m+1} \\ 2n-3 & : \quad Y = \mathbb{R}^{2m} \\ 2\#V^{\mathrm{ess}}(Y) & : \quad Y = \text{ a tree} \end{cases} . \tag{8.3}$$

Here, in the case of $Y$ a tree, $\#V^{\mathrm{ess}}$ stands for the number of **essential** vertices – vertices of degree strictly greater than two. These results are notable: (1) the lack of dependence on $m$ for $\mathrm{TC}(\mathcal{C}^n(\mathbb{R}^m))$ means that the ambient space is largely irrelevant to collision-avoidance complexity; (2) the lack of dependence on $n$ for $\mathrm{TC}(\mathcal{C}^n(Y))$ for $Y$ a tree means that the critical lack of room on a tree collapses the complexity of collision avoidance to what happens at the essential vertices. ⊚

## 8.10 Fibrations

There is a far-reaching generalization of fiber bundles befitting homotopy theory. In spirit, a fibration is a surjective map $p\colon E \to B$ with the property that (for $B$ path-connected) all fibers $p^{-1}(b)$ are homotopy equivalent, as opposed to homeomorphic. The proper definition does not reference the fibers at all but is founded in behavior of homotopies in manner not unlike covering spaces.

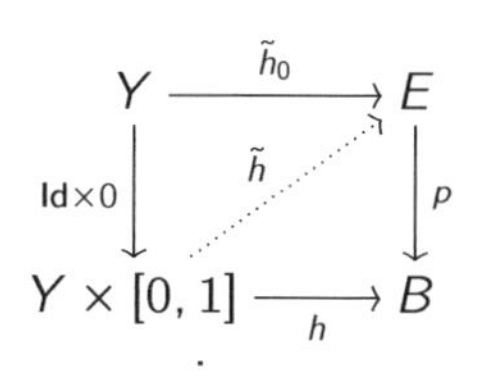


A [Hurewicz] **fibration** is an onto map $p\colon E \to B$ with the following **homotopy lifting property**: for any homotopy $h_t$ of a space $Y$ in $B$ and a lift $\tilde{h}_0\colon Y \to E$ of $h_0$ to $E$, there is a lifted homotopy $\tilde{h}_t\colon Y \to E$. In other words, in the appropriate commutative diagram, the dotted lift exists. At first glance, this definition seems obtuse: where is the fiber, $F$, and why does it have a well-defined homotopy type?

**Lemma 8.23.** *Fibers of a fibration over a path-connected base have constant homotopy type.*

**Proof.** Pick a basepoint $b_0 \in B$ and let $F = p^{-1}(b_0)$. Let $\beta\colon [0,1] \to B$ be a path in $B$ from $b_0$ to $b_1$. Define $h\colon F \times [0,1] \to B$ via $\beta$ on $[0,1]$ and via collapse-to-a-point on $F$. This has a lift $\tilde{h}_0 = \mathrm{Id}\colon F \times \{0\} \to p^{-1}(b_0)$. Thus, by homotopy lifting, $\tilde{h}_t = p^{-1}(\beta(t))$ is a homotopy from $F$ into $p^{-1}(b_1)$. Reverse the path and repeat to show a homotopy equivalence between fibers. ⊙

The idea of defining a fibration not in terms of explicit topological features of $F$, but rather in terms of implicit response of $p$ to homotopy-lifting, is deep and presages the use of homotopy testing as a means to define and extend other notions. Indeed, this is the pattern for defining **cofibrations** – a dual notion that characterizes maps $\iota\colon B \to E$ in terms of possessing the **homotopy extension property** – see [176] for details. These ingredients – fibrations generalizing projection and cofibrations generalizing inclusion – form the basis for the abstraction of homotopy theory to **model categories** [246, 101]. Upon first pass, the reader should note merely that such generalizations exist and flow from the use of commutative diagrams.

One simple example suffices to demonstrate the power of this approach. The reader may wonder why exact sequences have not made an appearance in the context of homotopy groups: they *are* very important, but come with concomitant subtlety (*cf.* the difference between the Mayer-Vietoris and Van Kampen theorems). The definition of a fibration allows for inference of fiber behavior via the long exact sequence associated to a fibration. Given $p\colon E \to B$ with fiber homotopy type $[F]$, the following sequence is exact:

$$\cdots \longrightarrow \pi_n(F) \xrightarrow{\pi(\iota)} \pi_n(E) \xrightarrow{\pi(p)} \pi_n(B) \xrightarrow{\delta} \pi_{n-1}(F) \xrightarrow{\pi(\iota)} \cdots \tag{8.4}$$

The maps are constructed as follows. Let $b_0 \in B$ be a basepoint, with $F = p^{-1}(b_0)$. The inclusion $\iota : F \hookrightarrow E$ sends a basepoint $f_0 \in F$ to $e_0 \in E$. The maps $\pi(\iota)$ and $\pi(p)$ are clear: it is the connecting homomorphism $\delta$ that is subtle. Consider the diagram whose top row is inclusion of the boundary $(n-1)$-sphere followed by collapse of same. The rightmost vertical arrow represents a class $[\alpha] \in \pi_n(B)$. Commutativity and the homotopy lifting property are used to generate the dotted vertical arrows and to show that the induced class $\delta([\alpha]) \in \pi_{n-1}(F)$ is well-defined.

$$\begin{array}{ccccc} \partial\mathbb{D}^n & \hookrightarrow & \mathbb{D}^n & \longrightarrow & \mathbb{S}^n \\ \vdots & & \vdots & & \downarrow \alpha \\ F & \xrightarrow{\iota} & E & \xrightarrow{p} & B \end{array}$$

As a simple example, this sequence yields a direct proof that covers induce isomorphisms on $\pi_n$ for $n > 1$: since the fiber is discrete, every third term of the sequence vanishes, yielding isomorphisms of the remaining pairs. More subtle is the example of the Hopf fibration $p : \mathbb{S}^3 \to \mathbb{S}^2$ with fiber $F = \mathbb{S}^1$; this induces the long exact sequence:

$$\cdots \longrightarrow \pi_n(\mathbb{S}^1) \longrightarrow \pi_n(\mathbb{S}^3) \longrightarrow \pi_n(\mathbb{S}^2) \longrightarrow \pi_{n-1}(\mathbb{S}^1) \longrightarrow \cdots$$

which, since $\pi_n(\mathbb{S}^1) = 0$ for all $n > 1$, implies that $\pi_n(\mathbb{S}^3) \cong \pi_n(\mathbb{S}^2)$ for all $n > 2$. Thus $\pi_3(\mathbb{S}^2) \cong \pi_3(\mathbb{S}^3) \cong \mathbb{Z}$.

## 8.11 Homotopy type theory

This text has largely avoided the (many and fruitful) applications of topology in the areas of logic and computer science, in part because of the significant overhead of definitions and formal structures required for a proper exposition. Some of the most recent activity in these domains is, however, too compelling not to limn. Let the reader beware that what follows is a severe redaction of a highly intricate and rapidly advancing research program.

Computer programmers use **types** to distinguish different classes of data: the reader may have seen $\lambda$-*calculus* or other systems that use formal rule systems to define functions from base or constant types. If this is not familiar, the reader may recall the difficulties stemming from *Russell's paradox* in set theory[3] and the resolution through distinguishing different types of objects (sets, classes, universes, *etc.*).

In (intensional) type theory, there are collections of *terms*, to which can be associated a *type* within a *universe* of types; these can range from variables to logical operators to functional types and more. There is a loose correspondence between type-theoretic constructs, logical constructs, and set-theoretic interpretations thereof. The novel ingredient in the recently-christened *homotopy type theory* is an injection of homotopy-theoretic perspectives. Beginning with the observation that a type is *something like* a space, as opposed to a set, and a term of a certain type is *something like* a point in said space, a homotopy-theoretic sensibility can inform and expand type theory. With the adjunction of one key axiom – the **univalence axiom**[4] of Voevodsky – one can construct a well-defined correspondence of the classical interpretations to the homotopy-theoretic constructs touched upon in this chapter. It is worth reproducing the table of correspondences from [291], which serves as a *Rosetta stone* for the

[3] *"The set of all sets that do not contain themselves..."* caused no small amount of trouble.

[4] The univalence axiom is about a universe $\mathcal{U}$ of types; it states that types which are formally equivalent in $\mathcal{U}$ are identical.

type-theoretic, logical, set-theoretic, and homotopy-theoretic views:

| TYPES | LOGIC | SETS | HOMOTOPY |
|---|---|---|---|
| $A$ | proposition | set | space |
| $a : A$ | proof | element | point |
| $B(x)$ | predicate | family of sets | fibration |
| $b(x) : B(x)$ | conditional proof | family of elements | section |
| $0, 1$ | $\bot, \top$ | $\varnothing, \{\varnothing\}$ | $\varnothing, \star$ |
| $A + B$ | $A \vee B$ | disjoint union | coproduct of spaces |
| $A \times B$ | $A \wedge B$ | set of pairs | product of spaces |
| $A \to B$ | $A \Rightarrow B$ | set of functions | function space |
| $\sum_{(x:A)} B(x)$ | $\exists_{x:A} B(x)$ | disjoint sum | total space |
| $\prod_{(x:A)} B(x)$ | $\forall_{x:A} B(x)$ | product | space of sections |
| $\mathrm{Id}_x$ | equality $=$ | $\{(x, x) : x \in A\}$ | path space $\mathcal{P}(A)$ |

This table is meant to inspire rather than to define: proper definitions are more involved and require the categorical language of Chapter 10 that is the focus of the remainder of this text.

# Notes

1. Knot theory began in earnest with the work of Kelvin and Tait as a problem in fluid dynamics. It was conjectured that atoms were knotted vortex tubes in the æther, and that a classification of knot types would reproduce and refine the periodic table. Poincaré's initial work on algebraic topology was likewise motivated by the desire to understand the dynamics of fluids. It is remarkable how much the field of topology owes to fluid dynamics. (The author, too, came to topology via fluids and dynamics.)
2. The Whitehead theorem has a computationally-friendly corollary: a map between *simply-connected* CW complexes that induces isomorphisms on all *homology* groups is a homotopy equivalence. Simply-connectivity is the key to this result.
3. Homotopy and homology are fundamentally entwined via configuration spaces. For a connected CW complex $X$, consider the unlabeled singular configuration space $X^n/S_n$ of $n$ unlabeled not-necessarily-distinct points on $X$. The **Dold-Thom theorem** states that in the limit as $n \to \infty$, the resulting configuration space has $\pi_n$ isomorphic to $H_n(X; \mathbb{Z})$. This is very deep and *seems* to presage a topological version of statistical physics.
4. Corollary 8.18 on topological social choice was discovered by Weinberger, [301] see also Chichilnisky et al. [66], and Baryshnikov [22]; however, in the process of publication, it was realized that the result is implicit in the 1954 paper of Eckmann, who was not motivated by social choice at all, but rather by problems of generalized means and homotopy theory.
5. Section 8.8 hints at the theory of **characteristic classes**. Given a vector bundle $\pi: E \to M$, a characteristic class is an element of $H^\bullet(E)$ carrying data about the bundle. Among the more interesting characteristic classes besides the Euler class are the **Stiefel-Whitney**, **Chern**, **Pontryagin**, and **Thom** classes [226]. The application of homotopy groups to classifying defects in liquid crystals in §8.5 is just the beginning of a number of exciting instances of algebraic topology in condensed matter physics, recent examples of which use characteristic classes to explain experimentally observed

phenomena. This text has skipped most of the applications of topology to physics and fields, not because of lack of interest, but because of the requisite depth: bundles and characteristic classes are the starting point for modern approaches to quantum field theory.

6. It is hard to overstate the importance of fibrations and cofibrations within homotopy theory. For a proper treatment, see, *e.g.*, the excellent text of May [221].
7. The antecedent to the work described in §8.9 is the insightful work of Blum, Shub, and Smale on topological complexity of computations [42, 278]. This work defines a topological complexity for computing roots of a complex polynomial of degree $k$ in terms of the section category of the bundle $\mathcal{C}^k(\mathbb{C}) \to \mathcal{UC}^k(\mathbb{C})$: see also, [17, 85, 295].
8. The cohomology computations hinted at in §8.9 go much deeper than explained in this text. See, *e.g.*, [112, 113, 114] for relations to cohomology operations, Steenrod squares, and the like. The topological complexity of the real projective spaces $\mathbb{P}^n$ is equal to its *immersion dimension* except for $n = 1, 3, 7$ [116] – this is a notoriously subtle and difficult to compute quantity.
9. The critique that TC does nothing to help with realistic motion-planning problems is perfectly true and perfectly ignorant of the illumination an obstruction theory brings.
10. The table in §8.11 is reproduced from [291] with some slight notational changes (legally, under their *Creative Commons* license). The reader is encouraged to see the source for more and better explanations.

Chapter 9

# Sheaves

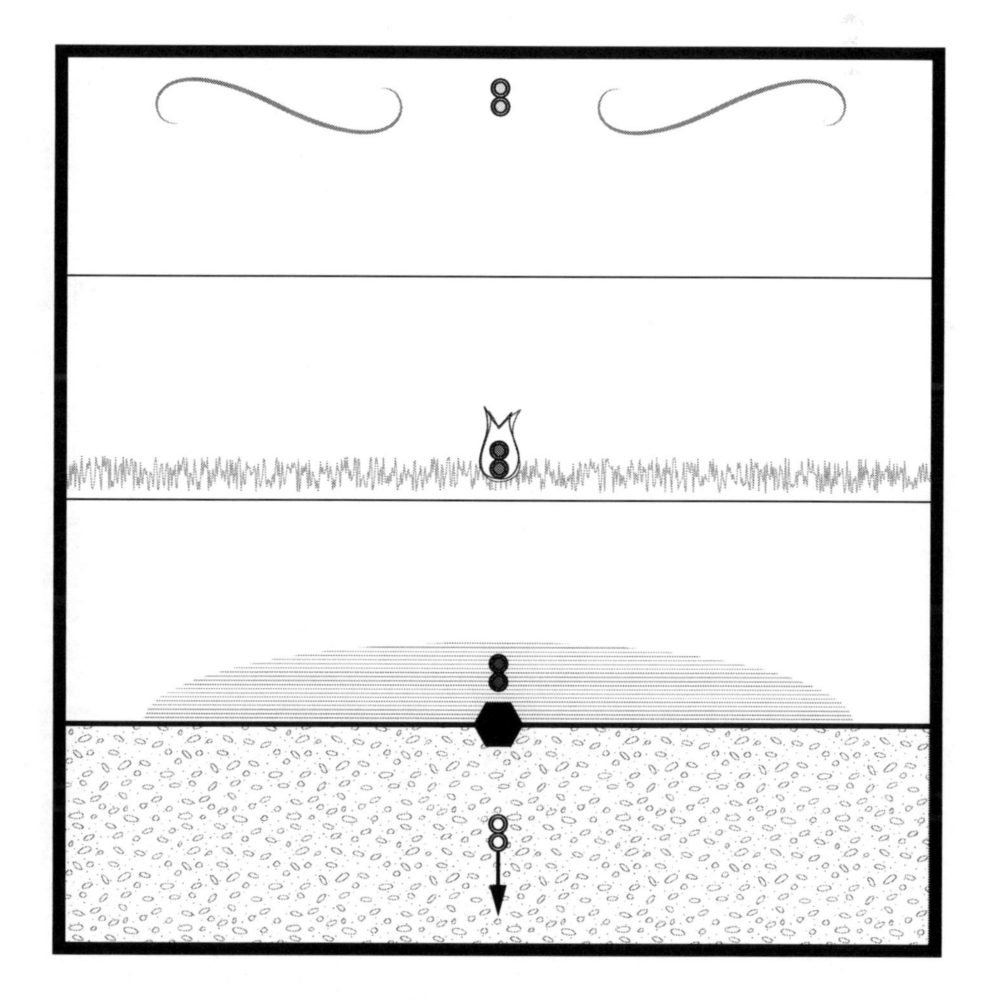

Data has, in most settings, ceased to be a scarce resource; the problem of how to get data has been eclipsed by how to manage its abundance and variety. Topology possesses several tools relevant to the aggregation and fusion of local data. Among the most powerful and flexible of these is the theory of sheaves, a structure for the collation of data parameterized by a space.

## 9.1 Cellular sheaves

A fiber bundle associates to sufficiently small open sets $U$ in the base space $B$ a product $U \times F$ with a fiber $F$. Very often, these fibers are vector spaces, modules, or groups, whose algebraic structure is respected by the ensuing topology of the total space $E$. Though useful and common, bundles (and even fibrations) are too restrictive to handle the general phenomenon of merging different forms of algebraic data over a base space.

Consider the following simple example. A robot locomotes through a system of narrow hallways. Using an omnidirectional laser scanner, it records real-valued sensor data – distance to the wall, say – along hallway directions. In the interior of a hallway, this directional data has rank two: forwards and backwards. At a branch point, where hallways meet, the number of feasible directions jumps instantaneously (in the conceptual limit where the hallways form a planar graph). The rank of data in the robot's *sensorium* changes. This idealized setting attaches to each point of a planar graph a real vector space whose dimension equals the number of directions along which one can move from that point. One can then imagine more complex sensors that store feature-detection, bearings, or other data in algebraic structures that can vary from place-to-place. Furthermore, it is possible to correlate data locally – as the robot moves along a hallway, there is consistent notion of *ahead* and *behind*. This amalgamation of algebraic data along a space is at the heart of the notion of a sheaf.

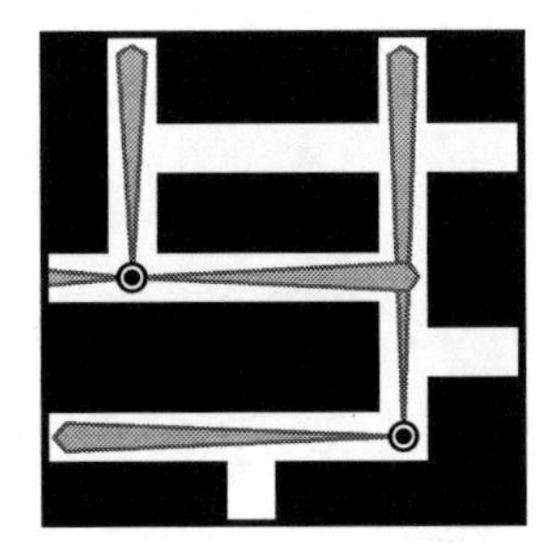

Though sheaf theory is a remarkably intricate language, the following treatment is, in keeping with the spirit of this text, elementary, emphasizing *cellular* sheaves; these possess computational and intuitive advantages reminiscent of cellular homology.

Fix $X$ a regular cell complex with $\trianglelefteqslant$ denoting the face relation: $\sigma \trianglelefteqslant \tau$ iff $\sigma \subset \overline{\tau}$. As a model for *data* over $X$, consider the setting of abelian groups and homomorphisms (or, if preferred, vector spaces and linear transformations). A **cellular sheaf** over $X$, $\mathcal{F}$, is generated by an assignment to each cell $\sigma$ of $X$ an abelian group $\mathcal{F}(\sigma)$ and to each face $\sigma \trianglelefteqslant \tau$ of $\tau$ a **restriction**[1] **map** – a homomorphism $\mathcal{F}(\sigma \trianglelefteqslant \tau)\colon \mathcal{F}(\sigma) \to \mathcal{F}(\tau)$ such that faces of faces satisfy the composition rule:

$$\rho \trianglelefteqslant \sigma \trianglelefteqslant \tau \quad \Rightarrow \quad \mathcal{F}(\rho \trianglelefteqslant \tau) = \mathcal{F}(\sigma \trianglelefteqslant \tau) \circ \mathcal{F}(\rho \trianglelefteqslant \sigma). \tag{9.1}$$

[1] Yes, it seems backwards to call this a restriction. The terminology comes from the topological perspective of §9.6, of which the cellular case acts as a nerve.

The *trivial* face $\tau \trianglelefteq \tau$ by default induces the identity isomorphism $\mathcal{F}(\tau \trianglelefteq \tau) = \text{Id}$. This simple definition of a sheaf as a representation of the face structure belies a powerful depth, one that is appreciated only later.

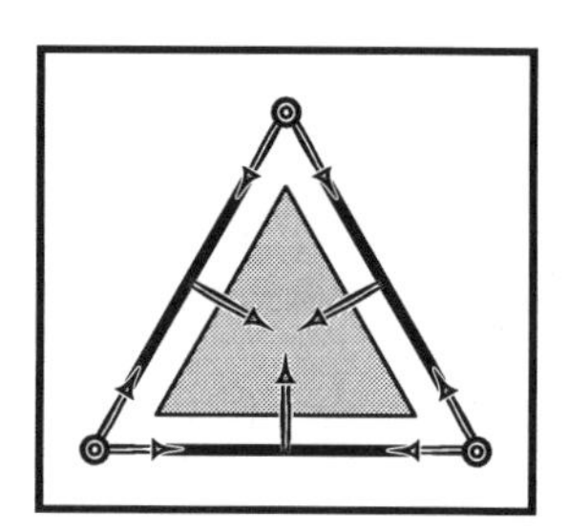

One says that the sheaf is *generated* by its values on individual cells of $X$: this data $\mathcal{F}(\tau)$ over a cell $\tau$ is also called the group of **local sections** of $\mathcal{F}$ over $\tau$: one writes $s_\tau \in \mathcal{F}(\tau)$ for a local section over $\tau$. Though the sheaf is generated by local sections, there is more to a sheaf than its generating data, just as there is more to a vector space than its basis. The restriction maps of a sheaf encode how local sections are continued into more global objects – sections defined over larger subsets of $X$. The value of the sheaf $\mathcal{F}$ on all of $X$ is defined to be collections of local sections that *continue* according to the restriction maps on faces:

$$\mathcal{F}(X) := \{(s_\tau)_{\tau \in X} : s_\sigma = \mathcal{F}(\rho \trianglelefteq \sigma)(s_\rho) \ \forall \rho \trianglelefteq \sigma\} \subset \prod_\tau \mathcal{F}(\tau). \tag{9.2}$$

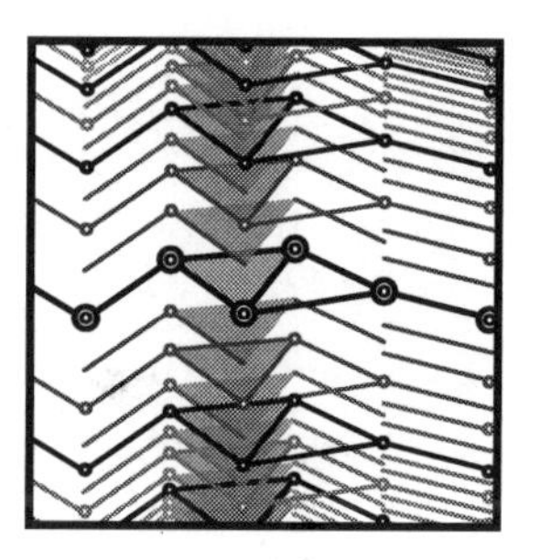

That is, $\mathcal{F}(X)$ consists of all choices of local data over cells which are compatible with respect to restriction maps: the **global sections**. In general, one thinks of $\mathcal{F}$ as a data structure that assigns to any subset $A \subset X$ the corresponding group $\mathcal{F}(A)$ of sections *over* $A$, where Equation (9.2) is modified to use the smallest collection of cells in $X$ containing $A$ and consistency is enforced on all faces within this $A$-containing subcomplex. In the case of a single point $x \in X$, one speaks of the **stalk** of $\mathcal{F}$ at $x$, $\mathcal{F}_x$. In the present setting, this is simply $\mathcal{F}_x = \mathcal{F}(\sigma)$, for $\sigma$ the unique cell in whose interior $x$ lies.

## 9.2 Examples of cellular sheaves

The simplest example of a cellular sheaf is the **constant** sheaf $\mathbf{G}_X$ on $X$, which assigns the group $\mathbf{G}$ to each cell of $X$ and the identity homomorphism to each face relation.

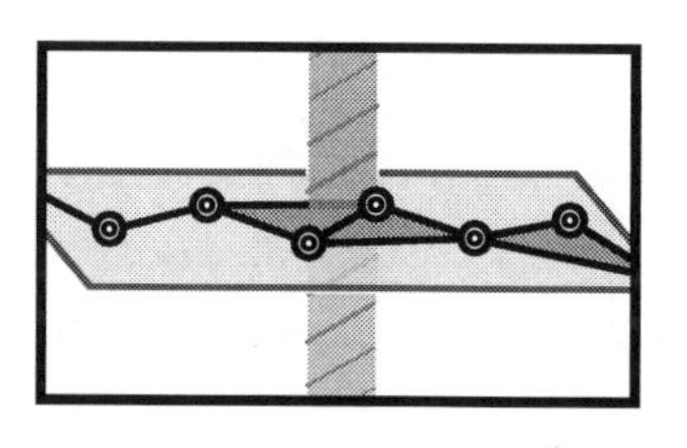

The data remain constant over $X$: every point has stalk $\mathbf{G}$ and the global sections match local sections, $\mathbf{G}_X(X) \cong \mathbf{G}$. The antipodal example to this is the **skyscraper** sheaf $\mathbf{G}_\sigma$ that assigns the group $\mathbf{G}$ to the cell $\sigma$ and zero to all other cells. In this case, all the restriction maps are the zero map (except the identity $\sigma \trianglelefteq \sigma$). The stalks vanish except on the cell $\sigma$. This sheaf has no nonzero global sections if $\dim \sigma > 0$, but for a vertex $v$, $\mathbf{G}_v(X) \cong \mathbf{G}$.

Note that the restriction maps are a crucial component of a sheaf. Given a pair of sheaves over $X$, $\mathcal{F}$ and $\mathcal{G}$, one defines the sum $\mathcal{F} \oplus \mathcal{G}$ to be the sheaf on $X$ whose data on $\sigma$ is $\mathcal{F}(\sigma) \oplus \mathcal{G}(\sigma)$ and whose restriction maps are likewise direct sums of the form $\mathcal{F}(\sigma\trianglelefteq\tau)\oplus\mathcal{G}(\sigma\trianglelefteq\tau)$. Note that summing up skyscraper sheaves over every cell of $X$ is *not* the same as the constant sheaf,

$$\bigoplus_{\sigma\in X}\mathbf{G}_\sigma \neq \mathbf{G}_X,$$

even though all the stalks agree: the restriction maps of this city full of skyscrapers are all zero. The value of the sum-of-skyscrapers sheaf on all of $X$ is $\oplus_v\mathbf{G}$ – an anarchic assignment of any $\mathbf{G}$-element to each vertex $v$ of $X$ agrees, via the restriction maps, to the null on higher-dimensional cells. On the other hand, $\mathbf{G}_X(X)\cong\mathbf{G}$, since all of the identity restriction maps force a perfect consensus among cells.

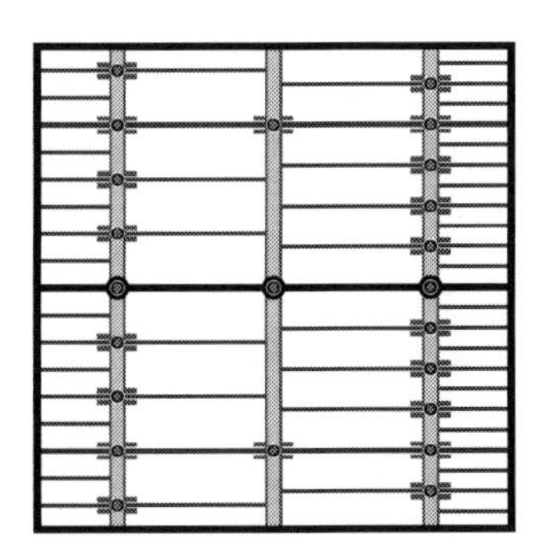

Nontrivial restriction maps lead to interesting situations vis-a-vis local versus global. Consider the case of a sheaf $\mathcal{F}$ whose stalks are all $\mathbb{Z}$. The restriction maps $\mathcal{F}(\sigma\trianglelefteq\tau)$ must therefore be homomorphisms $\mathbb{Z}\to\mathbb{Z}$; *i.e.*, multiplication by some constant. Note that these constants cannot be arbitrary, since by definition, composition must hold, meaning that the multiplication constants must factor according to the restriction maps. Then, assuming that none of these constants equals zero, and that the base complex $X$ is connected, then there are always nonzero global sections – in fact, a $\mathbb{Z}$'s worth. However, it is *not* the case that local sections can always be patched together to give a global section.

**Example 9.1 (Recurrence equations)** ⊚

Consider the simple linear recurrence $u_{n+1}=A_nu_n$, where $u_n\in\mathbb{R}^k$ is a vector of states and $A_n$ is a $k$-by-$k$ real matrix. These discrete-time analogues of nonautonomous linear ODE systems are important in everything from population models to audio filtering. Such a system can be thought of in terms of sheaves over the time-line $\mathbb{R}$ with the cell structure on $\mathbb{R}$ (or $\mathbb{R}^+$) having $\mathbb{Z}$ (or $\mathbb{N}$) as vertices. Let $\mathcal{F}$ be the sheaf that assigns to each cell the vector space $\mathbb{R}^k$. One can encode the dynamics of the recurrence relation as follows. Let $\mathcal{F}(\{n\}\trianglelefteq(n,n+1))$ be the update map $u\mapsto A_nu$ and let $\mathcal{F}(\{n+1\}\trianglelefteq(n,n+1))$ be the identity. Then global sections $u$ of $\mathcal{F}$ correspond precisely to solutions to the recurrence equation, since $u$ restricted to $\{n+1\}$ must equal $A_n$ times $u$ restricted to $\{n\}$ in order to be a consistent section. To be a sheaf taking values in $\mathbb{R}$-vector spaces, the dynamics have to be linear, so that the space of solutions is a linear subspace. ⊚

**Example 9.2 (Local cohomology)** ⊚

An excellent example for building intuition is to be found in local cohomology. Fix $X$ a cell complex and consider the cellular sheaf $\mathcal{F}$ that assigns to the cell $\sigma$ the (singular) $k^{\text{th}}$ **local cohomology** $H^k(X,X-\overline{\sigma})$, where $\overline{\sigma}$ is the closure of $\sigma$ in $X$ and the particular coefficient ring is left to taste. If $\sigma\trianglelefteq\tau$, then the inclusion $\iota\colon(X,X-\overline{\tau})\hookrightarrow(X,X-\overline{\sigma})$ induces on $H^k$ the homomorphism $\mathcal{F}(\sigma\trianglelefteq\tau)$, via

$$\mathcal{F}(\sigma\trianglelefteq\tau)=H(\iota)\colon H^k(X,X-\overline{\sigma})\to H^k(X,X-\overline{\tau}).$$

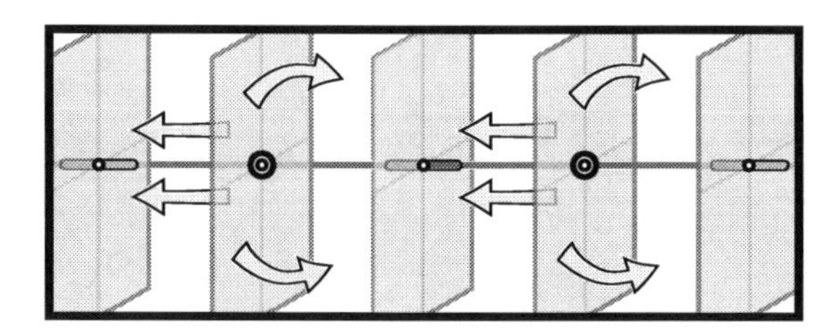

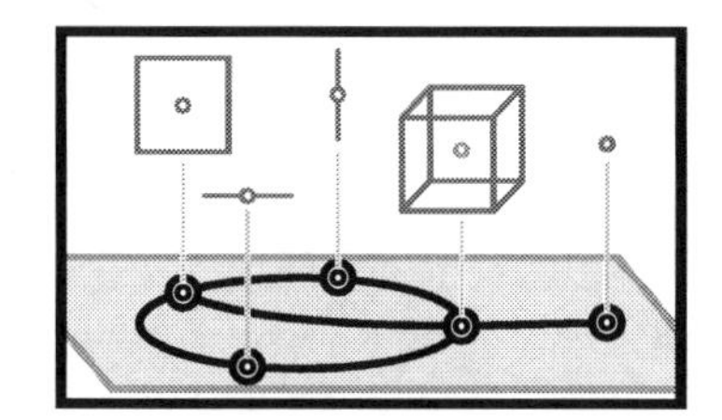

The stalk of this sheaf encodes local $H^n$. For example, on an $n$-dimensional manifold, this process yields a constant sheaf (of dimension one), called the **orientation sheaf**, *cf.* Example 4.18. The manifold is orientable if and only if the orientation sheaf has a global section. For a finite graph, the local $H^1$ sheaf has stalk dimension equal to 1 on edges and equal to $\deg(v) - 1$ on each vertex $v$. The restriction maps are projections for vertices of degree greater than two. ◎

**Example 9.3 (Fiber homology)** ◎

Consider a cellular map $f\colon X \to Y$ between cell complexes. There is an induced cellular sheaf on $Y$ that encodes $f$. Given $\sigma$ a simplex of $Y$, define the **fiber homology sheaf** $\mathcal{F}(\sigma) = H_k(f^{-1}\overline{\sigma})$. For $\sigma \trianglelefteq \tau$, one has $\overline{\sigma} \subset \overline{\tau}$ and likewise with the $f$-inverse image in $X$; hence the restriction map $\mathcal{F}(\sigma \trianglelefteq \tau)$ comes from the induced map $H_k(f^{-1}\overline{\sigma}) \to H_k(f^{-1}\overline{\tau})$. For example, if $h\colon X \to \mathbb{R}$ is a (cellular) height function, then the fiber homology sheaf on $\mathbb{R}$ records the homology of the level sets of $h$ as stalks over vertices of $\mathbb{R}$. ◎

**Example 9.4 (Logic gates)** ◎

Consider a simple XOR gate, with two binary inputs and one binary output given by exclusive conjunction. Topologize this gate as a directed Y-graph Y. Let $\mathcal{F}$ be the sheaf taking values in $\mathbb{F}_2$ vector spaces over Y with stalk dimension one everywhere except at the central vertex, where it equals two.

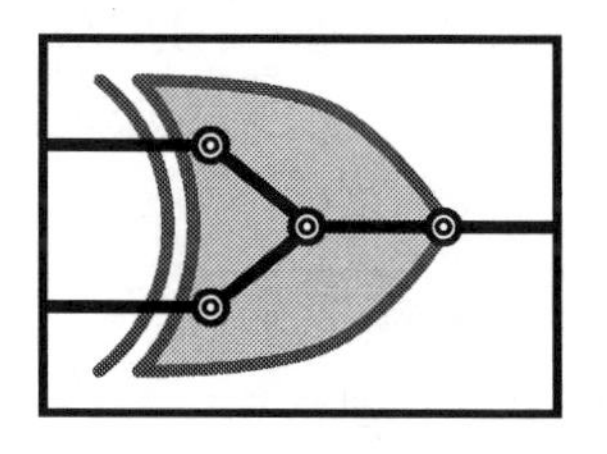

The restriction maps from the central vertex to the three edges are as follows. On the two input edges, the restriction map is projection to the first and second factors, respectively. The restriction map to the output edge is addition: $+\colon \mathbb{F}_2^2 \to \mathbb{F}_2$. This instantiates an exclusive-OR gate – the global sections correspond precisely to the truth table of inputs and outputs. A similar approach does *not* work for an AND gate, since the operation $\mathbb{F}_2^2 \to \mathbb{F}_2$ encoded by AND is no longer a homomorphism; neither is the involutive NOT, nor OR, nor NOR, nor NAND. See [159, 254] for other approaches to sheaf circuitry. ◎

## 9.3 Cellular sheaf cohomology

Sheaves are interpretable as an algebraic data structure over a space (cell complex, for the moment). Equally illustrative is the interpretation as a coefficient system for cohomology. A sheaf permits *local* coefficients that can change from cell-to-cell: $\mathcal{F}(\sigma)$ is the coefficient group over the cell $\sigma$ and the restriction maps $\mathcal{F}(\sigma \trianglelefteq \tau)$ encode the switching required to glue together local cochains into global cocycles.

The definition of cellular sheaf cohomology is straightforward and similar to that of ordinary cellular cohomology. Given $\mathcal{F}$ over a compact cell complex $X$, let $C^n(X;\mathcal{F})$ be the product of $\mathcal{F}(\sigma)$ over all $n$-cells $\sigma$: note that this aligns with the definition of cellular cohomology $C^n_{\text{cell}}(X;\mathbf{G})$ in the case of the constant sheaf $\mathbf{G}_X$. These cochains are connected by coboundary maps defined as follows:

$$0 \longrightarrow \prod_{\dim \sigma=0} \mathcal{F}(\sigma) \xrightarrow{d} \prod_{\dim \sigma=1} \mathcal{F}(\sigma) \xrightarrow{d} \prod_{\dim \sigma=2} \mathcal{F}(\sigma) \xrightarrow{d} \cdots \tag{9.3}$$

where, by analogy with (4.6), the coboundary map is given as

$$d(\sigma) := \sum_{\sigma \trianglelefteq \tau} [\sigma\colon \tau]\, \mathcal{F}(\sigma \trianglelefteq \tau). \tag{9.4}$$

Note that $d\colon C^n(X;\mathcal{F}) \to C^{n+1}(X;\mathcal{F})$, since $[\sigma\colon \tau] = 0$ unless $\sigma$ is a codimension-1 face of $\tau$. This gives a cochain complex: in the computation of $d^2$, the incidence numbers factor from the restriction maps, and the computation from cellular co/homology suffices to yield 0. The resulting **cellular sheaf cohomology** is denoted $H^\bullet(X;\mathcal{F})$. The cohomology of the constant sheaf $\mathbf{G}_X$ on $X$ is, clearly, $H^\bullet_{\text{cell}}(X;\mathbf{G})$. A skyscraper sheaf $\mathbf{G}_\sigma$ over a cell $\sigma$ has a (perfect) cochain complex with $d = 0$. As such, the cohomology is isomorphic to the cochain complex: $H^\bullet(X;\mathbf{G}_\sigma)$ vanishes except for grading $\dim \sigma$, at which the cohomology has rank equal to one.

One particularly useful set of interpretations for sheaf cohomology uses the terminology of local and global sections. Recall from Example 6.8 the interpretation of ordinary $H^0$ as connected components of a space; replacing the space with a sheaf over the space yields the following.

**Lemma 9.5.** *Sheaf cohomology in grading zero classifies global sections:*

$$H^0(X;\mathcal{F}) = \mathcal{F}(X). \tag{9.5}$$

It may seem puzzling to the beginner that $\mathcal{F}(X)$ is determined completely by the data on the vertices and edges of $X$. Is the data on the higher-dimensional cells ignored? No: the compatibility of data on higher-dimensional cells is built into the definition of a sheaf. In this, it is analogous to a planar tiling by square tiles with colored edges. Consider the plane as a cubical complex. One places square tiles at the vertices and demands that incident tiles have compatible edges. The global solutions (legal tilings of the plane) are classified completely by vertex and edge data: there are no additional constraints where the four corners of the tiles meet (on the 2-cells of the complex).

**Example 9.6 (Matrix equations)** 

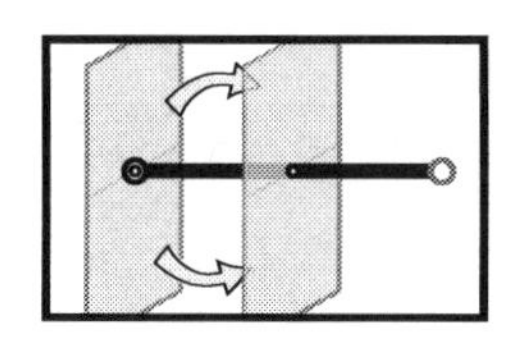

The elements of linear algebra recur throughout topology, including sheaf cohomology. Consider the following sheaf $\mathcal{F}$ over the closed interval with two vertices, $a$ and $b$, and one edge $e$. The stalks are given as $\mathcal{F}(a) = \mathbb{R}^m$, $\mathcal{F}(b) = 0$, and $\mathcal{F}(e) = \mathbb{R}^n$. The restriction maps are $\mathcal{F}(b \trianglelefteq e) = 0$ and $\mathcal{F}(a \trianglelefteq e) = A$, where $A$ is a linear transformation. Then, by definition, the sheaf cohomology is $H^0 \cong \ker A$ and $H^1 \cong \operatorname{coker} A$.

Cellular sheaf cohomology taking values in vector spaces is really a characterization of solutions to complex networks of linear equations. If one modifies $\mathcal{F}(b) = \mathbb{R}^p$ with $\mathcal{F}(b \trianglelefteq e) = B$ another linear transformation, then the cochain complex takes the form

$$0 \longrightarrow \mathbb{R}^m \times \mathbb{R}^p \xrightarrow{[A|-B]} \mathbb{R}^n \longrightarrow 0 \longrightarrow \cdots ,$$

where $d = [A|-B] : \mathbb{R}^{m+p} \to \mathbb{R}^n$. The zero$^{\text{th}}$ sheaf cohomology $H^0$ is precisely the set of solutions to the equation $Ax = By$, for $x \in \mathbb{R}^m$ and $y \in \mathbb{R}^p$. These are the global sections over the closed edge. The first sheaf cohomology measures the degree to which $Ax - By$ does not span $\mathbb{R}^n$. As before, all higher sheaf cohomology groups vanish. ⊚

As sheaves provide a means of doing cohomology, one anticipates that the methods of Chapters 5 *ff.* generalize as well. This is relatively straightforward, viewing a sheaf as a generalized coefficient system. By analogy with Lemma 5.5, consider a short exact sequence of sheaves over $X$,

$$0 \longrightarrow \mathcal{F} \xrightarrow{i} \mathcal{G} \xrightarrow{j} \mathcal{H} \longrightarrow 0 ,$$

where exactness is enforced cell-by-cell (or stalk-by-stalk). This leads to a connecting homomorphism and long exact sequence on sheaf cohomology:

$$\longrightarrow H^{n-1}(X;\mathcal{H}) \xrightarrow{\delta} H^n(X;\mathcal{F}) \xrightarrow{H(i)} H^n(X;\mathcal{G}) \xrightarrow{H(j)} H^n(X;\mathcal{H}) \longrightarrow . \tag{9.6}$$

For example, consider a sheaf $\mathcal{F}$ on $X$ and a closed subcomplex $A \subset X$. Let $\mathcal{F}_A$ be the restriction of $\mathcal{F}$ to $A$ (so that stalks are zero off of $A$ and all restriction maps $\mathcal{F}(\sigma \trianglelefteq \tau)$ are set to zero unless $\sigma, \tau \subset A$), and let $\mathcal{F}_{X-A}$ be the complementary restriction of $\mathcal{F}$ to $X-A$ (so that it vanishes on $A$). Then the following sequence of cellular sheaves on $X$ is exact:

$$0 \longrightarrow \mathcal{F}_{X-A} \xrightarrow{i} \mathcal{F} \xrightarrow{j} \mathcal{F}_A \longrightarrow 0 .$$

Here, $j$ is a projection map that sends stalks of $X-A$ to zero; $i$ is the inclusion of $\mathcal{F}$ restricted to $X-A$, the kernel of $j$. Equation (9.6) yields an analogue of the long exact sequence of $\mathcal{F}$ over the pair $(X, A)$:

$$\longrightarrow H^{n-1}(A;\mathcal{F}) \xrightarrow{\delta} H^n(X,A;\mathcal{F}) \xrightarrow{H(j)} H^n(X;\mathcal{F}) \xrightarrow{H(i)} H^n(A;\mathcal{F}) \longrightarrow .$$

Here, a convenient notation is adopted: $H^\bullet(A;\mathcal{F}) := H^\bullet(X;\mathcal{F}_A)$ and $H^\bullet(X,A;\mathcal{F}) := H^\bullet(X;\mathcal{F}_{X-A})$. This relative sheaf cohomology is significant. The reader may take the following as an exercise in definitions and the long exact sequence:

**Corollary 9.7.** *For $A$ a subcomplex of $X$, $H^0(X,A;\mathcal{F})$ classifies global sections of $\mathcal{F}$ on $X$ which vanish on $A$.*

This interpretation is in keeping with the notion of cohomology as an obstruction to solving certain problems of extension and restriction.

## 9.4 Flow sheaves and obstructions

Even the simple setting of a cellular sheaf over a network (a 1-d cell complex) has interesting applications. Consider a flow network, as per Example 6.5, in which a directed acyclic network $X$ from a source node s to a target node t has an assignment of capacity constraints, $\text{cap}(e) \in \mathbb{N}$ for each edge $e \in E(X)$. A flow $\varphi\colon E(X) \to \mathbb{N}$ assigns to edges a flow rate $0 \le \varphi(e) \le \text{cap}(e)$ in a conservative manner – at each node $v \in V(X)$,

$$\sum_{e \to v} \varphi(e) = \sum_{v \to e'} \varphi(e') =: \varphi_v, \tag{9.7}$$

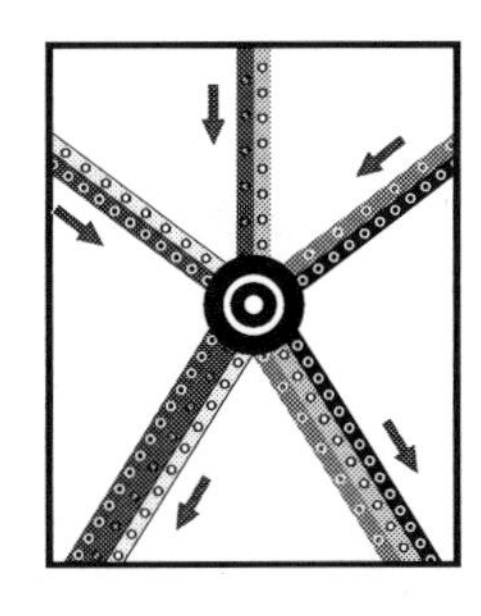

where $e \to v$ denotes an input edge and $v \to e'$ denotes an output edge at $v$. The **flow value**, $\text{val}(\varphi)$, equals the net flow out of s (or equivalently into t).

It is possible to interpret flows in terms of the cohomology of a **flow sheaf**, following the ideas of Hiraoka (with additional input from Robinson). Fix a capacity-respecting flow $\varphi$ on $X$ and choose a consistent **routing protocol** at each vertex. Namely, for each vertex $v$, choose a binary matrix $\Psi_v$ that encodes which portions of the incoming flow are sent to which portions of the outgoing flow at $v$,

$$\Psi_v\colon \bigoplus_{e\to v} \mathbb{R}^{\text{cap}(e)} \longrightarrow \bigoplus_{v\to e'} \mathbb{R}^{\text{cap}(e')},$$

where $(\Psi_v)_{i,j} = 1$ means that the $j^{\text{th}}$ unit of incoming flow is sent to the $i^{\text{th}}$ unit of outgoing flow; conservation is imposed by insisting that $\Psi_v$ is a permutation matrix padded with extra zero rows/columns. For purposes of building the appropriate cellular sheaf, subdivide $X$ to $\tilde{X}$ as follows:

1. Each edge of $X$ is split in two with an additional vertex in the middle.
2. Each such vertex $v$ added has trivial routing: $\Psi_v = \text{Id}$.
3. There is a single *feedback edge* from t to s with sufficiently large capacity.

Let $\mathcal{F}$ be the sheaf on $\tilde{X}$ taking values in $\mathbb{R}$-vector spaces whose stalk on each edge $e$ is $\mathcal{F}(e) = \mathbb{R}^{\mathrm{cap}(e)}$. At a vertex $v$, the stalk is $\mathcal{F}(v) = \mathbb{R}^{\mathrm{cap}(v)}$, where $\mathrm{cap}(v)$ is defined to be the sum of $\mathrm{cap}(e')$ for all incoming edges $e' \to v$. The sheaf restriction maps enforce conservation:

$$\mathcal{F}(v \trianglelefteq e) : \bigoplus_{e' \to v} \mathbb{R}^{\mathrm{cap}(e')} \longrightarrow \mathbb{R}^{\mathrm{cap}(e)}$$

where for $e \to v$, the map $\mathcal{F}(v \trianglelefteq e)$ is projection to the $\mathbb{R}^{\mathrm{cap}(e)}$-factor in $\mathcal{F}(v)$, and for $v \to e$, the map $\mathcal{F}(v \trianglelefteq e)$ is projection of the image of $\Psi_v$ to the $\mathbb{R}^{\mathrm{cap}(e)}$-factor. This flow sheaf satisfies a type of Poincaré duality:

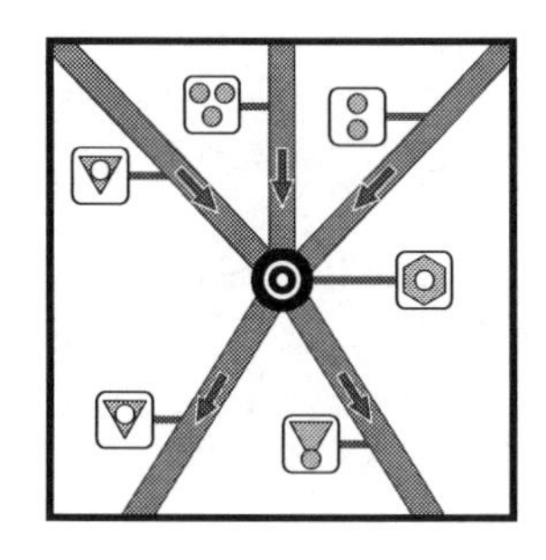

**Lemma 9.8.** *For a flow-sheaf $\mathcal{F}$ as above, $H^0(\tilde{X}; \mathcal{F}) \cong H^1(\tilde{X}; \mathcal{F})$ and $\dim H^0(\tilde{X}; \mathcal{F})$ equals the net flow value* $\mathrm{val}(\varphi)$.

**Proof.** The simplest proof of this duality is by counting: $C^0(\tilde{X}; \mathcal{F}) \cong C^1(\tilde{X}; \mathcal{F})$, since each vertex has stalk isomorphic to the sums of stalks of incoming edges, and this enumeration exhausts all cells of $\tilde{X}$. Passing to cohomology, the isomorphism comes from exactness (*i.e.*, Lemma 1.10): $\dim \ker d = \dim \mathrm{coker}\, d$. By Lemma 9.5, $\dim H^0(\tilde{X}; \mathcal{F})$ equals the number of global sections of $\mathcal{F}$; by conservation, each such section counts a unit of flow that circulates from source to target. ⊙

For a fixed flow $\varphi$ on $\tilde{X}$ there is a way to view the max-flow-min-cut theorem of Example 6.5 in the language of the long exact sequence of a pair in sheaf cohomology. Let $C \subset E(X)$ be a collection of edges in $X$, thought of as a putative cut-set. Since $C$ is not a subcomplex of $X$, consider instead $\tilde{C} \subset \tilde{X}$, the corresponding subcomplex of refined midpoint vertices of $C$ in $\tilde{X}$. The long exact sequence of the pair $(\tilde{X}, \tilde{C})$ in $\mathcal{F}$-cohomology yields:

$$0 \longrightarrow H^0(\tilde{X}, \tilde{C}) \xrightarrow{H(j)} H^0(\tilde{X}) \xrightarrow{H(i)} H^0(\tilde{C}) \xrightarrow{\delta} H^1(\tilde{X}, \tilde{C}) \xrightarrow{H(j)} H^1(\tilde{X}) \xrightarrow{H(i)} 0\,.$$

Consider the zero$^{\text{th}}$ cohomology $H^0(\tilde{C}; \mathcal{F})$ of $\tilde{C}$. By definition, this cohomology has dimension equal to the net capacity of the edge set $C$:

$$\dim H^0(\tilde{C}; \mathcal{F}) = \sum_{e \in C} \mathrm{cap}(e) =: \mathrm{val}(C).$$

By Lemma 9.8, the terms $H^1(\tilde{X}; \mathcal{F}) \cong H^0(\tilde{X}; \mathcal{F})$ each have dimension $\mathrm{val}(\varphi)$ the flow value. What connects these is the long exact sequence. By Corollary 9.7, the relative cohomology $H^0(\tilde{X}, \tilde{C})$ measures how much of the flow does *not* pass through $C$. Thus, by definition, if $C$ is a *cut*, then $H^0(\tilde{X}, \tilde{C}) = 0$, and one can say that this relative cohomology is the *obstruction* to being a cut. For $C$ a cut, this relative $H^0$ vanishes, and $H(i)$ above is injective by exactness, meaning that for any flow and cut,

$\text{val}(\varphi) \leq \text{val}(C)$: *the flow value is bounded by cut values* – the weak form of the max-flow min-cut theorem [146].

The crucial term in the long exact sequence is the connecting homomorphism $\delta$. Assuming $C$ a cut, then, by exactness, $H^0(\tilde{X}) \cong \ker \delta$ and $H^1(\tilde{X}) \cong \text{coker}\ \delta$. There is a splitting $H^1(\tilde{X}, \tilde{C}; \mathcal{F}) \cong H^1(\tilde{X}; \mathcal{F}) \oplus \text{im}\ \delta$. If a nonzero class is in the image of $\delta$, then it corresponds to a local section of $C$ that does not come from any global flow on $\tilde{X}$. Otherwise, it corresponds to a class in $H^1$ – a flow, by Lemma 9.8. Summarizing this discussion:

1. $\dim H^0(\tilde{X}; \mathcal{F}) = \text{val}(\varphi)$ is the flow value;
2. $\dim H^0(\tilde{C}; \mathcal{F}) = \text{val}(C)$ is the cut value;
3. $H^0(\tilde{X}, \tilde{C}; \mathcal{F})$ is the obstruction to being a cut; and
4. $\delta$ is the obstruction to $\text{val}(\varphi) = \text{val}(C)$.

When $\delta \neq 0$, the cut is not minimal or the flow is not maximal. When the two obstructions, $\delta$ and $H^0(\tilde{X}, \tilde{C}; \mathcal{F})$ both vanish, then the given flow and cut satisfy $\dim H^0(\tilde{X}; \mathcal{F}) = \dim H^0(\tilde{C}; \mathcal{F})$, or, maxflow-equals-mincut.

This construction of flow sheaves may seem an over-complicated restatement of conservation: net inputs equals net outputs. By adapting sheaves and cohomology to this elementary setting, one may properly generalize.

## 9.5 Information flows and network coding

The following is an application to **network coding** [146] – a branch of network information flow problems in which algebraic coding can be performed. This is motivated by communications: when sending data over a network, the data is split into packets that are individually routed from source to target(s) and then re-integrated. Packets of data are sent over edges at a particular bit rate, and packets are routed and/or coded at nodes to be broadcast along other edges.

Fix a communications network modeled as a finite directed acyclic graph $X$ from a fixed source (or sender) node $\mathsf{s}$ to multiple target (or receiver) nodes $\mathsf{t}_i$ in $X$. The data sent by $\mathsf{s}$ lies in a vector space over $\mathbb{F}_q$, a fixed finite field of $q$ elements. Data is transmitted along edges of $X$ and acted upon at nodes via linear transformations and rebroadcast, respecting directedness of $X$. The fundamental problems of network coding concern data throughput at the targets $\mathsf{t}_i$, given constraints on $X$, on $q$, on codings, and on bit-rate capacities of edges.

**Example 9.9 (Butterfly network)** ⊙

The classical *butterfly network* demonstrates that network coding can improve transmission rates [127, 196]. In this network, there is a single source $\mathsf{s}$ and two targets $\mathsf{t}_1$, $\mathsf{t}_2$. Each edge has unit capacity (only one packet in $\mathbb{F}_q$ can be sent per unit time). The goal is to send two packets of data, $a, b \in \mathbb{F}_q$ from the source to both targets as quickly as possible, assuming each edge can carry one bit of data per unit time. If the nodes act as routers – data are redirected along edges unchanged – then it requires at least five units of time to transmit $a$ and $b$ to both targets, since one central node must switch from transmitting $a$ to $b$. If, however, the central node receives both $a$

and $b$ and transmits $a+b$, then, in four time steps, $t_1$ receives $a$ and $a+b$ and $t_2$ receives $b$ and $a+b$. From this, the original signals can be *decoded* algebraically at the targets. ⊚

The general situation is similar. Fix $X$ a directed network with source s and multiple targets $t_i$ as above. The network data consists of vectors over a sufficiently large finite field $\mathbb{F} = \mathbb{F}_q$. Each edge $e$ in $X$ has a capacity $\mathrm{cap}(e) \in \mathbb{N}$, representing the bit-rate transmission capacity of $e$. The network coding is given in terms of $\mathbb{F}$-linear transformations at the nodes of $X$. To each node $v \in X$ is associated a **local coding map** ,

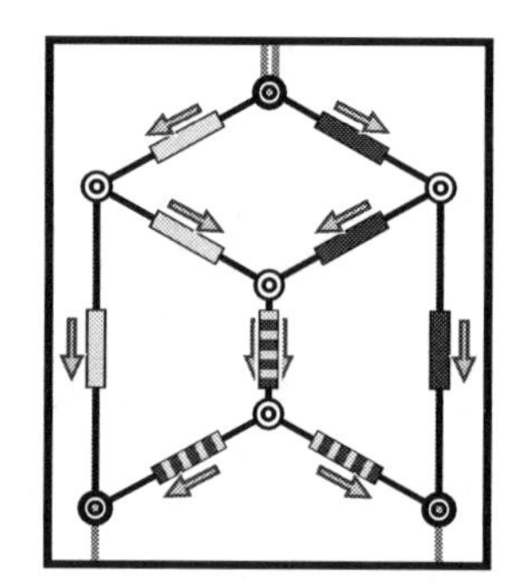

$$\Psi_v \colon \bigoplus_{e \to v} \mathbb{F}^{\mathrm{cap}(e)} \longrightarrow \bigoplus_{v \to e'} \mathbb{F}^{\mathrm{cap}(e')},$$

that maps the vector space of data entering $v$ to the vector space of data exiting $v$. For example, routing protocols are the (zero-padded) permutation matrices of §9.4. The central node $v$ of the butterfly network of Example 9.9 has $\Psi_v = [1|1]$ to encode via addition of signals. The decoding of messages received at $t_i$ is given by a decoding edge from $t_i \to s$ with corresponding decoding map $\Psi_{t_i}$ that disentangles coded messages at the targets. Let $\tilde{X}$ be the network obtained from $X$ by subdividing each edge (with inherited capacities and routings as in §9.4) and adding the decoding edges.

The following construction is due to Hiraoka and parallels his flow sheaf construction from §9.4. Given the local coding maps $\{\Psi_v\}$, build a **network coding sheaf** $\mathcal{F}$ on $\tilde{X}$ as follows. On each edge $e$, $\mathcal{F}(e) = \mathbb{F}^{\mathrm{cap}(e)}$. At a vertex $v$, the stalk is $\mathcal{F}(v) = \mathbb{F}^{\mathrm{cap}(v)}$, where $\mathrm{cap}(v) = \sum_{e' \to v} \mathrm{cap}(e')$. For $e \to v$, the map $\mathcal{F}(v \trianglelefteq e)$ is projection to the $\mathbb{F}^{\mathrm{cap}(e)}$-factor in $\mathcal{F}(v)$. For $v \to e$, the map $\mathcal{F}(v \trianglelefteq e)$ is projection of the image of $\Psi_v$ to the $\mathbb{F}^{\mathrm{cap}(e)}$-factor. The following results are straightforward generalizations of those of §9.4:

**Theorem 9.10 ([146]).** *With a network coding sheaf $\mathcal{F}$ for single-source-multi-target networks with decoding edges as above, $H^\bullet(\tilde{X}; \mathcal{F})$ has the following interpretations:*

1. *The net information flow rate equals* $\dim H^0(\tilde{X}; \mathcal{F})$.
2. *The net information flow rate which persists if a subnetwork $A$ becomes inoperative equals* $\dim H^0(\tilde{X}, \tilde{A}; \mathcal{F})$.
3. *The obstruction to extending an information flow on a subnetwork $A$ to all of $X$ respecting the coding and capacity constraints is $\delta \colon H^1(\tilde{X}, \tilde{A}; \mathcal{F}) \to H^1(\tilde{X}; \mathcal{F})$. If $\delta = 0$, all information flows on $A$ are extendable to $X$.*

## 9.6 From cellular to topological

Recall from Chapters 4-6 that it is relatively easy to define, visualize, and compute cellular co/homology, whereas it is of limited use in proving theorems such as homotopy invariance; the more powerful methods implicate the less-computable less-intuitive singular theory. A similar dichotomy exists in sheaf theory: all the standard texts [47, 95, 183] present definitions in the setting where the base space is a topological space with no fixed cell structure. Instead of stalks changing from cell-to-cell, they can change from point-to-point.

As an interpolative step, consider a space $X$ outfitted with a locally finite cover $\mathcal{U}$ by open sets $\{U_\alpha\}$. A sheaf $\mathcal{F}$ on $X$ *subordinate to the cover* $\mathcal{U}$ is, precisely, a cellular sheaf on the nerve $\mathcal{N}(\mathcal{U})$ of the cover. That is, there is an assignment to each cover element $U_\alpha$ of an abelian group $\mathcal{F}(U_\alpha)$, and for every non-empty intersection of cover elements, *e.g.*, $U_{\alpha\beta\gamma} := U_\alpha \cap U_\beta \cap U_\gamma$, there is the corresponding data $\mathcal{F}(U_{\alpha\beta\gamma})$. In addition, there are restriction homomorphisms according to the pattern of intersections in $\mathcal{U}$. For example, if $U_\alpha \cap U_\beta \neq \varnothing$, then $\mathcal{F}(U_\alpha \trianglelefteq U_{\alpha\beta})\colon \mathcal{F}(U_\alpha) \to \mathcal{F}(U_{\alpha\beta})$. Note that these restriction homomorphisms are indeed induced by restrictions of the cover elements, and the direction of the homomorphism is in the direction of the restriction. In the language of this chapter, one thinks of $\mathcal{F}$ as an assignment of data to $\mathcal{U}$ and the restriction maps as encoding how data changes when one narrows the field of view.

The cohomology of a sheaf subordinate to a cover is often called the **Čech cohomology**, $\check{H}^\bullet(\mathcal{U};\mathcal{F}) := H^\bullet(\mathcal{N}(\mathcal{U});\mathcal{F})$, of the sheaf. This has a natural parallel with the Čech co/homology of Chapter 4, and is stable under refinement of covers. The progression from sheaves on cellular complexes to topological spaces is clear in principle: one takes a *limit* of finer and finer covers. Since one can glue together local sections on cover elements, then, assuming a limit is possible, one anticipates the ability to assign *data* in the form of a group of local sections to *any* open subset of a topological space $X$. It would be confusing/redundant to call this a *singular sheaf* or even a *topological sheaf.* Perhaps *sheaf over a topology* would be the most descriptive term.

This discussion motivates the following definition. A **presheaf**, $\mathcal{P}$, on a space $X$, is an assignment to each open set $U \subset X$ of an abelian group $\mathcal{P}(U)$ of local data (or *sections*) along with restriction homomorphisms $\mathcal{P}(U \trianglelefteq V)\colon \mathcal{P}(U) \to \mathcal{P}(V)$ for every $V \subset U$. These restriction homomorphisms must satisfy composition and identity relations: $\mathcal{P}(U \trianglelefteq U) = \mathrm{Id}$ and

$$W \subset V \subset U \quad \Rightarrow \quad \mathcal{P}(U \trianglelefteq W) = \mathcal{P}(V \trianglelefteq W) \circ \mathcal{P}(U \trianglelefteq V),$$

*cf.* Equation (9.1). It is easy to confuse the directions of the inclusions in switching from celllular sheaves to sheaves on a topology. The reader will rightly suspect that a presheaf is *not quite* a sheaf. There are a number of subtleties associated with passing from a presheaf over a topological space to a sheaf. As a first example, note that stalks – pointwise data assignments – require a limiting process. For $x \in X$ and $\{U_i\}$ a nested sequence of open neighborhoods converging to $x$ (that is, any neighborhood of $x$ contains all $U_i$ for $i$ sufficiently large), one has a sequence of groups of sections and restriction homomorphisms

$$\cdots \longrightarrow \mathcal{P}(U_{i-1}) \longrightarrow \mathcal{P}(U_i) \longrightarrow \mathcal{P}(U_{i+1}) \longrightarrow \cdots \tag{9.8}$$

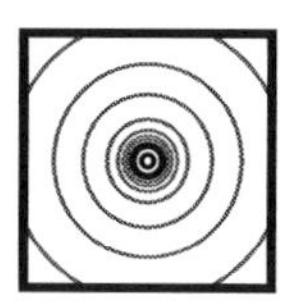

The **stalk** of $\mathcal{P}$ at $x$, $\mathcal{P}_x$, is the group of equivalence classes of sequences $(s_i)_i$ with $s_{i+1} = \mathcal{P}(U_i \trianglelefteq U_{i+1})(s_i)$ for all $i$; and where two such sequences are equivalent if they eventually agree: $[(s_i)] = [(s_i')]$ if and only if $s_i = s_i'$ for all $i$ large. One shows that $\mathcal{P}_x$ is a group and is independent of the system of neighborhoods chosen to limit to $x$. Examples of presheaves and their stalks include the following:

1. **Skyscrapers:** Fix a point $p \in X$ and consider the presheaf that sends an open set $U$ to 0 unless $p \in U$, for which it sends $U$ to $\mathbf{G}$, with all restriction homomorphisms being either Id or 0 depending on the presence of $p$; this presheaf has stalk 0 everywhere except for $\mathbf{G}$ at $p$.
2. **Constant functions:** The presheaf that assigns to open sets $U$ the group of constant functions $U \to \mathbf{G}$ is a presheaf with all restrictions being literal restrictions and all stalks isomorphic to $\mathbf{G}$.
3. **Smooth functions:** The presheaf that assigns to open sets $U$ of an $n$-manfiold $M$ the group $C^k(U;\mathbb{R})$ of $k$-times-differentiable real-valued functions (for $0 \leq k \leq \infty$) is a presheaf whose restriction homomorphisms are again restrictions; however, the stalks are the *germs*[2] of functions $\mathbb{R}^n \to \mathbb{R}$.
4. **Local homology:** The presheaf that assigns to open sets $U \subset X$ the local $k^{\text{th}}$ homology $H_k(X, X-U)$ has restriction homomorphisms induced via functoriality: for $V \subset U$, the induced homomorphism $H_k(X, X-U) \to H_k(X, X-V)$ behaves appropriately. The stalk of this presheaf at $p \in X$ is the local homology $H_k(X, X-p)$, which can vary greatly from point-to-point.
5. **Orientations:** For an $n$-manifold $M$, the (homology) **orientation presheaf** $\mathcal{O}$ assigns to $U$ open the local homology $H_n(M, M-U;\mathbb{Z})$. When this is a constant sheaf, the manifold is orientable (*cf.* Examples 4.18 and 9.2).
6. **Fibers:** Given $h\colon X \to Y$, the $k^{\text{th}}$ cohomology fiber presheaf over $Y$ assigns to $U \subset Y$ the cohomology $H^k(h^{-1}(U))$ of the inverse image. This presheaf can vary dramatically with change in $U$. For a sufficiently tame $h$, the stalk at $y \in Y$ is equal to $H^k(h^{-1}(y))$.

A sheaf is a presheaf that respects *gluings* as well as restrictions. In the cellular setting, this is accomplished by *fiat* as per Equation (9.2). In the topological setting, gluing, as with stalks, requires a limiting process. The definition is this: a presheaf $\mathcal{F}$ is a **sheaf** if and only if for any open $U$ and $\mathcal{U} = \{U_i\}$ a finite open cover of $U$, and for any local sections $s_i \in \mathcal{F}(U_i)$, which agree on all pairwise overlaps $U_{ij} = U_i \cap U_j$ via restriction maps, there *exists* a *unique* global section $s_U \in \mathcal{F}(U)$ which restricts to $s_i$ on each $U_i$. Note the double criterion of existence and uniqueness of global sections from local. This **gluing axiom** can be succinctly summarized as follows – for each such cover of $U = \cup_i U_i$, the following sequence is exact:

$$0 \longrightarrow \mathcal{F}(U) \xrightarrow{\mathcal{F}(U \trianglelefteq U_k)} \prod_k \mathcal{F}(U_k) \xrightarrow{d} \prod_{i,j} \mathcal{F}(U_{ij}) \ . \tag{9.9}$$

[2] Such germs are generalizations of Taylor series consisting of equivalence classes of functions that locally agree.

Here, the map $d$ is the coboundary map for cellular sheaf cohomology on the nerve $\mathcal{N}(\mathcal{U})$, *cf.* Equation (9.4): the $U_i$ correspond to vertices, the $U_{ij}$ to edges, and an orientation is chosen to determine the correct $\pm$ sign. Exactness means that $\mathcal{F}(U) = \ker d$. Thus, a good way to remember the gluing axiom is in terms of cohomology: to be a sheaf, $\mathcal{F}(U)$ must agree with the cellular sheaf cohomology $H^0(\mathcal{N}(\mathcal{U}); \mathcal{F})$ for any finite open cover $\mathcal{U}$ of $U$.

Every example of a presheaf in the list above is a sheaf *except* one. The canonical example of a non-sheaf presheaf is that of *constant* functions. This violates the existence property of gluing as follows. Given a disjoint union $U = U_1 \sqcup U_2$ and the cover $\mathcal{U} = \{U_1, U_2\}$, any constant functions on $U_1$ and $U_2$ trivially agree on the (empty) overlap. However, these local sections do not extend to a global section – a constant function on $U$ – if the values on $U_1$ and $U_2$ differ. In contrast, the presheaf of *locally* constant functions *is* a sheaf: the **constant sheaf**. Bounded real-valued maps are also examples of presheaves that are not sheaves, since maps that are locally bounded may not be globally bounded: *locally* bounded maps do form a sheaf.

The definition of cohomology $H^\bullet(X; \mathcal{F})$ for a sheaf over a topological space $X$ is more implicit than in the cellular setting. One approach takes $\mathcal{U}$ an open cover of $X$ and computes the Cech cohomology $\check{H}^\bullet(\mathcal{U}; \mathcal{F})$ of the sheaf subordinated to the cover. This clearly works for $H^0$ by dint of the gluing axiom, above. One can show that for sufficiently fine covers over sufficiently well-behaved sheaves, the higher cohomology stabilizes as well, and the limit is well-defined. The details of this and parallel constructions for sheaf cohomology can be found in [157, 183]. Suffice to say that most of the forms of cohomology seen in this text are expressible as a sheaf cohomology – either cellular or topological.

**Example 9.11 (Differential equations)** ◎

Many sheaves on manifolds can be generated by partial differential equations. Consider for example the sheaf of holomorphic functions on $\mathbb{C}$ that assigns to $U \subset \mathbb{C}$ open the functions $f \colon U \to \mathbb{C}$ satisfying $\partial f / \partial \bar{z} = 0$, where $\partial / \partial \bar{z}$ is the linear first-order partial differential operator which, when written out in real/imaginary components yields the more familiar *Cauchy-Riemann* PDEs. The condition of being holomorphic is local in nature, and the restriction of a holomorphic function is clearly holomorphic. Crucially, extension of holomorphic functions that agree on overlaps – *analytic continuation* – also holds, guaranteeing that holomorphic functions form a sheaf. Much of the impetus for sheaf cohomology came from problems of analytic continuation, viewed as producing global sections of this sheaf. Sheaf cohomology provides obstructions to analytic continuation, not just on $\mathbb{C}$, but on complex manifolds locally modelled on $\mathbb{C}^n$.

Other systems of PDEs arise in calculus on manifolds. The sheaf that associates to open sets $U$ of a manifold $M$ the $p$-forms $\Omega^p(U)$ is well-defined – restriction and extension of forms is sensible. These $p$-forms are linear objects that satisfy the partial differential equations $d^2 = 0$. The resulting sheaf cohomology is identical to the de Rham cohomology ${}_{dR}H^\bullet(M)$ of §6.9. ◎

## 9.7 Operations on sheaves

After working with enough examples of sheaves, certain common features resolve into canonical constructions. These steps are the beginnings of sheaf *theory* as opposed to working with sheaves one-at-a-time. This theory – a means of working with data over spaces in a platform-independent manner – quickly demands the development of tools and language beyond the scope of this chapter. A few basic definitions and examples will have to suffice until the next chapter brings that language to bear in §10.9.

**Morphisms:** A **sheaf morphism** $\eta\colon \mathcal{F} \to \mathcal{F}'$ between two sheaves on $X$ is a transformation between local sections that respects restrictions. That is, for each cell $\tau$ (or open set $U$) there is a homomorphism $\eta\colon \mathcal{F}(\tau) \to \mathcal{F}'(\tau)$ (resp. $\eta\colon \mathcal{F}(U) \to \mathcal{F}'(U)$) and for each $\sigma \trianglelefteq \tau$ (resp. $U \supset V$) the sheaf restriction maps commute with $\eta$:

$$\begin{array}{ccc} \mathcal{F}(\sigma) & \xrightarrow{\mathcal{F}(\sigma\trianglelefteq\tau)} & \mathcal{F}(\tau) \\ \eta\downarrow & & \downarrow\eta \\ \mathcal{F}'(\sigma) & \xrightarrow[\mathcal{F}'(\sigma\trianglelefteq\tau)]{} & \mathcal{F}'(\tau) \end{array} \qquad \begin{array}{ccc} \mathcal{F}(U) & \xrightarrow{\mathcal{F}(U\trianglelefteq V)} & \mathcal{F}(V) \\ \eta\downarrow & & \downarrow\eta \\ \mathcal{F}'(U) & \xrightarrow[\mathcal{F}'(U\trianglelefteq V)]{} & \mathcal{F}'(V) \end{array} \tag{9.10}$$

Sheaf morphisms are ways of transforming or evolving data over the same base space. Other sheaf operations answer the question of how to translate data on one space over to data on another by way of a mapping of base spaces.

**Direct image:** Assume a map $h\colon X \to Y$ of spaces or a cellular map on cell complexes. For $\mathcal{F}_X$ a sheaf over $X$, the **direct image** of $h$ is the sheaf $h_*\mathcal{F}_X$ on $Y$ that *pushes* data from $X$ to $Y$ by pulling back space via $h^{-1}$. In the topological setting, this means $h_*\mathcal{F}_X(U) = \mathcal{F}_X(h^{-1}(U))$, for $U \subset Y$ open. In the cellular setting, this means $h_*\mathcal{F}_X(\sigma) = \mathcal{F}_X(h^{-1}(\sigma))$ for $\sigma$ a cell of $Y$. For $h$ injective, one visualizes the direct image as embedding the data over $X$ into $Y$; for $h$ surjective, one might think of the direct image sheaf as being the data on $X$, *folded* atop $Y$.

**Inverse image:** For $h\colon X \to Y$ as before, one can *pull back* a sheaf from $Y$ to $X$. The **inverse image** of a sheaf $\mathcal{F}_Y$ on $Y$ is the sheaf on $X$ defined neatly in the cellular setting by $h^*\mathcal{F}_Y(\sigma) = \mathcal{F}_Y(h(\sigma))$ for $\sigma$ a cell of $X$. For $h$ cellular, this is well-defined. The difficulty arises in the topological setting. One wants to define $h^*\mathcal{F}_Y(U) = \mathcal{F}_Y(h(U))$ for $U \subset X$ open; however, $h(U)$ may not be open. Therefore, one takes a limit. Let $V_i$ be a system of nested open sets in $X$ with $\cap_i V_i = \overline{h(U)}$. Then the restriction maps give a sequence of abelian groups $\mathcal{F}_Y(V_i) \to \mathcal{F}_Y(V_{i+1})$. In the same manner that stalks over a point were defined in terms of equivalence classes of sequences of group elements, the larger subset $\overline{h(U)}$ has as its inverse image the 'limit' $h^*\mathcal{F}_Y(U) := \lim_{i\to\infty} \mathcal{F}_Y(V_i)$ consisting of equivalence classes of sequences of elements in $\mathcal{F}_Y(V_i)$.

In the case of direct and inverse images, one shows that the definitions lead to obvious restriction maps and the resulting presheaves are in fact sheaves. At first, these operations serve to increase the precision of language in defining and characterizing sheaves. Only later is it clear that these operations are at the heart of sheaf theory.

**Example 9.12 (Compact cohomology, redux)** ◎

Recall from §9.3 that for $A$ a subcomplex of $X$, the computation of cellular sheaf cohomology of $A$ proceeds by *"restricting"* the sheaf $\mathcal{F}$ to $A$, yielding $\mathcal{F}_A$, then computing $H^\bullet(X; \mathcal{F}_A)$. This *ad hoc* definition can now be specified and improved. Let $\iota: A \hookrightarrow X$ be the inclusion. Then one defines the cohomology of $\mathcal{F}$ on $A$ to be $H^\bullet(A; \iota^*\mathcal{F})$. This not only has the benefit of providing a sheaf on $A$, it easily extends to the setting where $A$ is not a subcomplex of $X$. Given any collection of (open) cells $A \subset X$, the **sheaf cohomology with compact supports** on $A$ is defined to be $H_c^\bullet(A; \mathcal{F}) := H^\bullet(X; \iota_*\iota^*\mathcal{F})$.

The notation $H_c^\bullet$ denotes, as in §6.4, compact supports. The need for this distinction is easily discerned with, say, a constant sheaf. If $\mathcal{F} = \mathbf{G}_X$ is a constant sheaf, then the cohomology $H_c^\bullet(A; \mathcal{F})$ is isomorphic to the singular compactly supported cohomology $H_c^\bullet(A; \mathbf{G})$ of $A$ as a topological space, so that for an open $n$-simplex $\sigma$ in $X$, $H_c^\bullet(\sigma; \mathbf{G}_X)$ vanishes except in grading $n$. Of course, if $A$ is a closed subcomplex of $X$, then $H_c^\bullet(A; \mathcal{F}) \cong H^\bullet(A; \mathcal{F})$. ◎

One strategy in sheaf theory is to manipulate data structures over spaces by pushing or pulling sheaves in a way that the impact on sheaf cohomology is controlled. In certain instances, simple general conditions guarantee such control.

**Theorem 9.13 (Vietoris Mapping Theorem).** *Assume $\mathcal{F}$ a sheaf on $X$ and $h: X \to Y$ a proper map with fibers satisfying $H^n(h^{-1}(y); \mathcal{F}) = 0$ for all $n > 0$. Then $h$ induces an isomorphism on sheaf cohomology: $H^\bullet(X; \mathcal{F}) \cong H^\bullet(Y; h_*\mathcal{F})$.*

In particular, if the fibers of $h$ are all zero-dimensional, then the acyclicity hypothesis of the theorem is always satisfied. The theorem holds for cellular sheaves also with obvious modifications.

**Example 9.14 (Flow sheaves, redux)** ◎

The flow sheaves of §9.4 can be represented as pushforwards of simple component sheaves of two types. To model a single unit of flow that circulates over a network (with feedback edge), consider first $S$, a circle, subdivided into a directed graph (with a single cycle), and let $\mathcal{F}_S = \mathbb{R}_S$ be the constant sheaf on $S$. Next, let $I = [0, 1]$ be an interval, subdivided into a directed linear graph with (potentially many) edges, and let $\mathcal{F}_I = \mathbb{R}_{(0,1]}$ denote the constant sheaf on $I$ that is set to $\mathbb{R}$ everywhere except zero at the left endpoint. With a bit of insight, one sees that every flow sheaf is precisely the pushforward of some finite number of copies of $\mathcal{F}_S$ and $\mathcal{F}_I$ over a disjoint union of circles and intervals, with the projection map $p$ gluing together subintervals in an orientation-preserving manner.

This clarifies the computation of flow sheaf cohomology. By Poincaré duality, $\mathcal{F}_S$ has cohomology $H^0 \cong H^1 \cong \mathbb{R}$. It is clear that $H^\bullet(I; \mathcal{F}_I) = 0$, since there are no global sections and the coboundary map $d$ is surjective. The projection map $p$ has discrete fibers; hence, by Theorem 9.13, flow sheaves have $H^0 \cong H^1$ and the dimension of this cohomology is precisely the flow value: *cf.* Lemma 9.8. ◎

**Example 9.15 (Euler characteristic, redux)** ⊙

There are other operations that apply to sheaves. Following the perspective that a sheaf is an algebraic enhancement of a topological space, one repeats standard topological constructs. Among the most primal is that of the Euler characteristic. For a cellular sheaf $\mathcal{F}$ on a cell complex $X$ taking values in vector spaces, one defines the Euler characteristic in the obvious way:

$$\chi(\mathcal{F}) := \sum_{\sigma} (-1)^{\dim \sigma} \dim \mathcal{F}(\sigma) = \sum_{n=0}^{\infty} (-1)^n \dim H^n(X; \mathcal{F}). \tag{9.11}$$

How does the sheaf-theoretic Euler characteristic relate to the familiar $\chi$ of previous chapters? Recall that the combinatorial definition of Euler characteristic used in the o-minimal setting of Chapter 3 is *not* the homological definition of Corollary 5.18 in general: the homological formula holds only for compact spaces, on which $\chi$ is, like homology, a homotopy invariant. The combinatorial $\chi$ has an interpretation in terms of compactly-supported cohomology. For $A$ a locally-compact definable space,

$$\chi(A) = \sum_{n=0}^{\infty} (-1)^n \dim H_c^n(A). \tag{9.12}$$

This can be interpreted at the level of sheaves. For $A$ a definable set represented as a subcollection $\iota \colon A \hookrightarrow X$ of open cells in a definable triangulation of $X$, the Euler characteristic of $A$ can be expressed as the Euler characteristic of the constant sheaf $\mathbb{R}_A$ on $A$:

$$\chi(A) = \chi(A; \mathbb{R}_A) = \chi(X; \iota_* \iota^* \mathbb{R}_X) = \sum_{n=0}^{\infty} (-1)^n \dim H_c^n(A; \mathbb{R}_A). \tag{9.13}$$

⊙

## 9.8 Sampling and reconstruction

Sheaf morphisms play a role in Robinson's recent sheaf-theoretic reinterpretation of the Nyquist-Shannon sampling theorem [255]. The classical result in sampling theory says the following. Consider a signal $f \colon \mathbb{R} \to \mathbb{R}$ whose values are known only on some periodic sampling, say $f(n)$ for $n \in \mathbb{Z}$. Under what conditions can $f$ be reconstructed from the samples? A little Fourier analysis yields the classical result: if $f$ is **bandlimited** – that is, the Fourier transform $\hat{f} \colon \mathbb{R} \to \mathbb{C}$ of $f$ has compact support on the interval $[-\frac{1}{2}, \frac{1}{2}]$ – then reconstruction of $f(t)$ from $f(\mathbb{Z})$ is (constructively) possible. This is useful in knowing, *e.g.*, how to encode and transmit human voice for maximum clarity over the telephone/internet; how to filter signals for transmission; and how to compress images and video. The Nyquist rate – that one must sample at a frequency at least twice the signal bandwidth – is iconic in signal processing, and all manner of attempts to generalize it abound in the literature on signal and image processing.

This venerable barrier to reconstruction has a modern topological interpretation [255]. Let $X$ be a cell complex and $\mathcal{F}$ a cellular sheaf on $X$ taking values in abelian

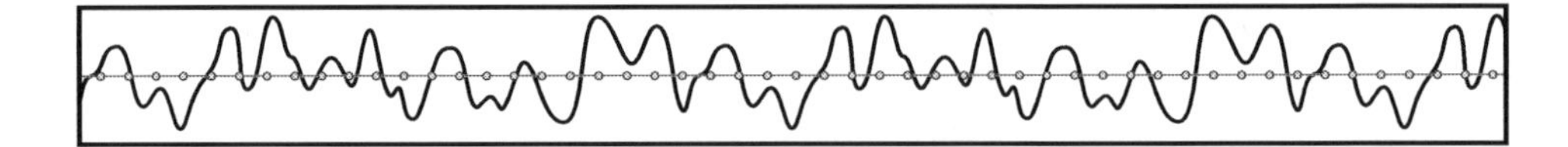

groups. One considers the global sections, $\mathcal{F}(X)$, to be the *signals*: in this section, the desiderata. Sampling will be interpreted not necessarily as a restriction to a subcomplex, but rather as a potentially more complex sheaf morphism, supported on a subcomplex. Define a **sampling** of $\mathcal{F}$ on a (closed) subcomplex $X_0 \subset X$ to be a surjective sheaf morphism $\mathcal{S}\colon \mathcal{F} \to \mathcal{F}_0$ from $\mathcal{F}$ to a **sampling sheaf** $\mathcal{F}_0$ on $X$, supported on $X_0$. Surjectivity means that on each cell of $X_0$, $\mathcal{S}$ is surjective onto that stalk. Such a sampling induces an **ambiguity sheaf**, $\mathcal{A} := \ker \mathcal{S} = \ker (\mathcal{F} \to \mathcal{F}_0)$, on $X$; the restriction maps of $\mathcal{A}$ are inherited from $\mathcal{F}$. Because of surjectivity of $\mathcal{S}$ and the definition of $\mathcal{A}$ as a kernel, one has the short exact sequence of sheaves on $X$: $0 \to \mathcal{A} \to \mathcal{F} \xrightarrow{\mathcal{S}} \mathcal{F}_0 \to 0$. This generates as per Equation (9.6) the long exact sequence

$$0 \longrightarrow H^0(X;\mathcal{A}) \longrightarrow H^0(X;\mathcal{F}) \xrightarrow{H(\mathcal{S})} H^0(X_0;\mathcal{F}_0) \xrightarrow{\delta} H^1(X;\mathcal{A}) \longrightarrow \cdots$$

One has the following immediate interpretation. The given sampling is a global section of $\mathcal{F}_0$. The desired signal is a corresponding global section of $\mathcal{F}$. The ability to reconstruct the original signal from the sample is equivalent to having $H(\mathcal{S})$ invertible. By exactness, one observes the following: (1) $H(\mathcal{S})$ is injective iff $H^0(X;\mathcal{A}) = 0$; (2) $H(\mathcal{S})$ is surjective iff $\delta = 0$. Otherwise said, $H^0(X;\mathcal{A})$ is the obstruction to sampling reconstruction, whereas the connecting homomorphism $\delta$ indicates the degree of redundancy in the sampling.

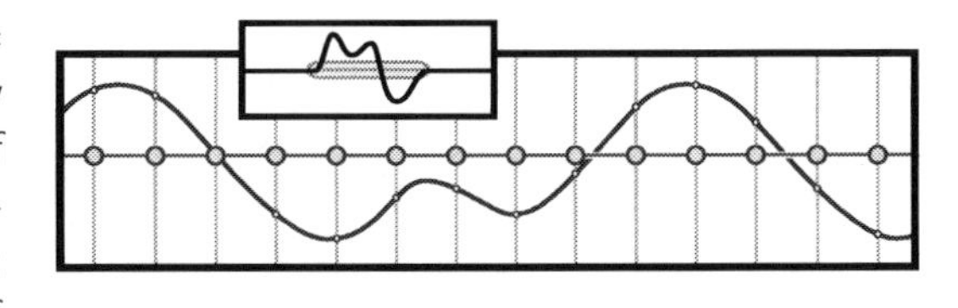

For the example of the classical Nyquist-Shannon theorem: $X = \mathbb{R}$, with the cell structure induced by $X_0 = \mathbb{Z}$; $\mathcal{F}$ is the constant sheaf of continuous $\mathbb{C}$-valued functions supported on a fixed interval $[-B, B]$; $\mathcal{F}_0 = \mathbb{C}_{X_0}$ is the constant $\mathbb{C}$-sheaf over $X_0$; and the sampling morphism $\mathcal{S}\colon \mathcal{F} \to \mathcal{F}_0$ is the inverse Fourier transform $\mathcal{S}(n, \hat{f}) = \int_{-B}^{B} \hat{f}(\xi)e^{-2\pi i n\xi}d\xi$. One thinks of $X$ as the frequency domain, with a section of $\mathcal{F}$ *not* as the original signal $f(x)$, but as its Fourier transform $\hat{f}(\xi)$. Then, the sampling morphism, being the inverse Fourier transform, yields the values of $f(n)$ for $n \in \mathbb{Z}$ – these are the samples of the original signal $f(x)$.

In this instantiation, the ambiguity sheaf, $\mathcal{A} = \ker \mathcal{S}$, has stalk over $n \in \mathbb{Z}$ equal to the subgroup of all band-limited functions whose $n^{\text{th}}$ Fourier coefficient vanishes. The global sections of $\mathcal{A}$, elements of $H^0(X;\mathcal{A})$, therefore consist of bandlimited functions $\hat{f}\colon [-B, B] \to \mathbb{C}$ for which $\int_{-\infty}^{\infty} \hat{f}(\xi)e^{-2\pi i n\xi}d\xi = 0$ for all $n$. Basic Fourier theory says that for $B \leq \frac{1}{2}$, all the Fourier coefficients of $\hat{f}$ vanish, as must $\hat{f}$. Thus, $H^0(X;\mathcal{A}) = 0$ and signal reconstruction is possible. When $B < \frac{1}{2}$, $\delta \neq 0$, meaning that oversampling has occurred: the $B = \frac{1}{2}$ case is sharp.

This is simply a reinterpretation of Nyquist-Shannon into sheaf-theoretic terms, and neither replaces nor improves the harmonic-analysis proof; as well, using the constant sheaf on $\mathbb{R}$ does not reveal what the long-exact sequence on sheaf cohomology is really capable of. However, the benefits of this more general interpretation are worth noting: the result holds for other kinds of samplings which are not amenable to harmonic analysis, as well as to other base spaces than Euclidean $\mathbb{R}^n$ [255].

**Example 9.16 (Cut samples; Flow signals)** ◎
If $\tilde{X}$ is taken to be the subdivision of a directed network $X$ as in §9.4 and $\mathcal{F}$ a network flow sheaf, then a valid sampling morphism is $\mathcal{S}\colon \mathcal{F} \to \mathcal{F}_{\tilde{C}}$, where $\mathcal{F}_{\tilde{C}}$ is the restriction of $\mathcal{F}$ to an edge-cut $C$, suitably subdivided so that $\tilde{C}$ is a closed subcomplex of $\tilde{X}$. In this case, $H^\bullet(\tilde{X}; \mathcal{F}_{\tilde{C}}) = H^\bullet(\tilde{C}; \mathcal{F})$. The ambiguity sheaf in this case is $\ker \mathcal{S}$, which yields the relative cohomology $H^\bullet(\tilde{X}; \mathcal{A}) = H^\bullet(\tilde{X}, \tilde{C}; \mathcal{F})$. One recovers the results from §9.4: $H^0(\tilde{X}, \tilde{C}; \mathcal{F})$ classifies the obstruction to signal reconstruction (*is $C$ a cut?*) and the redundancy of the sampling (*is the cut minimal?*) is measured by $\delta$. The expressiveness of sheaf-theoretic language allows one to think of both the Max-Flow-Min-Cut and the Nyquist-Shannon theorems as expressions of the same principle, wherein the cut is a *sampling* of the flow sheaf *signal*. ◎

## 9.9 Euler integration, redux

The Euler integration of Chapter 3, though simple enough to define combinatorially, has its origins in sheaf theory, through independent work of Kashiwara [190] and MacPherson [215] in the 1970s. The interpretation as a calculus was envisaged and promoted by Schapira [269] and Viro [297] independently years later. The subject was independently rediscovered as combinatorics [262, 268] in a more limited form. The sheaf-theoretic version has impacted algebraic geometry [172] and motivic integration [67, 92] independent of its newfound utility in signal/data processing.

Euler calculus is built on a basis of constructible sheaves – these differ slightly from the cellular sheaves introduced thus far in that cellular sheaves begin with a fixed cell structure, but constructible sheaves do not fix the cell structure in advance, but use tameness to imply a cell structure. For simplicity, fix an o-minimal structure as in §3.5 and let $X$ be a definable space. A **constructible sheaf** on $X$ is a cellular sheaf for some definable triangulation of $X$.

The idea behind the Euler integral is that one converts a constructible function $h\colon X \to \mathbb{Z}$ into a sheaf $\mathcal{F}_h$ whose Euler characteristic *is* the integral of $h$ with respect to $d\chi$. This procedure is simple in the case where $h\colon X \to \mathbb{N}$. On each simplex $\sigma$ in a tame triangulation of $X$, define $\mathcal{F}_h$ to be the $\mathbb{R}$-vector space of dimension $h(\sigma)$. All the restriction maps vanish. Put simply, $\mathcal{F}_h$ is a sum of skyscraper sheaves over the simplices that converts $h$-values to dimensions. Then, by construction,

$$\chi(\mathcal{F}_h) = \chi\left(\bigoplus_\sigma \mathbb{R}^{h(\sigma)}\right) = \sum_\sigma (-1)^{\dim \sigma} h(\sigma) = \int_X h\, d\chi.$$

The only difficulty is in how to adapt the definition to integrands $h\colon X \to \mathbb{Z}$ that can take on both positive and negative values. This is accomplished by switching from individual sheaves $\mathcal{F}_h$ to complexes of sheaves $\mathcal{F}_h^\bullet$ of the form

$$\mathcal{F}_h^\bullet = \cdots \longrightarrow 0 \longrightarrow \mathcal{F}_h^- \longrightarrow \mathcal{F}_h^+ \longrightarrow 0 \longrightarrow \cdots$$

where $\mathcal{F}_h^+$ the sum of skyscraper sheaves for the restriction of $h$ to $\{h > 0\}$ and $\mathcal{F}_h^-$ the sum of skyscraper sheaves for the restriction of $-h$ to $\{h < 0\}$. The grading in the sequence above is chosen so that

$$\chi(\mathcal{F}_h^\bullet) = \chi(\mathcal{F}_h^+) - \chi(\mathcal{F}_h^-) = \int_X h\, d\chi.$$

The sheaf-theoretic perspective provides motivation for some of the otherwise esoteric-seeming results of Chapter 3. The Fubini-type result (Theorem 3.11) is nothing more than the commutativity of $\chi$-on-sheaves with the direct image operation. Likewise, the Fourier-Sato transform, $\mathcal{F}_\mathcal{S}$, of Example 3.18, is a particular instance of the general **Fourier-Sato transform** of sheaves [191] in the setting of constructible functions. The student who progresses in sheaf theory may find it helpful to translate as much of the general machinery as possible into specific examples from Euler calculus.

## 9.10 Cosheaves

Sheaf theory is built for cohomology; one suspects that the homological counterpart should be and be by no means unimportant. This dual is called a **cosheaf** and has its own set of emerging applications.

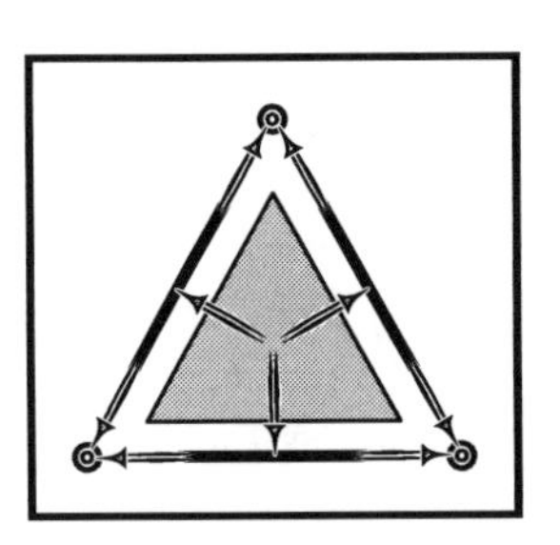

A cosheaf is a data structure over a space that encodes extension rather than restriction. For a cell complex $X$, a cosheaf $\hat{\mathcal{F}}$ assigns to each cell $\sigma$ an abelian group $\hat{\mathcal{F}}(\sigma)$ and to face pairs $\sigma\trianglelefteq\tau$ an **extension map** – a homomorphism $\hat{\mathcal{F}}(\sigma\trianglelefteq\tau)$ which sends $\hat{\mathcal{F}}(\tau) \to \hat{\mathcal{F}}(\sigma)$ and composes properly: for $\rho\trianglelefteq\sigma\trianglelefteq\tau$, the extension maps satisfy $\hat{\mathcal{F}}(\rho\trianglelefteq\tau) = \hat{\mathcal{F}}(\rho\trianglelefteq\sigma) \circ \hat{\mathcal{F}}(\sigma\trianglelefteq\tau)$.

One imagines extending data from cells to faces. Note how this flips directions in the definition of a sheaf. Continuing the pattern of appending the *co-* prefix to connote this contravariance, one calls the data $\hat{\mathcal{F}}(\sigma)$ over a cell the **costalk**. From this basis of costalks, one assembles the full cosheaf structure on $X$. In analogy with Equation (9.2), one defines:

$$\hat{\mathcal{F}}(X) := \bigoplus_\tau \hat{\mathcal{F}}(\tau) \Big/ \sim \quad : \quad s_\rho \sim \hat{\mathcal{F}}(\rho\trianglelefteq\sigma)(s_\sigma). \tag{9.14}$$

This is much harder to parse than the global-sections intuition of a sheaf over $X$: the cosheaf over $X$ is an equivalence class of data assignments to cells of $X$, where the equivalence is up to compatible extensions that fall from higher-dimensional to lower-dimensional cells.

As a simple example, consider the skyscraper cosheaf $\hat{\mathbf{G}}_\sigma$ that assigns an abelian group $\mathbf{G}$ to a cell $\sigma$ and 0 to all other cells. The value of this cosheaf over all of $X$ is zero *unless* $\sigma$ has no cofaces (*i.e.*, it is locally top-dimensional), in which case the global sections are $\hat{\mathcal{F}}(X) \cong \mathbf{G}$. Recall: global nonzero sections of a skyscraper *sheaf* exist only if the cell is minimal in the face poset (a vertex); for a skyscraper *cosheaf*, this is turned upside down and only maxima in the face poset contribute globally.

**Example 9.17 (Sensing and inference)** ⊙

The duality between sheaves and cosheaves is expressed in applications to sensing. Consider the problem of integrating sensor data into an inference. Assume that for, say, a finite collection of sensors $x_i \in \mathcal{Q}$, each returns some sensed attributes $S(x_i) \subset \mathcal{S}$ about a fixed object $z \in \mathcal{O}$. The set of nodes defines a simplex $X$ (of dimension $|\mathcal{Q}|-1$) that will act as a base space. Consider the sheaf on $X$ taking values in vector spaces defined as follows. To each vertex $x_i$ is assigned the vector space whose basis is the finite set $S(x_i)$ of attributes: for example, *the target is grey and large*; *the target has a long nose*. The restriction maps act as inclusions to vector spaces obtained by taking the unions over the vertices of the associated simplex. The global sections $\mathcal{F}(X)$ give the full sensorium. *Sensing yields a sheaf.*

Dually, there is a cosheaf defined in terms of what the sensors allow one to conclude. Consider the assignment of a vector space $\hat{\mathcal{F}}(x_i)$ to each vertex $x_i \in \mathcal{Q}$ whose basis is all objects in $\mathcal{O}$ consistent with $S(x_i)$: *rhinoceros, elephant, whale*; *anteater, elephant*. The extension maps are projections to the subspace defined by a subbasis generated by intersection: *elephant*. The global sections of this cosheaf $\hat{\mathcal{F}}(X)$ return constraint satisfactions consistent with sensing. *Inference is a cosheaf.* One suspects this cartoonish example can be greatly generalized. ⊙

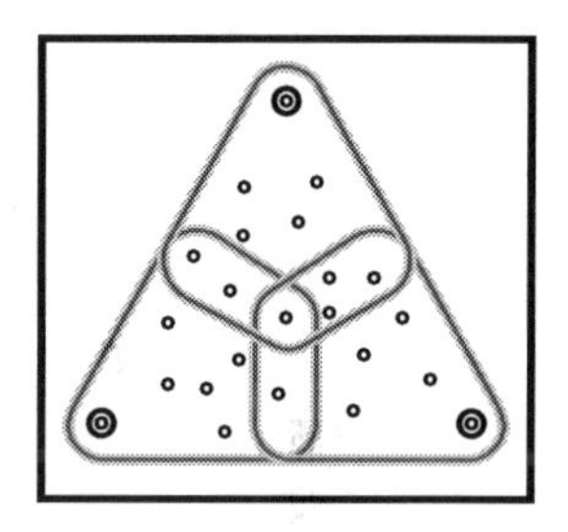

As foreshadowed, cosheaves are built for homology. The obvious cellular cosheaf chain complex is given as,

$$\cdots \xrightarrow{\partial} \bigoplus_{\dim \sigma=2} \hat{\mathcal{F}}(\sigma) \xrightarrow{\partial} \bigoplus_{\dim \sigma=1} \hat{\mathcal{F}}(\sigma) \xrightarrow{\partial} \bigoplus_{\dim \sigma=0} \hat{\mathcal{F}}(\sigma) \longrightarrow 0 \tag{9.15}$$

where the boundary map is:

$$\partial(\tau) = \sum_{\sigma \lhd \tau} [\sigma\colon \tau]\, \hat{\mathcal{F}}(\sigma \lhd \tau). \tag{9.16}$$

Notice that $\partial$ *decreases* the grading by one and that $\partial^2 = 0$, as a boundary operator must. The resulting **cosheaf homology**, $H_\bullet(X; \hat{\mathcal{F}})$ gives a collation and classification of homological features in the data structure $\hat{\mathcal{F}}$.

If $\mathbf{G}_X$ is the constant cosheaf on $X$, then the cosheaf homology agrees with the cellular homology with $\mathbf{G}$ coefficients. For a skyscraper cosheaf $\hat{\mathcal{F}}_\sigma$ the cosheaf homology vanishes unless $\sigma$ is a maximal cell in the face poset (a cell of locally top dimension), in which case $H^\bullet(X; \hat{\mathcal{F}}_\sigma) \cong \mathbf{G}$ in grading dim $\sigma$. This duality between dimensions and sheaf cohomology / cosheaf homology on skyscrapers is reminiscent of the delicacies of duality between homology and compactly supported cohomology from Chapter 6.

**Example 9.18 (Precosheaves)** ◎

In the topological setting, a **precosheaf** assigns data to open sets $U \subset X$ and for a subset $V \subset U$ the extension map $\hat{\mathfrak{F}}(U \trianglelefteq V)$ acts as $\hat{\mathfrak{F}}(V) \to \hat{\mathfrak{F}}(U)$ with the corresponding composition in the case of $W \subset V \subset U$. To be a cosheaf, a precosheaf must behave well under limits of covers: The analogue of Equation (9.9) for cosheaves is the dual condition, that

$$\bigoplus_{i,j} \hat{\mathfrak{F}}(U_{ij}) \xrightarrow{\partial} \bigoplus_k \hat{\mathfrak{F}}(U_k) \xrightarrow{\hat{\mathfrak{F}}(U \trianglelefteq U_k)} \hat{\mathfrak{F}}(U) \longrightarrow 0 . \tag{9.17}$$

is exact for any finite open cover $\mathcal{U} = \{U_i\}$ of $U$. In other words, $\hat{\mathfrak{F}}(U)$ agrees with the cellular cosheaf homology $H_0(\mathcal{N}(\mathcal{U}); \hat{\mathfrak{F}})$ for any cover $\mathcal{U}$. In practice, this agreement *cannot* be assumed and can be challenging to confirm.

Many of the examples of sheaves over a topology come from functions: continuous functions, smooth functions, vector fields, differential forms, etc. The dual cosheaf notions are generated (taking a hint from Poincaré duality) by restricting to functions of compact support. For example, let $M$ be a manifold. Then the precosheaf on $M$ of compactly supported continuous functions $\hat{\mathfrak{F}}(U) = C_c(U, \mathbb{R})$ with the extension map from $V$ to $U \supset V$ being extension-by-zero is a cosheaf. Likewise, the sheaf of differential $p$-forms $\Omega^p$ on $M$ is complemented by the cosheaf of $p$-currents $\Omega_p$ (see §6.11). ◎

**Example 9.19 (Homology fiber cosheaves)** ◎

One simple example of a cosheaf comes from a map $f\colon X \to Y$ where $Y$ is a locally connected space. The precosheaf that assigns to $U \subset Y$ the homology $H_k(f^{-1}(U))$ is in fact a cosheaf on $Y$, with extension arising from the induced homomorphism on $H_k$. The cellular case is analogous (*cf.* Example 9.3). ◎

## 9.11 Bézier curves and splines

The problem of patching together local data over a cell complex is commonly encountered by architects and designers working with polyhedral representations of objects. The subject is classical, as exemplified by the simple problem of how to specify a polynomial planar curve $\gamma$ between two endpoints with global control over the degree and local control at the endpoints. This is precisely the context in which a **Bézier curve** is appropriate.

For example, a planar Bézier curve is specified by the locations of the two endpoints, along with additional **control points**, each of which may be interpreted as a *handle* (or tangent vector at the endpoint) specifying derivative data of the resulting curve at each endpoint. The reader who has used any modern drawing software will have a visceral understanding of the control that these handles give over the resulting smooth curve. Most programs use a cubic Bézier curve in the plane – the image of a cubic polynomial $\vec{p}(t)$ for $0 \leq t \leq 1$. In these programs, the specification of the endpoints and the endpoint handles (tangent vectors) completely determines the curve uniquely.

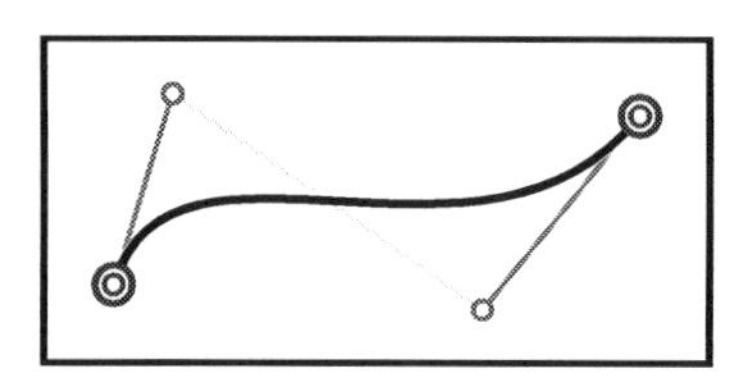

This can be viewed from the perspective of a cosheaf $\hat{\mathcal{S}}$ over the closed interval $I = [0,1]$. The costalk over the interior $(0,1)$ is the space of all cubic polynomials from $[0,1] \to \mathbb{R}^2$, which is isomorphic to $\mathbb{R}^4 \oplus \mathbb{R}^4$ (one cubic polynomial for each of the $x$ and $y$ coordinates). If one sets the costalks at the endpoints 0, 1 to be $\mathbb{R}^2$, the physical locations of the endpoints, then the obvious extension maps to the endpoints are nothing more than evaluation at 0 and 1 respectively. The corresponding cosheaf chain complex is:

$$\cdots \longrightarrow 0 \longrightarrow \mathbb{R}^4 \oplus \mathbb{R}^4 \xrightarrow{\partial} \mathbb{R}^2 \oplus \mathbb{R}^2 \longrightarrow 0.$$

Here, the boundary operator $\partial$ computes how far the cubic polynomial (edge costalk) 'misses' the specified endpoints (vertex costalks). It is clear that $H_0(\hat{\mathcal{S}}) = 0$, since $\partial$ is surjective. It is also clear that the space of global solutions is $H_1(\hat{\mathcal{S}}) \cong \mathbb{R}^2 \oplus \mathbb{R}^2$, meaning that there are four degrees of freedom available for a cubic planar Bézier curve with fixed endpoints: these degrees of freedom are captured precisely by the pair of handles, each of which is specified by a (planar) tangent vector. Note the interesting duality: the global solutions with boundary condition are characterized by the top-dimensional homology of the cosheaf, instead of the zero-dimensional cohomology of a sheaf.

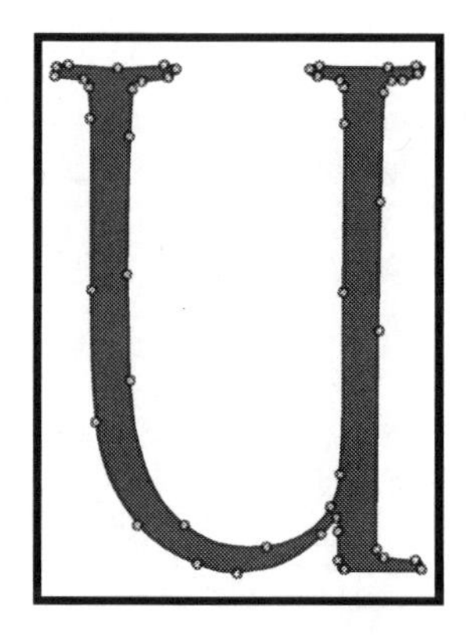

In practice, a compound curve drawn in any vector graphics software is a **spline** obtained by patching together segments of Bézier curves, usually cubic. Such curves are specified by handles at each endpoint of each segment. For a $C^1$ (differentiable) curve, one aligns the directions of the handles at segment endpoints; corners can also be specified, yielding a $C^0$ curve. This finite-dimensional representation of complex systems of curves is of foundational importance: all modern font systems[3] are based on such splines.

A cosheaf for a polynomial Bézier surface over a polygonal 2-cell with constraints on boundary points would be more challenging to write out, but would have meaningful homology in degree two. Patched together, such higher-dimensional polynomial splines are crucial in architecture and design, contributing immensely to both aesthetic and technical aspects of what can be built. This prompts the study and classification of splines with various constraints.

The following is a classification of simpler, $\mathbb{R}$-valued splines over simplicial manifolds, based on independent work of Billera [36] and Yuzvinsky [304] reinterpreted in

[3] True type fonts use quadratic Bézier splines; Postscript and Metafont use cubic Bézier splines.

the context of cellular cosheaves. Given $X \subset \mathbb{R}^n$ a simplicial complex realized with convex simplices, consider $P_m$ the ring of $\mathbb{R}$-valued polynomials in $n$ variables which have degree $\leq m$. By $S^r_m$ denote the (vector) space of degree $\leq m$ $\mathbb{R}$-valued splines on $X$ which are of global smoothness class $C^r$, for fixed $m > r \geq 0$. For example, a dome over a triangulated disc might be specified as an element of $S^1_4$ if it were built of quartic surface patches which must meet preserving first derivatives.

The following cosheaf captures the spline constraints. Let $\hat{\mathcal{S}}^r_m$ denote the cellular cosheaf on $X$ whose costalk on a simplex $\sigma$ is the quotient

$$\hat{\mathcal{S}}^r_m(\sigma) = P_m/(P_m \cap I^r(\sigma)) ,$$

where $I^r(\sigma)$ is the subspace of polynomials in $n$ variables which vanish on $\sigma$ (and thus on the affine space spanned by this convex simplex) but which have complementary degree at least $r+1$. The extension map is given as follows: for $\sigma \trianglelefteq \tau$, then $I^r(\tau) < I^r(\sigma)$ and there is a natural map on the quotient. This satisfies the cosheaf composition law and thus yields a well-defined cosheaf homology.

**Theorem 9.20 ([36]).** *If $X$ as above is an $n$-dimensional simplicial manifold-with-boundary, then $S^r_m \cong H_n(X - \partial X; \hat{\mathcal{S}}^r_m)$.*

Note that $\hat{\mathcal{S}}^0_1$ is precisely the setting of continuous piecewise-linear functions. In this case, it can be shown that $\dim S^0_1 = |V|$, the number of vertices of $X$. One can imagine the use of a long exact sequence to characterize constraints, the use of Euler characteristic to bound dimensions, and the use of exact sequences of cosheaves to decompose $S^r_m$: all this and more appear (implicitly) in [36].

## 9.12 Barcodes, redux

The full story of cosheaves and their applications has yet to be written. This chapter closes with an example from Curry [77] related to topological data analysis. Recall the setting of §5.13-5.15, in which a linear sequence of homologies and induced homorphisms forms, via indecomposables, infographics of parameterized homologies: barcodes. For example, consider the *zigzag* sequence of the form

$$\cdots \hookleftarrow X_i \hookrightarrow X_i \cup X_{i+1} \hookleftarrow X_{i+1} \hookrightarrow \cdots .$$

Passing to $k^{\text{th}}$ homology $H_k$ in field-$\mathbb{F}$ coefficients gives a sequence of $\mathbb{F}$-vector spaces with alternating linear transformations. This $k^{\text{th}}$ homology of the sequence is precisely a cosheaf over a base space $\mathbb{R}$, discretized by $\mathbb{Z} \subset \mathbb{R}$. Costalks over edges, $H_k(X_i; \mathbb{F})$, are mapped to costalks over vertices, $H_k(X_i \cup X_{i-1}; \mathbb{F})$ and $H_k(X_i \cup X_{i+1}; \mathbb{F})$. There are no higher-dimensional cells, and so composition is trivially satisfied.

It is not necessary to begin with alternating zigzag sequences. For example, the monotone sequences of §5.13 can be recast as a cosheaf over $\mathbb{R}$ by interweaving backward-pointing identity maps, as done with recurrence relations in Example 9.1. By such means, one can recast the theory of barcodes into the language of cosheaves. *A homology barcode is a cosheaf over $\mathbb{R}$.*

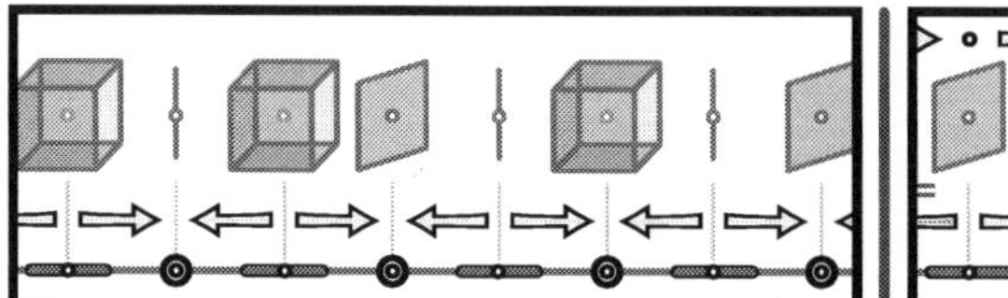

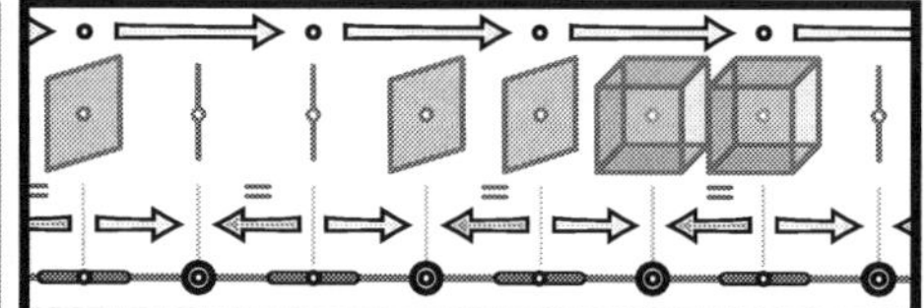

Homology cosheaves with field coefficients are classified completely by the Structure Theorem 5.21 into **cosheaf interval indecomposables** which are the constant cosheaves with costalk $\mathbb{F}$ over some interval. Note, however, that because of the decomposition of $\mathbb{R}$ into vertices and edges, these indecomposable intervals can have endpoints corresponding either to vertices or edges of the cell structure on $\mathbb{R}$. There are exactly four different types of bars that can arise as cosheaf interval indecomposables: a closed interval, an open interval, or (left/right) half-open intervals. Since cosheaves form coefficients for homology, one can sensibly speak of the homology of a cosheaf interval indecomposable. Over a base space of $\mathbb{R}$, the only homology that can be nonvanishing is $H_0$ and $H_1$. It is a simple exercise to compute:

**Lemma 9.21 ([77]).** *Cosheaf interval indecomposables and their homologies are:*

| interval | type | $H_0$ | $H_1$ |
|---|---|---|---|
| *closed bar* | $\hat{\mathfrak{F}} = \bullet --- \bullet$ | $H_0(\mathbb{R};\hat{\mathfrak{F}}) \cong \mathbb{F}$ | $H_1(\mathbb{R};\hat{\mathfrak{F}}) \cong 0$ |
| *open bar* | $\hat{\mathfrak{F}} = \circ --- \circ$ | $H_0(\mathbb{R};\hat{\mathfrak{F}}) \cong 0$ | $H_1(\mathbb{R};\hat{\mathfrak{F}}) \cong \mathbb{F}$ |
| *left-open bar* | $\hat{\mathfrak{F}} = \circ --- \bullet$ | $H_0(\mathbb{R};\hat{\mathfrak{F}}) \cong 0$ | $H_1(\mathbb{R};\hat{\mathfrak{F}}) \cong 0$ |
| *right-open bar* | $\hat{\mathfrak{F}} = \bullet --- \circ$ | $H_0(\mathbb{R};\hat{\mathfrak{F}}) \cong 0$ | $H_1(\mathbb{R};\hat{\mathfrak{F}}) \cong 0$ |

This is neither frivolous nor inconsequential, as evidenced in the following cosheaf interpretation of level-set persistence. Let $h\colon X \to \mathbb{R}$ be a cellular map from a compact cell complex $X$ to the reals, outfitted with a cell structure whose vertex set $\{v_i\}$ includes critical values of $h$. Let $\hat{\mathfrak{B}}_k$ be the $k^{\text{th}}$ homology fiber cosheaf of $h$ with coefficients in a field $\mathbb{F}$. Specifically, given the vertices $v_0 < v_1 < \cdots < v_N \subset \mathbb{R}$, define the $k^{\text{th}}$ homology cosheaf $\hat{\mathfrak{B}}_k$ as the cosheaf of $h$-level-set preimage homologies whose extension maps are induced on homology by inclusions as in the following diagram:

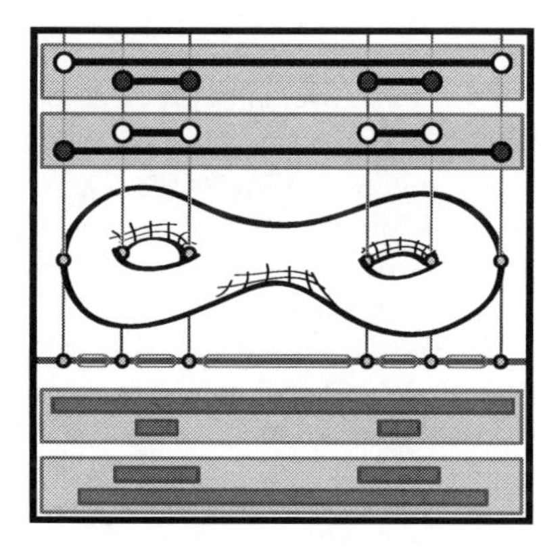

$$\cdots \longleftarrow H_k(v_{i-1}<h<v_i) \longrightarrow H_k(v_{i-1}<h<v_{i+1}) \longleftarrow H_k(v_i<h<v_{i+1}) \longrightarrow \cdots .$$

The following theorem blends homology, persistence, cosheaves, and Morse theory into a compact package:

**Theorem 9.22 ([77], cf. [78]).** *For $h\colon X \to \mathbb{R}$ and $\hat{\mathcal{B}}_k$ as above, the homology of $X$ in $\mathbb{F}$-coefficients can be computed in terms of the homology of $\mathbb{R}$ with coefficients in the homology cosheaf: for any $n$,*

$$H_n(X;\mathbb{F}) \cong H_0(\mathbb{R};\hat{\mathcal{B}}_n) \oplus H_1(\mathbb{R};\hat{\mathcal{B}}_{n-1}).$$

As a simple example, consider the canonical Morse function $h\colon S \to \mathbb{R}$ on a compact genus-2 surface with one minimum, four saddle critical points, and one maximum. The corresponding critical values induce a cell structure on $\mathbb{R}$ and the nonzero homology cosheaves are $\hat{\mathcal{B}}_0$ and $\hat{\mathcal{B}}_1$. These have interval decompositions as follows: $\hat{\mathcal{B}}_0$ has one closed bar and two open bars; $\hat{\mathcal{B}}_1$ has one open bar and two closed bars. Theorem 9.22 allows one to read off $H_\bullet(S;\mathbb{F})$ trivially by counting these intervals. For example, dim $H_1(S;\mathbb{F})$ equals the number of closed intervals in $\hat{\mathcal{B}}_1$ plus the number of open intervals in $\hat{\mathcal{B}}_0$. The reader will note that this graphical language also expresses Poincaré duality as a neatly visceral symmetry on the level of cosheaf barcodes.

## Notes

1. Sheaf theory has had an enormous impact within mathematics, in algebraic geometry, number theory, complex analysis, logic, and more. The agricultural terminology that suffuses this subject is entwined with France and the fascinating history of sheaves.
2. This chapter pretends to survey sheaf theory; in reality, the subject is so broad and deep as to evade compression. Experts in sheaf theory will find this chapter ridiculously elementary. This, however, is how mathematics transitions from pure to applied: slowly and through simple examples. It is the author's hope that the themes of elementary homological algebra, persistence, and sheaves will make their way via applications into the undergraduate linear algebra curriculum, where they belong.
3. The perspective of using cellular sheaves is unorthodox, but hopefully edifying. Cellular sheaves and cosheaves have developed in fits and starts. Their first explicit instantiation (so far as the author can tell) was by Fulton, Goresky, MacPherson, and McCrory in their famous 1977-78 seminar. Relevant references include the theses of Shepard [276], Vybornov [298] and Curry [77].
4. Sheaves, like simplicial complexes, are defined abstractly but have a geometric realization. To any sheaf $\mathcal{F}$ is associated an **étale space**: a topological space (also denoted $\mathcal{F}$) and a projection $\pi\colon \mathcal{F} \to X$ that, like a covering space, is a local homeomorphism with discrete fibers. Sections of the sheaf are precisely sections to the map $\pi$. This perspective can be useful; however, as the topology on the étale space is almost always non-Hausdorff, it is very unenlightening from the point of view of visualizing sheaves and their cohomology.
5. Sheaves taking values in vector spaces or abelian groups are convenient for doing cohomology; if one is willing to give up a simply-defined $H^k$ for $k > 0$, then it is possible to work with sheaves taking values in sets, spaces, or other categories (see §10.9). Such non-abelian data types are certainly natural and useful, though not covered in this chapter.
6. Most of the examples of this chapter are intentionally elementary, with networks as base spaces. Even so, it is a harbinger that such simple sheaves have clean applications. A number of applications of sheaves to data have recently emerged, using networks as a base space [146, 150, 201, 254, 255, 256].

7. Several applications of sheaves have been suggested in the computer science literature: see, *e.g.*, [159] and subsequents. Structural aspects seem to have displaced applications as the focus of this line of inquiry. This author believes that sheaf cohomology is the missing ingredient to make sheaf theory applied as opposed to applicable.
8. The full statement and proof of the Lefschetz formula of Goresky-MacPherson in §7.7 uses sheaf theory extensively.
9. The treatment of flow sheaves in §9.4 does not provide an independent proof of the classical Max-Flow/Min-Cut theorem, since the interpretations of the relative sheaf cohomology as obstructions subtly entail the theorem, and a careful proof of the interpretation and computation of obstructions would re-prove the result. There is a wholly sheaf-theoretic proof of the max-flow min-cut theorem by Krishnan [201]: see Example 10.25.
10. Constructible functions on a definable $X$ are locally defined, and thus form a sheaf $\mathsf{CF}_X$. Euler integration, the Fubini Theorem, and more follow directly from canonical sheaf operations [270].
11. The use of cosheaves in splines is not in the literature, but it is an obvious reformulation of existing work. The initial work of Billera [36] presented a chain complex by fiat and argued that the homology was meaningful. Yuzvinsky [304] reformulated the problem as a sheaf over the dual poset to a certain hperplane arrangement associated to the cell complex. Schenck and collaborators [223, 271] applied methods from spectral sequences and more classical commutative algebra to push the theory further. This subject blends softly into algebraic geometry; the treatment in this text is intended to emphasize topological aspects of the problem.
12. The reformulation of persistent homology and barcodes in terms of cosheaves is from the thesis of Curry [77]. Robinson [256] likewise recently interpreted persistent cohomology as a sheaf over $\mathbb{R}$. Theorem 9.22 can be derived from earlier works [57, 52] without using cosheaves: both of these, and the dual sheaf-theoretic version [78] are consequences of the **Leray spectral sequence**.
13. It is to be hoped that the language of sheaves and cosheaves provides the key to understanding and computing multi-dimensional persistence by means of generalized barcodes. This ambitious project will likely require nontrivial contributions from both sheaf theory and representation theory.
14. It is not an exaggeration to say that in sheaf theory, individual sheaves are of secondary importance. The real power in sheaf theory comes from dextrous use of complexes of sheaves and sheaf morphisms. This text has too little room to demonstrate this fully, as well as to unfold the rest of the six canonical operations of which direct and inverse image are two.

Chapter 10

# Categorification

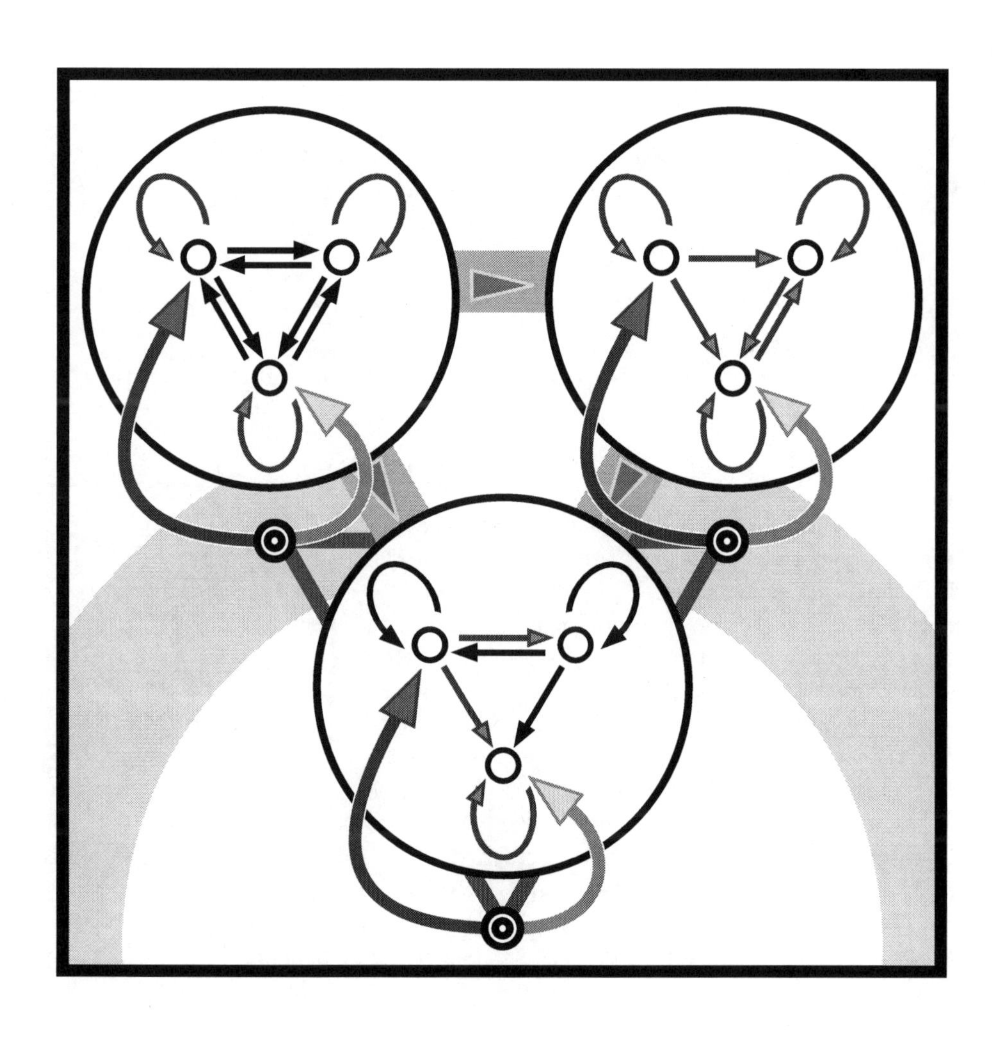

Mathematics is the science of patterns. As mathematicians have unraveled the patterns comprising algebra, analysis, topology, and more, certain meta-patterns have emerged. Notions of equivalence, limits, duality, and transformation have taken shape and precipitated a unified theory. It has been noted repeatedly that the power of topological invariants lies not only in their ability to characterize *spaces*, but *maps* as well. The oft-invoked *functoriality* and *naturality* of co/homology and homotopy groups are crucial ingredients of topology, pure and applied. The study of functoriality and its generalizations leads to **category theory**.

## 10.1 Categories

A **category** C consists of: (1) a collection[1] of **objects** $\mathcal{O}$; (2) for each ordered pair $(a, b)$ in $\mathcal{O}$, a set of **morphisms** $\mathcal{M}(a, b)$; and (3) for each ordered triple $(a, b, c)$ in $\mathcal{O}$ a **composition** operation $\circ\colon \mathcal{M}(a, b) \times \mathcal{M}(b, c) \to \mathcal{M}(a, c)$. In addition, these satisfy the following:

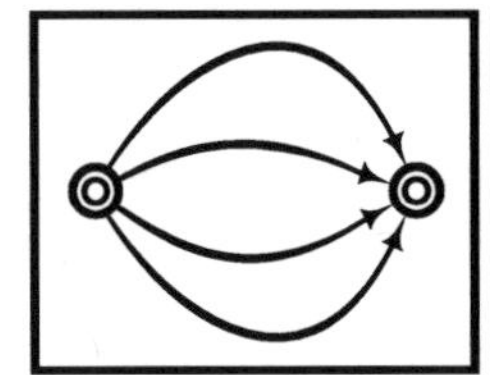
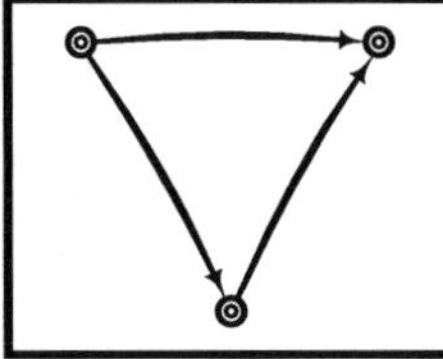
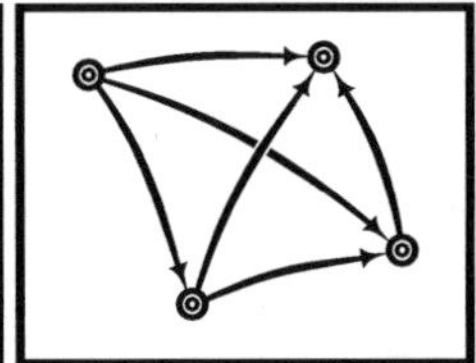
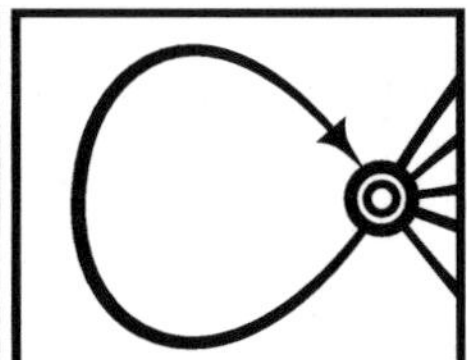

1. **Associativity:** Composition of morphisms is associative.
2. **Identity:** For each $a \in \mathcal{O}$, there exists an identity morphism $\mathrm{Id} \in \mathcal{M}(a, a)$ with $f \circ \mathrm{Id} = f$ for all $f \in \mathcal{M}(a, b)$ and $\mathrm{Id} \circ g = g$ for all $g \in \mathcal{M}(b, a)$.

The word *category* is so generic and ubiquitous as to be uninformative. The above definition is unrelated to the *LS category* of Chapter 7; this precise word was chosen to evoke an Aristotelian organization. This definition is, like all else Aristotelian, deceptively unexciting. A more transparently beautiful definition is possible when the category is sufficiently small, say, when objects and morphisms are countable. It is possible to represent such a category as a diagram of points (objects) and arrows (morphisms). A category C is visualized as a directed graph of vertices ($\mathcal{O}$) and, for each oriented pair of vertices $a, b$, a set $\mathcal{M}(a, b)$ of arrows **from** $a$ **to** $b$ (the direction is important). To each vertex is attached a loop-like identity arrow. Composition of arrows can be visualized by a 2-simplex whose boundary is the commutative triangle. The associativity of composition likewise has an arrow diagram best represented as the wireframe of a 3-simplex. This encapsulates an *Apollonian*[2] approach to cate-

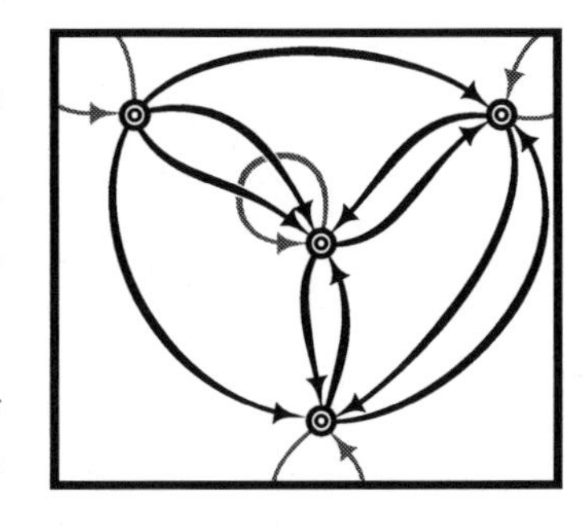

[1] A **class** as opposed to a set, but the initiate should not worry about such things.

[2] This deity connotes reason and order, and he keeps a full quiver of arrows.

gories. Very small categories are easily visualized (*cf.* the chain complex of a finite cell complex); categories with large sets of objects and morphisms (*e.g.*, singular chain complexes) demand too much of the Apollonian seer. Viewing the algebraic laws of composition and associativity in their simplicial guise is this subject's first hint at its relevance to topology.

The most common examples of categories are not small enough to be so illustrated; these include:

1. Vector spaces (over a field $\mathbb{F}$) and linear transformations Vect;
2. Groups and homomorphisms Grp;
3. Abelian groups and homomorphisms Ab;
4. Graded abelian groups and graded homomorphisms GrAb;
5. Topological spaces and continuous maps Top;
6. Topological spaces and homotopy classes of maps hTop.
7. Manifolds and smooth maps Man;
8. Sets and functions Set;
9. Finite sets and functions FinSet;
10. Chain complexes and chain maps ChCo; and
11. Posets and order-preserving functions Pos.

Other examples are less sweeping, though still useful:

1. Given a topological space $X$, the category $\mathsf{Op}_X$ has objects equal to the open sets of $X$ (including the empty set!), with inclusions $V \subset U$ defining morphisms $V \to U$.
2. A regular cell complex $X$ is a category $\mathsf{Face}_X$ whose objects are cells with morphisms $\sigma \to \tau$ iff $\sigma$ is a face of $\tau$, $\sigma \trianglelefteq \tau$.
3. Any poset (partially ordered set) $(P, \trianglelefteq)$ is a category with objects elements of $P$ and morphisms $a \to b$ iff $a \trianglelefteq b$. The previous two examples are special cases of a poset category.
4. A group **G** can be defined as a category with one object $\star$ and with morphisms $\star \to \star$ corresponding to elements of **G**, the composition being the group operation. The identity of **G** is the identity morphism, and each morphism must be invertible (where invertible hopefully means what you think it means: see §10.2).
5. A **groupoid** is, despite the grotesque name, easily defined: it is a category with invertible morphisms (a group being a single-object groupoid). A **monoid** is a category with a single object (a group being a monoid-with-inverses).

The reasons for wanting to use categorical language take time and space to fully unfurl. It is perhaps best to internalize some examples of categories that have appeared implicitly in other portions of this text.

**Example 10.1 (Simplices)** ⊙

The definition of the standard $n$-simplex in Chapter 2 was explicitly geometric. A categorical $n$-simplex is the category

$$[n] \quad := \quad 0 \longrightarrow 1 \longrightarrow 2 \longrightarrow \cdots \longrightarrow n-1 \longrightarrow n,$$

whose objects are the ordinals through $n$ and whose morphisms are induced by the total order $\lhd$. Of course, there are more morphisms than the *generators* displayed above, as composition must be applied: for example, drawing the picture associated to [3] (without drawing the identity morphisms) should reveal a familiar picture. In §10.3 it will be shown how to build complexes (in a category) from such simplices by building a larger category, Simp, whose objects are the $n$-simplices above. ⊚

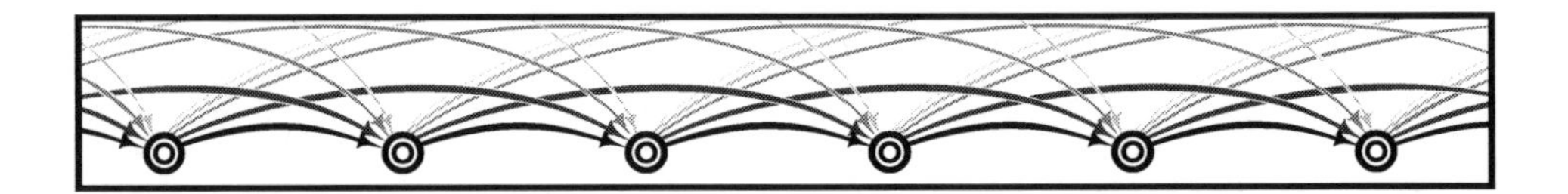

**Example 10.2 (Temporal dynamics)** ⊚

Dynamics bifurcates into continuous-time and discrete-time. Continuous-time dynamics (such as, solutions to differential equations) involves a flow on a space $X$ – an action of $\mathbb{R}$ on $X$ via homeomorphisms. In contrast, discrete-time dynamics involves an iterated homeomorphism $h : X \to X$, interpretable as an action of $\mathbb{Z}$ on $X$. Both settings are interpretable as a space $X$ acted on by a category – in these cases, the totally ordered groups $\mathbb{R}$ and $\mathbb{Z}$, with objects interpretable as *time* and morphisms determined by the total ordering. With a categorical perspective, it becomes clear how to investigate dynamics with more subtle temporal features, such as irreversibility, multi-dimensional time, locally-orderable but not globally-orderable time, and branching timelines: one simply replaces *time* with a different category. ⊚

**Example 10.3 (Boolean logics)** ⊚

Logic gates and circuits are built on a Boolean foundation, expressed as a ring $\{\bot, \top\}$ (or, often, $\{0, 1\}$) connoting false/true, along with the operations of disjunction ($\wedge$, *i.e.*, AND), conjunction ($\vee$, *i.e.*, OR), and negation ($\neg$, *i.e.*, NOT). These are greatly generalizable. A **Boolean algebra** is a set $B$ with a pair of distinguished members: 0, the minimum; and 1, the maximum; having commutative, associative, and distributive binary operations $\wedge$, $\vee$, as well as a unary operation $\neg$ which relate to the min/max values as follows:

$$b \vee 0 = b = b \wedge 1 \quad ; \quad b \vee \neg b = 1 \quad ; \quad b \wedge \neg b = 0.$$

A Boolean algebra forms the objects of a category whose morphisms are $a \to b$ iff $a \vee b = b$ (or, equivalently, $a \wedge b = a$). It helps to read this out loud using logical terminology. Boolean algebras form the objects of a category, Bool, whose morphisms are functions preserving 0, 1, $\wedge$, $\vee$, and $\neg$: $f(a \wedge b) = f(a) \wedge f(b)$, *etc.* Boolean algebras are a first hint at the utility of categories in logic, as commutative diagrams in Bool yield equations inside each Boolean algebra. ⊚

**Example 10.4 (Flow category)** ⊚

It is not the case that morphisms need to encode something explicitly algebraic or set-theoretic: dynamics is another source of morphisms. Consider the following category $C_h$ associated to a smooth function $h\colon M \to \mathbb{R}$ on a manifold $M$ [69]. The objects of $C_h$ are the critical points of $h$. Morphisms are flowlines of the gradient flow of $h$, up to time-reparametrization.

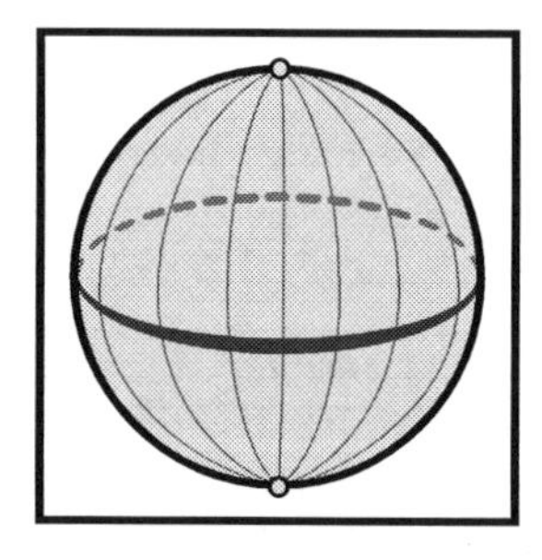

To allow for composition, flowlines are interpreted as **broken flowlines** with critical points in the interior allowed. This forms a category with some additional structure – each set of morphisms $\mathcal{M}(p, q)$ can be given the structure of a topological space. For example, the round sphere $\mathbb{S}^2 \subset \mathbb{R}^3$ with the simple linear height function yields a category with two objects (the two critical points $a$ and $b$). The morphisms are spaces: $\mathcal{M}(a, a)$ and $\mathcal{M}(b, b)$ are each a single point (the identity, corresponding to the invariant fixed point as a flowline); $\mathcal{M}(a, b)$ is homeomorphic to $\mathbb{S}^1$ and represents all possible flowlines from top-to-bottom; $\mathcal{M}(b, a)$ is empty. ⊚

## 10.2 Morphisms

The beauty of category theory is the ability to work with mathematical operations in a platform-independent manner. As a sample of what is possible with these very basic definitions, consider the following constructs, interpretable as lifting basic ideas from the algebraic to the categorical. All of the following demonstrate the difficulties and opportunities implicit in working with morphisms.

**Monic:** What does it mean to have a morphism in a category C that is *injective*? One is not permitted to discuss the inverse image of an arrow; kernels, images, and all other linear-algebraic thinking requires a reformation. The categorical analogue of an injection is a **monic** morphism: a morphism $f$ is monic if it is left-cancellative, *i.e.*, whenever $f \circ g = f \circ h$, then $g = h$. This is perhaps best digested in diagrammatic form. It is a delightful exercise to convince oneself that this is, indeed, the proper definition of injective when one has no recourse to sets, but only to morphisms, identities, and composition.

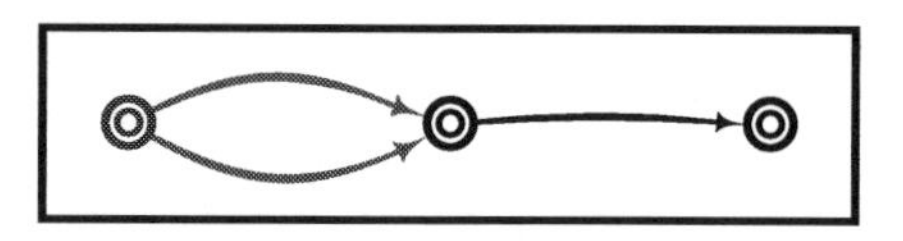

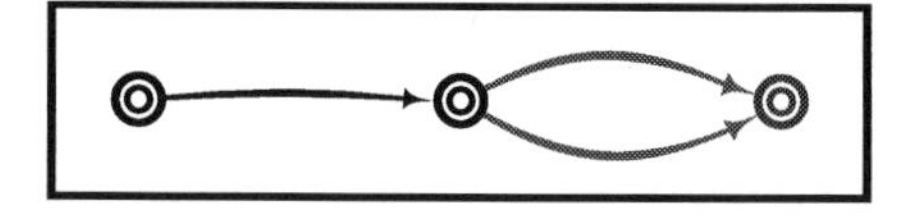

**Epic:** The associated notion of an onto morphism – **epic** – is pleasantly symmetric to the monic case. A morphism $f$ is epic if whenever $g \circ f = h \circ f$, then $g = h$ (right-cancellative). The symmetry between epic and monic is manifest and is not the last time that initially disconnected notions (into, onto) are revealed as dual under the appropriate categorical lift.

**Iso:** The reader might guess that an iso-morphism would be any morphism that is both epic and monic. This is *not* equivalent to the true definition of an isomorphism – a morphism $f$ that has an inverse $\overline{f}$ such that $f \circ \overline{f}$ and $\overline{f} \circ f$ are both identities (on potentially different objects). For example, the inclusion $\mathbb{N} \hookrightarrow \mathbb{Z}$ is epic and monic in the category of monoids[3], but is not an iso. In sufficiently nice categories, such as Set, iso *is* equivalent to epic-plus-monic. Isomorphic *objects* in a category are ones which have an isomorphism between them. In Set, isomorphic objects are equicardinal (via a relabelling of elements); in Top, isomorphic objects are homeomorphic spaces; in hTop, isomorphic objects are homotopic spaces; in Grp, isomorphic means isomorphic.

**Initials and terminals:** The simplex category $[n]$ has two distinguished objects: 0 and $n$, at the *beginning* and *end* of the category. In a general category, an **initial** object is one with a *unique* morphism from it to *every* object in the category (*cf.* 0). A **terminal** object is one with a unique morphism from *every* object in the category to it (*cf.* $n$). Initials and terminals are easily shown to be unique up to isomorphism. Not all categories possess initial or terminal objects; even those derived from a total order (like $(\mathbb{Z}, \leq)$) do not necessarily contain their *limits* in this sense. Examples of categories with initials or terminals include the following:

1. $\mathsf{Op}_X$ has $\varnothing$ as initial and $X$ as terminal objects.
2. The empty set $\varnothing$ is the initial object of Set: any 1-point set is a terminal object (unique up to isomorphism).
3. In Grp and Vect, the singleton object is both initial and terminal.
4. In a given Boolean algebra, 0 is initial and 1 is terminal.
5. In Bool, the 2-element algebra $\{0, 1\}$ is initial, and the single-element algebra $\{0\}$ is terminal.

**Equalizers:** In linear algebra, one cares about the kernel of a transformation; in calculus, one defines level sets as kernel-like objects. The category-theoretic analogue is an **equalizer**.

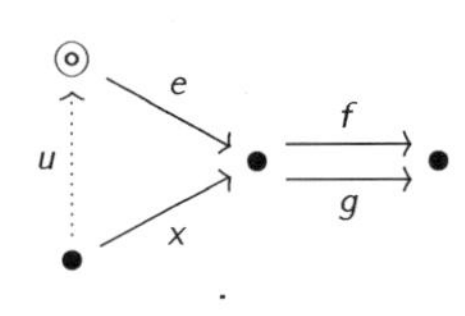


Given a pair of morphisms $f, g \in \mathcal{M}(a, b)$, the equalizer is defined to be the **universal** object-morphism pair $e : \circledcirc \to a$ such that $f \circ e = g \circ e$. That is, given any morphism $x$ with $f \circ x = g \circ x$, there exists a *unique* morphism $u$ by which $x$ factors through $e$. Note how everything is defined in terms of morphisms – the objects are implicit. The connection to kernels and level sets is discernable when working in the appropriate categories. The diagram implies that equalizer is the solution to the equation "$f - g = 0$" when such is sensible.

[3]A category whose objects are monoids and whose morphisms are monoid homomorphisms that respect multiplication and identities.

## 10.3 Functors

By this point, the reader will not be surprised to learn that in category theory, it is not the *objects* that matter so much as the *morphisms*. However, the delimited purview of a single category is too weak: vim resides in structure-respecting transformations between categories. A **functor** is an assignment $\mathcal{F}\colon \mathsf{C} \to \mathsf{C}'$ taking objects to objects and morphisms to morphisms in a manner that respects composition and identities: identity morphisms are sent to identity morphisms and likewise with composed morphisms. Some readers will find it helpful to write out all the equations implicit in this formulation.

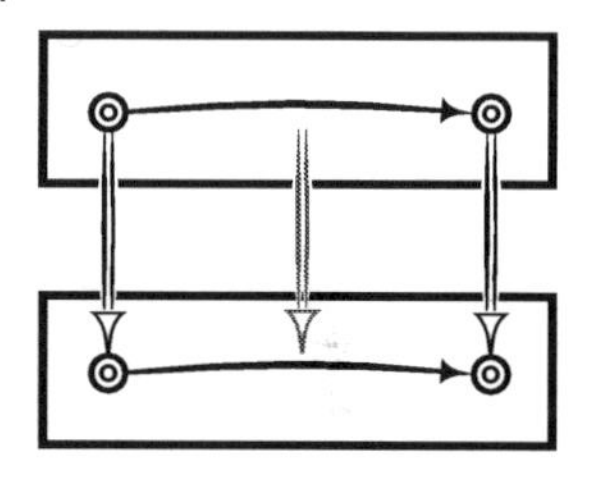

The simplest functors are the **forgetful functors**, which simply remove structure from a category. For example, the removal of a (topological, group, differential, order) structure from a (space, group, manifold, poset) is a forgetful functor from the category (Top, Grp, Man, Pos) to Set. Homology and homotopy groups are at the heart of algebraic topology: *these are functors* and convert topological data to algebraic data in a manner that is *functorial*. The reason for that particular word choice should now be clear. Homology $H_\bullet$ is a functor from Top to GrAb, since a map $f\colon X \to Y$ has induced homomorphism $H(f)\colon H_\bullet(X) \to H_\bullet(Y)$. Cohomology $H^\bullet$ is not *quite* a functor from Top $\to$ GrAb, since the induced homomorphism goes from $H^\bullet(Y)$ to $H^\bullet(X)$: however, the induced maps behave as a functor would, but backwards. The language of an older dispensation distinguished **covariant** and **contravariant** functors – homology is covariant; cohomology, contravariant. Contemporary fashion keeps functors face-forward and expresses cohomology as a functor on a flipped category.

**Example 10.5 (Duality & opposites)** ⊚

Duality is seen to live in many forms throughout mathematics, hinting at a general construct. Given a category C, the **opposite category** $\mathsf{C}^\circ$ has the same objects, but reverses the direction of all arrows: $\mathcal{M}^\circ(A, B) := \mathcal{M}(B, A)$.

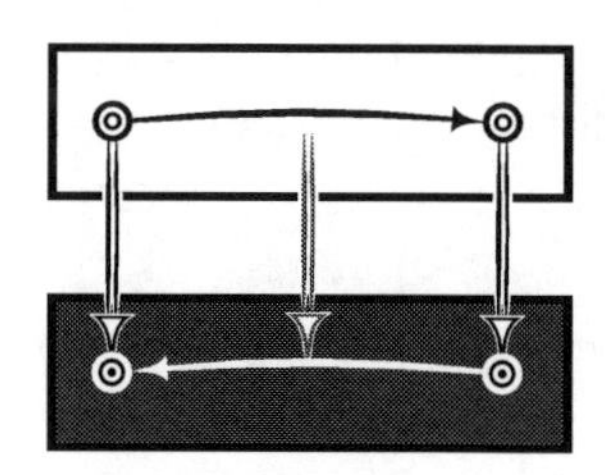

For example, given a poset $(P, \trianglelefteq)$, thought of as a category with a unique morphism $a \to b$ iff $a \trianglelefteq b$, the dual category is the poset $(P, \trianglerighteq)$ which reverses the partial order. Opposites can also be used to set aright those functors that do things *in reverse*. While homology $H_k(\cdot)$ gives a functor from Top $\to$ Ab, cohomology $H^k(\cdot)$ yields a functor from $\mathsf{Top}^\circ \to$ Ab. The operation of passing to the opposite category is, like other forms of duality, involutive: $(\mathsf{C}^\circ)^\circ = \mathsf{C}$. ⊚

**Example 10.6 (Simplicial sets)** ⊚

The treatment of simplicial complexes in Chapter 2 conflated geometric things (spaces, gluings) with algebraic data (simplices, orderings, faces). This mixture of the geometric and the algebraic is best viewed via functors. Recall the definition of an $n$-simplex

$[n]$ from Example 10.1. One can build a larger category, the **simplex category**, Simp, whose objects are $[n]$ for $n \in \mathbb{N}$ and whose morphisms are functors $[n] \to [m]$, that is, order-preserving functions. These morphisms are generated by the **face maps**, $D_k\colon [n] \to [n-1]$, and **degeneracy maps**, $S_k\colon [n] \to [n+1]$, which respectively skip or repeat the $k^{\text{th}}$ index:

$$D_k(0 \to 1 \to \ldots \to n) = (0 \to 1 \to \ldots \to k{-}1 \to k{+}1 \to \ldots \to n)$$
$$S_k(0 \to 1 \to \ldots \to n) = (0 \to 1 \to \ldots \to k{-}1 \to k \to k \to k{+}1 \to \ldots \to n)$$

These do not freely generate the category – there is a list of relations that these morphisms must satisfy in order to mimic the network of faces of simplices [158, 220]. As indicated by the terminology, there are **degenerate simplices** in this theory. For example, a morphism $[3] \to [1]$ given by $(0,1,2,3) \to (0,0,1,1)$ resembles a degenerate 3-simplex with projected image a 1-simplex.

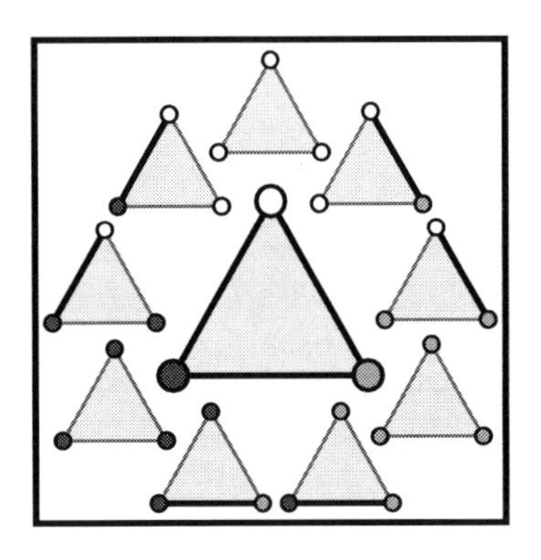

It is easy to use such a structure as a *model* on which to build representations that capture the combinatorics of oriented simplices. For example, a **simplicial set** is a functor $X\colon \mathsf{Simp}^{\mathsf{o}} \to \mathsf{Set}$. This functor associates a set to each object in Simp and *glues* them together along all faces. Simplicial sets are ideal for keeping track of the combinatorics of an oriented simplicial complex in a unified package, and are an especially nice class of structures on which to do homotopy theory: one has the freedom of working in infinite dimensions (note that Simp contains simplices of all dimensions) while maintaining an efficient bookkeeping. One thinks of a simplicial set as a single infinite-dimensional Platonic simplex outfitted with a list of folding instructions sufficient to produce the (potentially finite) output. To complete the intuition back to the topological, any simplicial set can be converted into a topological space by means of another functor, the **geometric realization** functor, $|\bullet|$, which, in one instantiation, results in a CW complex with one $n$-cell for each nondegenerate $n$-simplex [158, 220]. These are not the best spaces for computation, since the data structure is prodigal: each nondegenerate 2-simplex comes with 9 degenerate 2-simplices and an infinite number of higher-dimensional degenerate cousins. ⊚

**Example 10.7 (Nerves, redux)** ⊚

The complexity implicit in the definition of simplicial sets induces elegance elsewhere. Nerves of covers can be lifted to more general settings, to great effect. The **nerve** of a small category, $\mathcal{N}(\mathsf{C})$, is the simplicial set whose $n$-simplices are ordered sequences of $n$ composable morphisms in C – a chain of $n$ arrows. Thus, the objects of C are the vertices of $\mathcal{N}$; the arrows of C form its edges; pairs $\to\to$ of incident arrows are 2-simplices, *etc.* Such a formal 2-simplex really does *look* like a 2-simplex, since the two arrows can be composed to obtain a third *edge*. This hints at how to specify the face and degeneracy maps, as must be done for a simplicial set. The degeneracy map $D_k$ removes the $(k+1)^{st}$ arrow and replaces the $k^{th}$ arrow with the composition (or eliminates the first/last arrow if $k = 0, n$ resp.). The face map $S_k$ inserts an additional arrow by using the identity arrow at that object in C. One checks that the relations hold and that the result is a simplicial set. In the case of $\mathsf{C}(\mathcal{U})$ the poset induced by intersections of elements of a finite cover $\mathcal{U} = \{U_i\}$ of a space, the nondegenerate simplices of $\mathcal{N}(\mathsf{C}(\mathcal{U}))$ are precisely the simplices of the classical nerve complex $\mathcal{N}(\mathcal{U})$ as in §2.6. The categorical nerve, being a simplicial set, has a great many more degenerate simplices. These all collapse out, and the geometric realization of the categorical nerve is homotopic to the classical nerve. ⊚

## 10.4 Clustering functors

Categorical language has found its way into a few disciplines outside of Mathematics, with Computer Science being chief among them. At first glance, such applications might seem like a translation to a foreign language: intricate, symbol-ridden, and unreadable. The following application to statistics should convince an otherwise sceptical reader of the utility of categorical thinking.

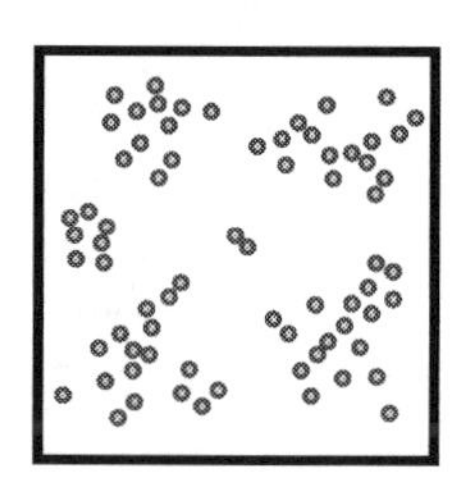

Sections §2.2, §2.5, and §5.14 have discussed methods for approximating the topology of a cloud of data points via homology. The first-order term of this sequence is the computation of the number of connected components. Though this problem of **clustering** is easily stated, its importance in statistics and the natural and engineering sciences is immense. The subtlety of partitioning a discrete set $\mathcal{Q} \subset \mathbb{R}^n$ into clusters is evidenced both by the enormous literature and by results like the following.

Consider a clustering algorithm as a function which takes as input a finite metric space $\mathcal{Q}$ (thus, pairwise distances between points are known, but placement up to rigid Euclidean motions is irrelevant) and returns a partition into **clusters**. Desirable properties for a clustering algorithm to possess would seem to include the following:

1. **Scale-invariance:** Clusters are invariant under rigid rescaling of the metric.
2. **Surjectivity:** For any partition of a finite set $\mathcal{Q}$, there exists a metric on $\mathcal{Q}$ which yields that partition as the clusters.
3. **Consistency:** Given an input $\mathcal{Q}$ with resulting cluster, move the points of $\mathcal{Q}$ so that within clusters, distances between points do not increase, and between clusters, distances between points do not decrease. The resulting input $\mathcal{Q}'$ has clustering identical to that of $\mathcal{Q}$.

The consistency property, though hardest to state, is no less desirable than the others: indeed, it seems vital. The following theorem of Kleinberg asserts the mutual incompatibility of all three conditions, in a manner not unlike the Arrow impossibility theorem in voting.

**Theorem 10.8 ([194]).** *There does not exist a clustering algorithm which is scale-invariant, surjective, and consistent.*

The critical observation of Carlsson-Mémoli is this: *clustering can be functorial.* A classical clustering algorithm takes an object from a category of finite metric spaces and returns an object in the **cluster category**, Clust, whose objects are pairs $(\mathcal{Q}, P^{\mathcal{Q}})$ consisting of a finite set $\mathcal{Q}$ and a partition $P^{\mathcal{Q}}$ thereof. The morphisms in Clust consist of functions $f\colon \mathcal{Q} \to \mathcal{Q}'$ that send points-to-points and partition elements to partition elements in such a manner that $f^{-1}(P^{\mathcal{Q}'})$ is a *refinement* of $P^{\mathcal{Q}}$ – a cluster morphism can coalesce clusters but not break them up.

A **clustering functor** is a functor from a category of finite metric spaces to Clust. It not only assigns clusters to a point-cloud, it converts morphisms between point-clouds into correspondences between and refinements of the resulting clusters. The desired properties for a clustering algorithm – *e.g.*, consistency or scale-invariance – should be built into the morphisms of the categories chosen. Consider the category $\mathsf{FinMet}^{\leq}$ whose objects are finite metric spaces $\mathcal{Q}$ and whose morphisms are *distance non-increasing*. That is $f\colon \mathcal{Q} \to \mathcal{Q}'$ with $d(f(x), f(x')) \leq d(x, x')$. With this structure for $\mathsf{FinMet}^{\leq}$, a clustering functor to Clust must of necessity satisfy a property like consistency. The other conditions of Theorem 10.8 can be likewise programmed into the categories in such a manner that the proper interpretation is that there exist no nontrivial onto functors from $\mathsf{FinMet}^{\leq}$ to Clust: see [60] for details.

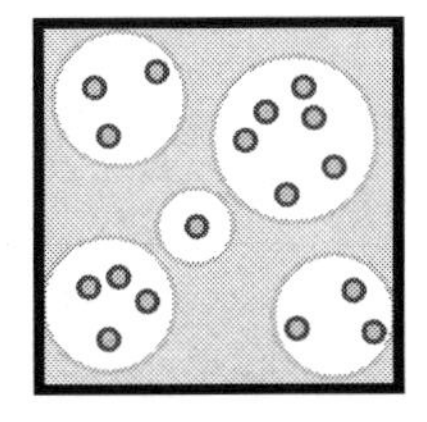

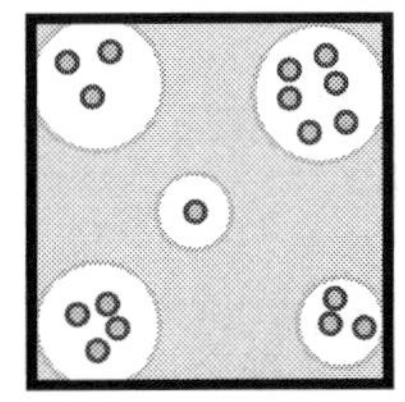

What is the good of this? Category theory is criticized as an esoteric language: formal and fruitless for conversation. *This is not so.* The virtue of reformulating (the negative) Theorem 10.8 functorially is a clearer path to a positive statement. If the goal is to have a theory of clustering; if clustering is, properly, a nontrivial functor; if no nontrivial functors between the proposed categories exist; then, naturally, the solution is to alter the domain or codomain categories and classify the ensuing functors. One such modification is to consider a category of persistent clusters.

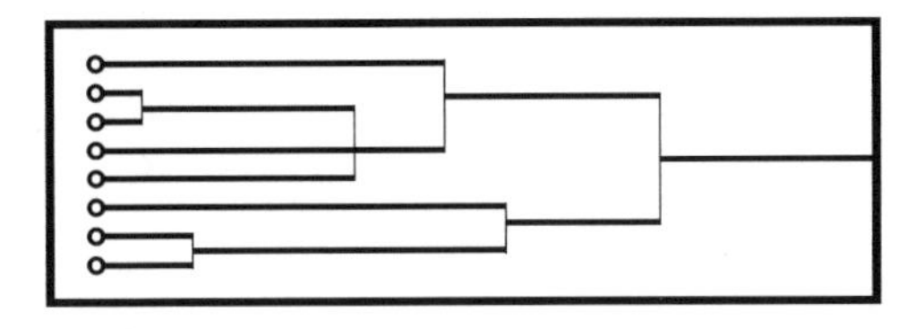

Define PClust, the category of persistent clusters, to be the category whose objects are pairs $(\mathcal{Q}, P_t^{\mathcal{Q}})$, where $P_t^{\mathcal{Q}}$ is a persistent partition: a family of partitions of $\mathcal{Q}$ depending on $t \in [0, \infty)$ such that $P_t^{\mathcal{Q}}$ is a refinement of $P_{t'}^{\mathcal{Q}}$ for $t \leq t'$. The morphisms are $t$-dependent morphisms from Clust – a $t$-dependent family of refinements of clusters. Such a persistent clustering is related to the notion of a **dendrogram**: one thinks of $t$ as something like a decreasing *resolution* of the clustering.

**Theorem 10.9 ([60]).** *There exists a unique functor* $\mathsf{FinMet}^{\leq} \to \mathsf{PClust}$ *which takes the input* $\{\bullet - \bullet\}$ *consisting of two points at distance* $R$ *to the persistent cluster having* *one cluster for $t \geq R$ and two clusters for $0 \leq t < R$.*

This provides a resolution to the conundrum of Theorem 10.8: it is surjective and consistent, and the persistent clustering scales with metric scaling in a clear manner. Not surprisingly, this clustering method is well-known: it is called **single linkage** clustering and is equivalent to saying that the clusters are given by $\pi_0(\mathrm{VR}_t(\mathcal{Q}))$ – the connected components of the distance-$t$ Vietoris-Rips complex of $\mathcal{Q}$. The appearance of the Vietoris-Rips complex here is not unexpected, but pleasant nonetheless. Functoriality of the clusters is captured in the appearance of the functor $\pi_0$.

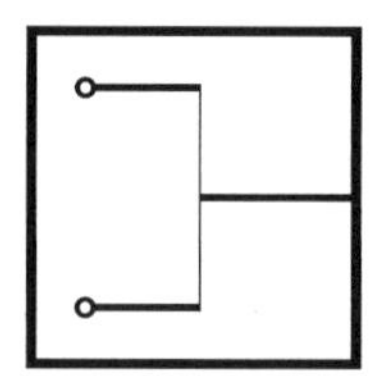

## 10.5 Natural transformations

The reader for whom this material is an introduction may suspect that a joke is being played when it is asserted that the *truly* interesting objects of study in category theory are neither morphisms nor functors, but rather correspondences between functors. Nevertheless, this is asserted in all seriousness: such a construction was indeed the true impetus for the creation of category theory [211]. A **natural transformation**, $\eta\colon \mathcal{F} \Rightarrow \mathcal{F}'$ connects a pair of functors $\mathcal{F}, \mathcal{F}'\colon \mathsf{C} \to \mathsf{C}'$ by sending each object $a \in \mathsf{C}$ to a morphism $\eta(a)\colon \mathcal{F}(a) \to \mathcal{F}'(a)$ so that for each morphism $h\colon a \to b$ in $\mathsf{C}$, there is a commutative square connecting $\mathcal{F}(h)$ and $\mathcal{F}'(h)$:

$$\begin{array}{ccc} \mathcal{F}(a) & \xrightarrow{\mathcal{F}(h)} & \mathcal{F}(b) \\ {\scriptstyle \eta(a)}\downarrow & \Downarrow & \downarrow{\scriptstyle \eta(b)} \\ \mathcal{F}'(a) & \xrightarrow[\mathcal{F}'(h)]{} & \mathcal{F}'(b) \end{array}$$

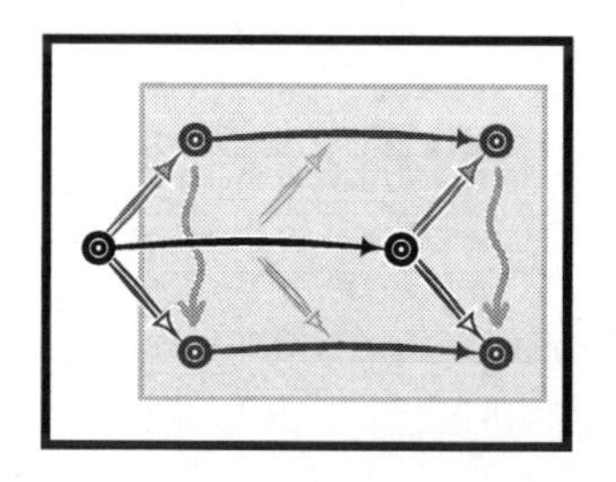

Said better, a natural transformation is a functor in the Category of categories. This is a deep idea – that categories themselves form the Objects of a Category whose Morphisms are functors (under functor composition and with the identity functor playing the obvious role). The Functors of this Category relate a pair of Morphisms: in the original category, this is precisely what a natural transformation is.

Natural transformations are, like functors, composable. Given $\eta\colon \mathcal{F} \Rightarrow \mathcal{G}$ and $\eta'\colon \mathcal{F}' \Rightarrow \mathcal{G}'$, there is the obvious natural transformation $\eta \circ \eta'\colon \mathcal{F} \cdot \mathcal{F}' \to \mathcal{G} \cdot \mathcal{G}'$. However, it is also possible to compose a natural transformation $\eta$ with a functor $\mathcal{H}$, either on the left, $\mathcal{H} \cdot \eta$, or on the right, $\eta \cdot \mathcal{H}$, obtaining a modified natural transformation.

**Example 10.10 (Translation)** ⊚

Consider the reals $(\mathbb{R}, \leq)$ with the total order as a category: there is one morphism $a \to b$ whenever $a \leq b$. Translation by $\epsilon$ is a functor $T_\epsilon\colon \mathbb{R} \to \mathbb{R}$ that sends $a \mapsto a + \epsilon$ and preserves the order (hence morphisms). The translation functor does very little: just a shift in objects. Thus, it comes as no surprise that there is a natural transformation $\tau_\epsilon\colon \mathrm{Id} \Rightarrow T_\epsilon$ from the identity that takes $(a \leq b)$ to $(a + \epsilon \leq b + \epsilon)$.

One can interpret the *naturality* in this setting as the indifference to whether one $\leq$-compares objects before or after the translation. ⊚

**Example 10.11 (Snakes)** ⊚

The Snake Lemma (Lemma 5.5) asserts the existence of the connecting homomorphism $\delta\colon H_\bullet(\mathcal{C}) \to H_{\bullet-1}(\mathcal{A})$, given a short exact sequence on $\mathcal{A} \to \mathcal{B} \to \mathcal{C}$. What makes the result so powerful is the so-called *naturality* of the resulting long-exact sequence, as per Equation (5.5). This is equivalent to saying that $\delta$ is a natural transformation as follows. Consider the category ChCoSES of short exact sequences of chain complexes. The Snake Lemma asserts that $\delta$ is a natural transformation on homology functors to GrAb:

$$\left(\begin{array}{c} 0 \to \mathcal{A} \to \mathcal{B} \to \mathcal{C} \to 0 \\ \downarrow \\ H_\bullet(\mathcal{C}) \end{array}\right) \overset{\delta}{\Rightarrow} \left(\begin{array}{c} 0 \to \mathcal{A} \to \mathcal{B} \to \mathcal{C} \to 0 \\ \downarrow \\ H_{\bullet-1}(\mathcal{A}) \end{array}\right).$$

⊚

**Example 10.12 (Co/homology equivalences)** ⊚

One of the great advantages of the multiple homology theories developed in Chapter 4 is that they are all isomorphic *and* that these isomorphisms are *natural*. This means that not only do the cellular and singular homologies of a cell complex agree, but maps between cell complexes induce the "same" homomorphism on homologies, as noted in §5.4. Of course, this naturality really means that there are natural transformations between the various homology functors: cellular, singular, Morse, Čech, *etc.*, restricted to the subcategory of spaces/maps on which both theories are defined. These are, specifically, **natural isomorphisms** between homology functors, meaning that, *e.g.*, between $H_\bullet^{\text{sing}}$ and $H_\bullet^{\text{cell}}$, there are an inverse pair of natural transformations whose compositions (both ways) yield the identity natural transformation on each homology functor. Natural isomorphisms are fundamental. ⊚

**Example 10.13 (Retraction to a cone point)** ⊚

Given any category C, there is a unique functor $\mathcal{R}\colon \mathsf{C} \to \{\bullet\}$ to the category with one object and one (identity) morphism. This $\mathcal{R}$ collapses all objects to $\bullet$ and all morphisms to the identity. For any fixed object $a \in \mathsf{C}$, the obvious functor $\mathcal{I}_a\colon \{\bullet\} \to \mathsf{C}$ that sends $\{\bullet\} \to a$ acts like an inclusion that satisfies $\mathcal{R}\cdot\mathcal{I}_a = \mathrm{Id}_\bullet$. In what sense could $\mathcal{I}_a \cdot \mathcal{R}$ be compared to the identity functor $\mathrm{Id}_{\mathsf{C}}$? In the case that C has an initial object $0 \in \mathsf{C}$, then there is a natural transformation $\eta\colon \mathcal{I}_0 \cdot \mathcal{R} \Rightarrow \mathrm{Id}_{\mathsf{C}}$ given by sending $(a \to \bullet \to 0) \to a$, using the unique morphism from the initial. This hints at the role played by an initial: it acts as the apex of a cone along which the category is, metaphorically, contractible.[4]

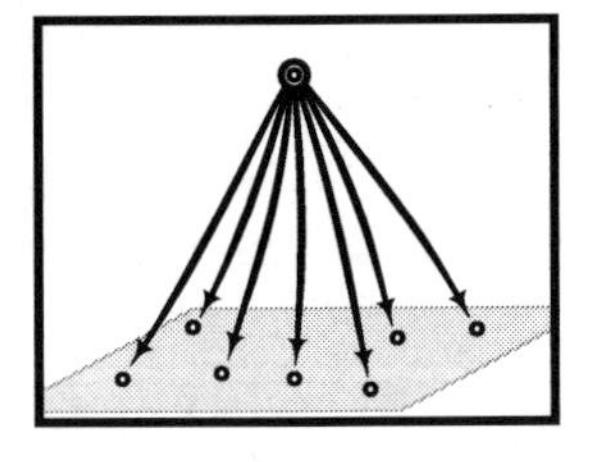

⊚

[4]More than metaphor: the nerve of any small category with an initial object has contractible geometric realization.

One says that categories are *isomorphic*, $\mathsf{C} \cong \mathsf{D}$, when there are functors $\mathcal{F}\colon \mathsf{C} \to \mathsf{D}$ and $\mathcal{G}\colon \mathsf{D} \to \mathsf{C}$ which are inverses in that $\mathcal{G} \cdot \mathcal{F} = \mathrm{Id}_{\mathsf{C}}$ and $\mathcal{F} \cdot \mathcal{G} = \mathrm{Id}_{\mathsf{D}}$. This is rarely satisfied, even for categories that seem very closely related. More common (and useful) is the case where there are not equalities but rather natural transformations $\mathcal{G} \cdot \mathcal{F} \Rightarrow \mathrm{Id}_{\mathsf{C}}$ and $\mathcal{F} \cdot \mathcal{G} \Rightarrow \mathrm{Id}_{\mathsf{D}}$ to the identities. One says that such categories $\mathcal{C} \approx \mathcal{D}$ are **equivalent categories**. This is analogous to the way in which homotopic spaces are topologically equivalent, though not necessarily homeomorphic.

**Example 10.14 (Duals and isomorphisms)** ⊚

Duality can be subtle. On the category FinVect of finite-dimensional vector spaces, there is the dual-space functor $\mathcal{D}\colon \mathsf{FinVect} \to \mathsf{FinVect}$ that sends $V \to V^{\vee}$. Although $V \approx V^{\vee}$ are isomorphic as vector spaces, this isomorphism is *not* natural. What this means *precisely* is that there is no natural isomorphism $\mathcal{D} \Rightarrow \mathrm{Id}$. It is true, however, that any $V$ in FinVect is *naturally* isomorphic to its double-dual $(V^{\vee})^{\vee}$: *i.e.*, there is a natural isomorphism $\eta\colon \mathcal{D}^2 \Rightarrow \mathrm{Id}$. ⊚

## 10.6 Interleaving and stability in persistence

One of the more widely-cited results in topological data analysis is the Stability Theorem for persistent homology [70], alluded to in §7.2 for sublevel set persistence. The treatment of persistence in §5.13 and §7.2 usedc a discretization of the parameter line: in practice, one may want to use a real parameter. Consider, therefore, the setting of persistence over $\mathbb{R}$, in which $\{X_t\colon t \in \mathbb{R}\}$ is a family of spaces with inclusion maps $X_a \subset X_b$ for $a \leq b$, thought of as (lower) excursion sets $X_t = \{h \leq t\}$ of a height function $h\colon X \to \mathbb{R}$.

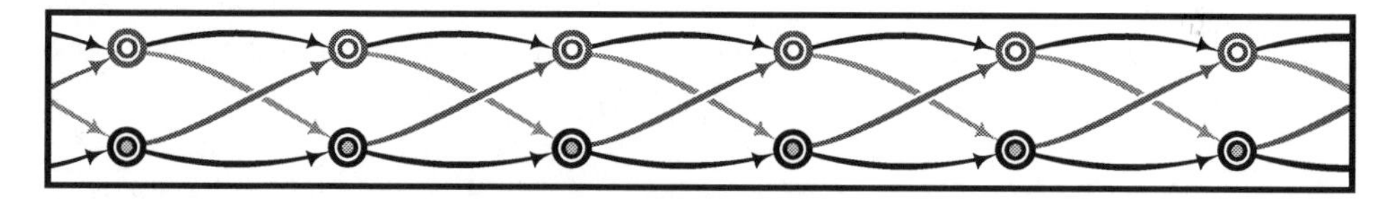

One question of stability is this: for $h'$ close to $h$, how much can the topology of the excursion sets $X_t$ change? Any individual $X_t$ can change dramatically with a small perturbation; the content of the Stability Theorem is that the impact on persistent homology is small. This requires some notion of proximity for persistent homology. Current practice uses the following definition. Given two $\mathbb{R}$-indexed homology sequences, say, $H_k(X_t)$ and $H_k(X'_t)$, they are said to be $\epsilon$-**interleaved** if there exist homomorphisms $\phi_t\colon H_k(X_t) \to H_k(X'_{t+\epsilon})$ and $\phi'_t\colon H_k(X'_t) \to H_k(X_{t+\epsilon})$ such that $\phi'_t \circ \phi_t$ and $\phi_t \circ \phi'_t$ are each the inclusion maps on homology induced by the shift $+2\epsilon$.

An $\epsilon$-interleaving implies that the homologies of the two sequences can only differ substantially over short parameter intervals. This motivates defining a pseudometric on persistent homologies by declaring the distance $d(H_k(X_t), H_k(X'_t))$ to be the infimum of $\epsilon$ over all $\epsilon$-interleavings. This is not quite a metric, since it may take on the value $+\infty$ (if no interleaving exists) and two persistence diagrams with interleaving distance zero are not necessarily identical (since the infimum may be 0 without a 0-interleaving). Nevertheless, reflexivity and the triangle inequality do hold for this pseudo-metric. In this language, the relevant result is:

**Theorem 10.15 ([70]).** *Given $h, h' \colon X \to \mathbb{R}$, the interleaving distance between the sublevel set persistent homology sequences is bounded by the $L_\infty$ norm of the difference of the height functions:*

$$d(H_k(X_t), H_k(X'_t)) \leq \|h - h'\|_\infty .$$

This result can be translated into categorical language. Consider a function $h \colon X \to \mathbb{R}$. Then for $a \leq b$, the inclusion $X_a \to X_b$ of sublevel sets of $h$ defines an excursion-set functor $\mathcal{E}_h \colon (\mathbb{R}, \leq) \to \mathsf{Top}$. The resulting sublevel set persistent homology is the composition of this functor with $H_\bullet$.

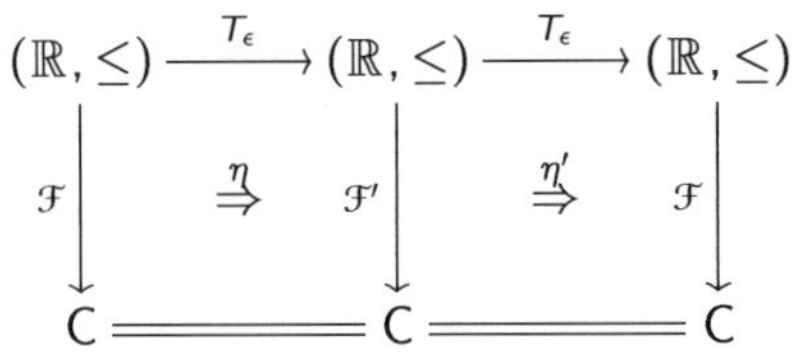

$$\begin{array}{ccccc} (\mathbb{R},\leq) & \xrightarrow{T_\epsilon} & (\mathbb{R},\leq) & \xrightarrow{T_\epsilon} & (\mathbb{R},\leq) \\ \mathcal{F}\downarrow & \overset{\eta}{\Rightarrow} & \downarrow\mathcal{F}' & \overset{\eta'}{\Rightarrow} & \downarrow\mathcal{F} \\ \mathsf{C} & = & \mathsf{C} & = & \mathsf{C} \end{array}$$

The prevalence of functors motivates the question: what is the interleaving distance of arbitrary functors $(\mathbb{R}, \leq) \to \mathsf{C}$ for $\mathsf{C}$ a category? The answer involves natural transformations [51]. By Example 10.10, translation in $(\mathbb{R}, \leq)$ is a functor $T_\epsilon$ isomorphic to the identity functor via the natural transformation $\tau_\epsilon$. An $\epsilon$-**interleaving** of two functors $\mathcal{F}, \mathcal{F}' \colon \mathbb{R} \to \mathsf{C}$ is a pair of natural transformations $\eta \colon \mathcal{F} \Rightarrow \mathcal{F}' \cdot T_\epsilon$ and $\eta' \colon \mathcal{F}' \Rightarrow \mathcal{F} \cdot T_\epsilon$ such that

$$(\eta' \cdot T_\epsilon) \circ \eta = \mathcal{F} \cdot \tau_{2\epsilon} \quad \text{and} \quad (\eta \cdot T_\epsilon) \circ \eta' = \mathcal{F}' \cdot \tau_{2\epsilon}.$$

The interleaving distance on functors, $d(\mathcal{F}, \mathcal{F}')$, is the infimal $\epsilon$ for an $\epsilon$-interleaving. As before, this is not a metric, but rather a pseudo-metric than can take on the value $\infty$ for non-interleavable functors.

**Proof.** *(of Theorem 10.15)* [51] Given $h, h' \colon X \to \mathbb{R}$, these define excursion set functors $\mathcal{E}_h, \mathcal{E}_{h'} \colon \mathbb{R} \to \mathsf{Top}$. Note that for $\epsilon = \|h - h'\|_\infty = \sup_x |h(x) - h'(x)|$,

$$\begin{aligned} \mathcal{E}_h(t) &= \{h \leq t\} \subset \{h' \leq t + \epsilon\} = \mathcal{E}_{h'}(t + \epsilon) \\ \mathcal{E}_{h'}(t) &= \{h' \leq t\} \subset \{h \leq t + \epsilon\} = \mathcal{E}_h(t + \epsilon). \end{aligned}$$

This implies an $\epsilon$-interleaving $\eta, \eta'$ of $\mathcal{E}_h$ and $\mathcal{E}_{h'}$. Note that for any $\mathcal{F}, \mathcal{F}'$ that are $\epsilon$-interleaved, so are $\mathcal{G} \cdot \mathcal{F}$ and $\mathcal{G} \cdot \mathcal{F}'$, for any functor $\mathcal{G}$, by functoriality. Applying the homology functor $H_k$ to $\mathcal{E}_h$ and $\mathcal{E}_{h'}$ yields an $\epsilon$-interleaving on homology for $\epsilon = \|h - h'\|_\infty$. ⊙

## 10.7 Limits

Let $\mathsf{J}$ be a (small) **index category**. A **diagram** is a functor $\mathcal{F} : \mathsf{J} \to \mathsf{C}$ from an index category to a representation category. A diagram can sometimes be thought of as a "picture" of J in C [20, 160]. Another interpretation is that a diagram is something akin to a "sequence" in C (as is the case when $J = \mathbb{N}$). This second interpretation prompts the notion of a limit of a diagram.

Every student of Mathematics eventually grasps that limits are as subtle as they are useful. A limiting process has two inputs: that which is converging to a limit, and the indexing family over which the convergence occurs. The most familiar examples of limits – a limit of a sequence of points in a metric space, or an intersection of nested open sets in a topological space – limit over $\mathbb{N}$ as a poset. For a categorical limit, the converging objects reside in a category C and the indexing family is an index category J. The limit of a diagram $\mathcal{F} : J \to C$ is a distinguished object $\lim_J \mathcal{F} \in C$ that is thought of as a *terminus* of $\mathcal{F}$.

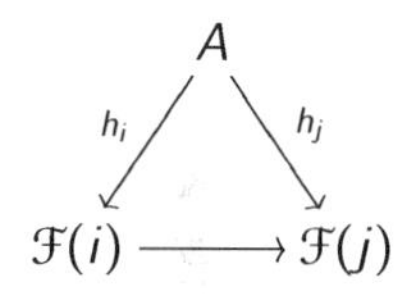


The definition is facilitated by an auxiliary construct. Fix $\mathcal{F} : J \to C$ a diagram. A **cone** over the diagram $\mathcal{F}$ is a J-indexed family of morphisms $h_j : A \to \mathcal{F}(j)$ from a fixed object $A$ in C to the image of J in C that respects composition (as per the diagram) for each morphism $i \to j$ in J. The collection of cones $(A, h_J)$ over $\mathcal{F}$ forms the objects of the **cone category**, $\text{Cone}_{\mathcal{F}}$, where a morphism between cones $(A, h_J) \to (A', h'_J)$ means that there is a morphism $A \to A'$ that makes the triangles with all $h_j$ and $h'_j$ pairs commute. One visualizes the cones over $\mathcal{F}$ as pyramid-like structures balanced atop the base image of $\mathcal{F}$.

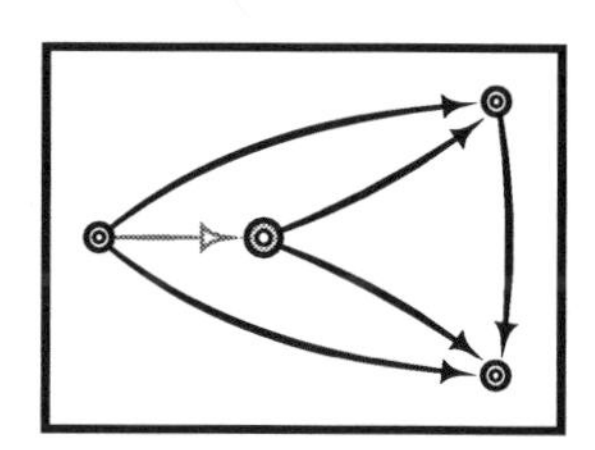

The **limit** of a diagram $\mathcal{F} : J \to C$ is the terminal object in the cone category $\text{Cone}_{\mathcal{F}}$. This has an interpretation as a universal property: such a limit gives a distinguished object $\lim_J \mathcal{F} = \lim_j \mathcal{F}(j) \in C$ and J-respecting morphisms $h_j : \lim_J \mathcal{F} \to \mathcal{F}(j)$ such that any other cone must factor through the limit. As terminals, limits carry a connotation that blends intersection, restriction, GCD, preimage, and gluing.

**Example 10.16 (Limit examples)** ◎

As in the case of calculus, limits may or may not exist. When a limit exists, it usually corresponds to something important. The following examples of limits reinforce the interpretation of a limit as being restrictive in nature.

**[intersection]** Consider an index category of the form $J = \bullet\ \bullet$ with no non-identity morphisms, and a diagram $\mathcal{F} : J \to \text{Op}_X$. This is, simply, a pair of open sets in a space $X$. The cone category $\text{Cone}_{\mathcal{F}}$ has as objects open sets in $X$ with inclusions into the two open sets defined by $\mathcal{F}$. The limit is the largest such open set (any other factors through it via inclusion): this is the intersection of the two open sets.

**[products]** The same index category $J = \bullet\ \bullet$ when sent via $\mathcal{F}$ to Top yields a different style of limit: the cartesian product of spaces. Let $\mathcal{F}$ have image objects topological spaces $X$ and $Y$. A cone over $\mathcal{F}$ is a space $Z$ and maps $Z \to X$, $Z \to Y$ such that any other space $Z'$ with maps to $X$ and $Y$ must factor through $Z$: this is the cartesian product $Z = X \times Y$. The same construction works to give the familiar products in

Vect and Grp as well: all are limits of this simple $\mathsf{J} = \bullet\ \bullet$.

**[AND]** Using again the same index category, but taking a diagram into a poset $(P, \trianglelefteq)$ gives as limit the meet, $\wedge$, of the image of $\mathcal{F}$ in $P$: the $\trianglelefteq$-largest object of $P$ that is $\trianglelefteq$-smaller that both terms in the image of $\mathcal{F}$. In a Boolean algebra, the limit corresponds to the logical AND of the two image objects.

**[equalizers]** Consider the index category $\mathsf{J} = \bullet \rightrightarrows \bullet$ with two objects and a pair of morphisms between them. For any diagram $\mathcal{F}$ of $\mathsf{J}$ in $\mathsf{C}$, one notes that a cone over $\mathcal{F}$ is precisely an object in $\mathsf{C}$ that factors through both morphisms of the image of $\mathcal{F}$. The limit is therefore the universal such object: the equalizer. Limits therefore encompass kernels in linear algebra.

**[cohomology]** This implies, in particular, that the simplest cohomology group is a limit. The zero$^{\text{th}}$ cohomology $H^0(\mathcal{C})$ of a cochain complex $\mathcal{C}$ is simply the kernel of $d : C^0 \to C^1$. Since a limit is a generalized kernel, $H^0$ should be expressible as a limit: it is. For a discrete example, let $X$ be a cell complex with face poset given by $\trianglelefteq$. Then

$$H^0(X; \mathbf{G}) = \lim_{\sigma \in X} \mathbf{G}, \tag{10.1}$$

where the limit is over the constant diagram that sends each simplex in the face poset of $(X, \trianglelefteq)$ to the group $\mathbf{G}$.

**[terminals]** The empty category is a valid choice for $\mathsf{J}$. The only diagram of this $\mathsf{J}$ in $\mathsf{C}$ is the trivial diagram. By definition, a cone is simply an object of $\mathsf{C}$, and the limit, if it exists, is precisely a terminal object in $\mathsf{C}$: thus limits generalize terminals.

**[pullbacks]**

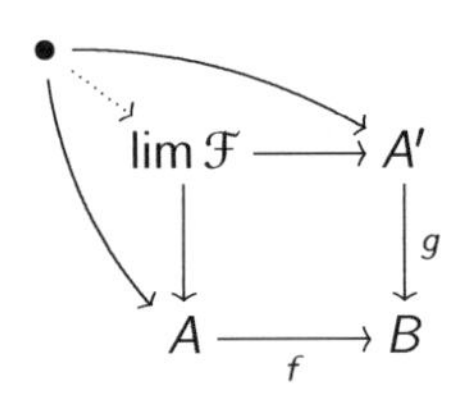


The diagram $\mathsf{J} = \bullet \to \bullet \leftarrow \bullet$ leads to an interesting type of intersection in the limit. Assume a diagram $\mathcal{F}$ that embeds $\mathsf{J}$ into $\mathsf{C}$ as $A \to B \leftarrow A'$. Then, a cone over $\mathcal{F}$ and the terminal limit is determined by commutative squares in the appropriate diagram. The colimit is called the **pullback** of the diagram and is sometimes denoted $A \times_B A'$, co-opting notation from Top. Examples of pullbacks in this category include: (1) the pullback of a fiber bundle $\pi: E \to B$ to a bundle over $X$ via a map $f: X \to B$; and (2) the preimage of a subset $A \subset Y$ (with inclusion map $\iota: A \to Y$) under the map $f: X \to Y$. ⊚

There are more interesting limits to be had with the use of complex, non-finite index categories, including the limits used in calculus. The terminology is intentionally suggestive: in the same way that a limit of a sequence in calculus is a single point that best approximates the $\mathbb{N}$-indexed sequence of points, $\lim \mathcal{F}$ is a single object in $\mathsf{C}$ that best approximates the image of the diagram $\mathcal{F}$. What distinguishes a categorical limit is its implicit uniqueness and its attendant morphisms to the diagram.

## 10.8 Colimits

The astute reader will note that the above examples possess parallel or dual notions which should likewise have a categorical formulation: direct sums, unions, disjoint unions, free products, and the like. Each is a **colimit** obtained by dualizing the definition as follows.

Given a diagram $\mathcal{F}\colon \mathsf{J} \to \mathsf{C}$, a **cocone** is a is an J-indexed family of morphisms $h_j\colon \mathcal{F}(j) \to A$ to a fixed object $A$ in C from the image of J in C that respects composition in J. These are objects in the corresponding **cocone category**, $\mathsf{CoCone}_{\mathcal{F}}$, where a morphism between cones $(A, h_{\mathsf{J}}) \to (A', h'_{\mathsf{J}})$ means that there is a morphism $A \to A'$ that makes the triangles with all $h_j$ and $h'_j$ pairs commute. The **colimit** of a diagram $\mathcal{F} : \mathsf{J} \to \mathsf{C}$ is the *initial* object in the cocone category of $\mathcal{F}$. This, too, has an interpretation as a universal property: the colimit gives a distinguished object $\operatorname{colim}_{\mathsf{J}} \mathcal{F} = \operatorname{colim}_j \mathcal{F}(j) \in \mathsf{C}$ along with J-respecting morphisms $h_j\colon \operatorname{colim}_{\mathsf{J}} \mathcal{F} \to \mathcal{F}(j)$ that factors through any other cocone.

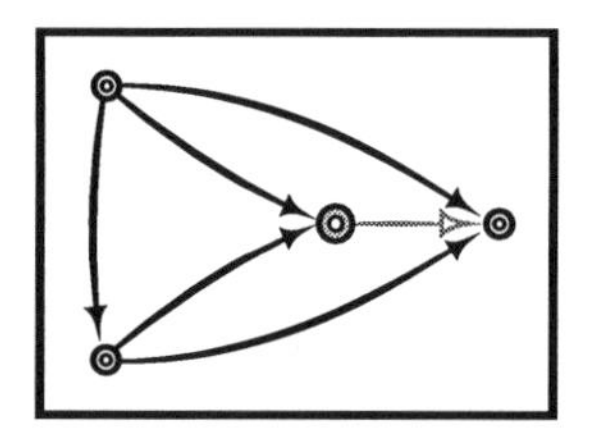

**Example 10.17 (Colimit examples)** ⊚

The colimit generalizes the initial (and indeed is the initial if J is empty). Other examples of colimits build an interpretation of a colimit as being agglomerative or disjunctive in nature.

**[union]** For the simple index category $\mathsf{J} = \bullet\ \bullet$ and a diagram $\mathcal{F}\colon \mathsf{J} \to \mathsf{Op}_X$, $\mathsf{CoCone}_{\mathcal{F}}$ has as objects open sets in $X$ containing the two open sets defined by $\mathcal{F}$. The colimit is the *smallest* such open set (it factors through any other via inclusion): this is the union.

**[coproducts]** A diagram of the same index category to Top has image spaces $X$ and $Y$. The colimit of $\mathcal{F}$ is a space $\operatorname{colim}_{\mathcal{F}}$ and maps from $X$ and $Y$ such that any other space with maps from $X$ and $Y$ must factor through the colimit: this is precisely the **coproduct** (or disjoint union) $X \sqcup Y$. The same colimit in algebraic categories like Vect or Grp is the **direct sum**, $\oplus$, of the objects. All of these colimits express a union and a disjunction.

**[OR]** In a poset $(P, \trianglelefteq)$, the colimit is the join, $\vee$, of the $\mathcal{F}$-image: the $\trianglelefteq$-smallest object of $P$ that is $\trianglelefteq$-larger that both terms in the image of $\mathcal{F}$. In a Boolean algebra, the colimit corresponds to the logical OR of the two image objects.

**[coequalizers]** The index category $\mathsf{J} = \bullet \rightrightarrows \bullet$ leads to a colimit that is dual to an equalizer. This is called the **coequalizer** of the diagram, and, in Vect, expresses the cokernel. This emphasizes that colimits are more like quotient objects than subobjects.

**[homology]** As with cohomology and limits, the zero$^{\text{th}}$ homology, $H_0(\mathcal{C})$ of a chain complex $\mathcal{C}$ is a colimit. For example, if $(X, \trianglelefteq)$ is a cell complex with face poset, then

$$H_0(X; \mathbf{G}) = \operatorname*{colim}_{\sigma \in X} \mathbf{G}, \tag{10.2}$$

where the limit is over the constant diagram $(X, \trianglelefteq) \to \mathbf{G}$.

**[pushouts]** The diagram $\mathsf{J} = \bullet \leftarrow \bullet \rightarrow \bullet$ leads to an amalgamation in the colimit. The colimit is called the **pushout** of the diagram and is sometimes denoted $A \cup_B A'$, co-opting notation from Top. An example of a pushout in this category is the wedge sum of pointed spaces $A \vee A'$, where $B$ is a singleton and $f$ and $g$ are inclusions. An algebraic example of a pushout in Grp is in the form of the Van Kampen Theorem (8.4), the statement of which is simplified greatly with categorical language: $\pi_1(U \cup V)$ is the pushout of the diagram

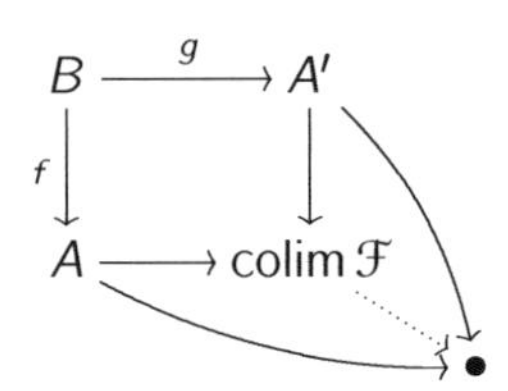


$$\pi_1(U) \leftarrow \pi_1(U \cap V) \rightarrow \pi_1(V).$$

**[infinite-dimensional spaces]** Throughout the text, invocations of certain infinite-dimensional cell complexes – such as $\mathbb{S}^\infty$ or $\mathbb{P}^\infty$ – have been made in ignorance of their precise definition. Colimits assist with this. Consider the diagram of $\mathbb{N}$ in Top given by the standard embeddings $\mathbb{R}^0 \hookrightarrow \mathbb{R}^1 \hookrightarrow \mathbb{R}^2 \hookrightarrow \cdots$. One can define $\mathbb{R}^\infty$ to be the colimit of this diagram. The reader can easily adapt this to define $\mathbb{S}^\infty$, $\mathbb{P}^\infty$, $\mathbb{T}^\infty$, and inductively built CW complexes. Fortunately, these colimits in Top exist. ⊙

In applications, one must work to show that [co]limits exist and to show how they behave under a given functor. A category for which [co]limits of *all* diagrams exist is called a **[co]complete** category. Functors that preserve [co]limits are called, of course, **[co]continuous** . A visceral comprehension of limits and colimits is essential to applications of category theory. Contemporary problems in data, networks, and sensing all involve localization of data and integration of local data into global: *limits and colimits.*

## 10.9 Sheaves, redux

The language of categories mirrors, expands, and simplifies greatly the many definitions of Chapter 9.

**Cellular Sheaves:** Let $X$ be a regular cell complex and $\mathsf{Face}_X$ be the poset category of the cells of $X$ under the face relation $\trianglelefteq$, so that objects are cells and there is a unique morphism $\sigma \to \tau$ for every face $\sigma \trianglelefteq \tau$ (with the identity face giving the identity morphism). By now the reader has probably observed that a cellular sheaf over $X$ is neither more nor less than a functor $\mathcal{F}\colon \mathsf{Face}_X \to \mathsf{C}$, where, in Chapter 9, Vect and Ab were used extensively for C. The first consequence of this extended language is that one may easily interpret what is meant by a sheaf of sets, or a sheaf of monoids, or any kind of categorical data: it is the composition that is key. One thinks of a cellular sheaf as being a *representation* of $\mathsf{Face}_X$ in C.

The wide variety of categories available as data types greatly expands the vector spaces used in Chapter 9 and yields a number of interesting objects. For example, a **complex of groups** is a sheaf on a cell complex $X$ taking values in Grp, a **complex**

**of spaces** is a Top-valued sheaf on $X$, *etc.* The important construct comes in gluing together local data over cells into a global object. The process for doing this gluing is specified in Equation (9.2): it is a choice of data on each cell that agrees according to faces and restriction maps. As an exercise in understanding concepts, the reader should show that the value of the sheaf $\mathcal{F}$ on all of $X$ is precisely the limit over the face poset category:

$$\mathcal{F}(X) = \lim_{\sigma \in X} \mathcal{F}(\sigma) = H^0(X; \mathcal{F}). \tag{10.3}$$

This equality is a slight abuse of notation. To wit: the explicit definition of $\mathcal{F}(X)$ from Equation (9.2) specifies local data on each cell. This is precisely a cone over the diagram $\mathcal{F}$: $\mathsf{Face}_X \to \mathsf{C}$. By the definition of a limit, there is thus a unique map from $\mathcal{F}(X)$ to the limit. By equality is meant that this unique map is a natural isomorphism, as can be shown by means of the compatibility condition for the assignment of local data. Note that for data taking values in more general categories than Vect or Ab, one must become concerned with the existence of limits. Fortunately, finite limits tend to be uncomplicated things. The same cannot be said for infinite limits.

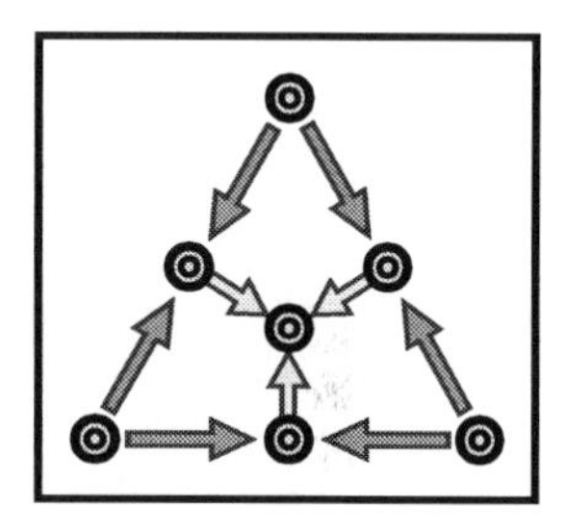

**Topological Sheaves:** The subtleties of sheaves over a tooplogy demand the refinements of categorical language. A presheaf on a space $X$ taking values in a category $\mathsf{C}$ is, precisely, a functor $\mathcal{F}$: $\mathsf{Op}_X^{\circ} \to \mathsf{C}$, where the preservation of composition of morphisms corresponds to the respecting of restriction maps. The stalk of a (pre)sheaf was defined via Equation (9.8) as an awkward sort of limiting equivalence. In truth, it is a colimit,

$$\mathcal{F}_x = \operatorname*{colim}_{U \ni x} \mathcal{F}(U),$$

where the $U$ are open sets containing $x$, partially-ordered by reverse inclusion ($U \to V$ for $V \subset U$) to provide a diagram over which the colimit is computed. It is an exercise to show that this categorical colimit gives the same answer as the more explicit mechanical process of (9.8).

As noted in Chapter 9, a presheaf alone does not make a sheaf. For this, an additional condition must be satisfied. This *gluing axiom* has several equivalent formulations in the language of this chapter. The most direct is a reinterpretation of the exact sequence in Equation (9.9). Namely, for any open cover $\mathcal{U} = \{U_i\}$ of $U$, the value of $\mathcal{F}$ on $U$ is an equalizer:

$$\mathcal{F}(U) \xrightarrow{\mathcal{F}(U \trianglelefteq U_i)} \prod_i \mathcal{F}(U_i) \underset{\mathcal{F}(U_j \trianglelefteq U_{ij})}{\overset{\mathcal{F}(U_i \trianglelefteq U_{ij})}{\rightrightarrows}} \prod_{i,j} \mathcal{F}(U_{ij}) . \tag{10.4}$$

While this formulation is correct and canonical, it is perhaps not optimal in its reliance on a mechanistic collation of pairwise gluings. A more elegant reformulation uses the full nerve complex $\mathcal{N}(\mathcal{U})$ of the cover $\mathcal{U}$ of $U$ and recapitulates the approach in §9.6

of approaching sheaves over a topology via nerves. The gluing axiom is equivalent to saying that

$$\mathcal{F}(U) = \lim_{U_J \in \mathcal{N}} \mathcal{F}(U_J), \tag{10.5}$$

independent of the cover $\mathcal{U}$. As in Equation (10.3), the equality is an abuse of notation, meaning that $\mathcal{F}(U)$ is naturally isomorphic to the limit of $\mathcal{F}$ over the cover. This limit is computed over the face poset of the nerve $\mathcal{N}$ of the cover $\mathcal{U}$, as in the cellular case above. A cell $U_J$ in the nerve $\mathcal{N}$ is indexed by a multi-index $J$ that determines a nonempty intersection of the open sets $U_i$, $i \in J$, making $\mathcal{F}(U_J)$, and the limit, well-defined. This condition is worth repeating: *a sheaf converts open covers into isomorphic limits.* Depending on one's predilection and mood, the equalizer or limit formulations of the gluing axiom will come more readily.

**Sheaf Operations:** The various manipulations of sheaves outlined in §9.7 provide excellent instances of categorical constructs. For example, a sheaf morphism is neither more nor less than a natural transformation $\eta\colon \mathcal{F} \Rightarrow \mathcal{F}'$ between sheaves-as-functors. In fact, depending one which subject one learns first – sheaves or categories – this example may assist in making concrete an otherwise opaque definition. Thinking of a natural transformation as a mapping between data structures over a category can be illuminating.

Sheaves provide a topological and data-centric view of categories that may assist in making the subject more visceral. One can think of any functor $\mathcal{F}\colon \mathsf{C} \to \mathsf{C}'$ as an assignment of data from the target category $\mathsf{C}'$ to the source category $\mathsf{C}$ in a sheaf-like manner. Natural transformations $\eta\colon \mathcal{F} \to \mathcal{F}'$ provide the means of transforming from one data structure to another. When the target category is sufficiently algebraic (say, Ab), then it is possible to define kernels and cokernels of data, leading to the cohomology of $\mathcal{F}$. This foreshadows one of the most powerful set of topological tools in category theory: **derived functors**, of which sheaf cohomology is a motivating precursor.

## 10.10 The genius of categorification

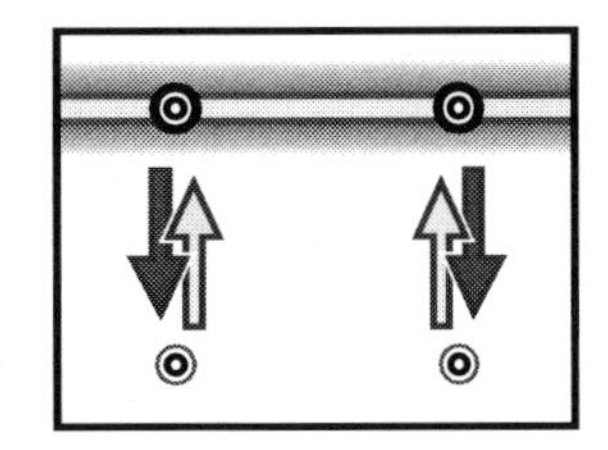

The goal of this chapter is to point the reader to **categorification**: the systematic lifting of, say, numerical data to a higher categorical structure, with a concomitant functoriality. This functoriality is key, and permits a lifting of numerical equality (*"I wonder why these two numbers happen to be equal?"*) to an algebraic equivalence (*"These two structures are isomorphic."*), with the ability to make additional high-level connections. A subsequent **decategorification** back to the numerical can, in the best cases, provide explanatory power and intuition for deeper results.

Categorification – like category theory – is not itself a branch of topology; however, it has been so influenced by and effective in topology that it is fitting to end this text with a gentle invocation. The focus will be on the sprit of the subject rather than on rigorous results.

**Example 10.18 (Arithmetic, simple)** ◎

Counting is the primal decategorification, so internal as to have been sublimated as such. To each finite collection of objects (apples, oranges, coins, cats, or czars), one associates an element of $\mathbb{N}$ – the cardinality of the set. This abstraction permits referencing a generic set of $N$ objects, without having to specify the membership thereof. To categorify this, the reader might first try the category of finite sets, FinSet. The *cardinality* function $|\cdot|$, sends objects of this category to $\mathbb{N}$ and descends to a bijection from isomorphism classes of objects in FinSet to $\mathbb{N}$. Note that FinSet has more structure than $\mathbb{N}$: an isomorphism between objects specifies identities as well as preserving cardinalities. Because FinSet is a category, one can apply categorical constructs and decategorify to see the numerical impact. For example, the colimit of a pair of sets (the disjoint union) decategorifies to addition; the limit of a pair (the cartesian product) decategorifies to multiplication. Certain basic laws – commutativity, associativity, distributivity – have higher equivalents, and the order relation $\leq$ on $\mathbb{N}$ is enriched to the language of injectives and projectives. Interesting though this may be, a reformulation of arithmetic in terms of category theory provides no new insight; rather, it is an elementary example of how the most primal bits of mathematics are the shadows of emanations from higher-up the hierarchy of structure. ◎

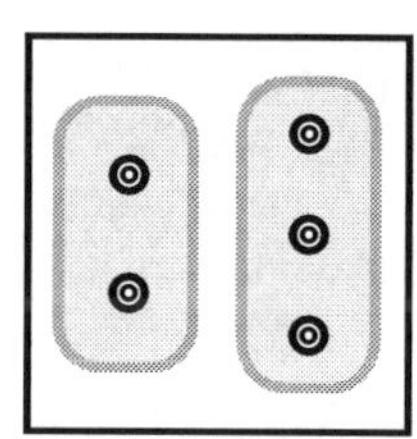
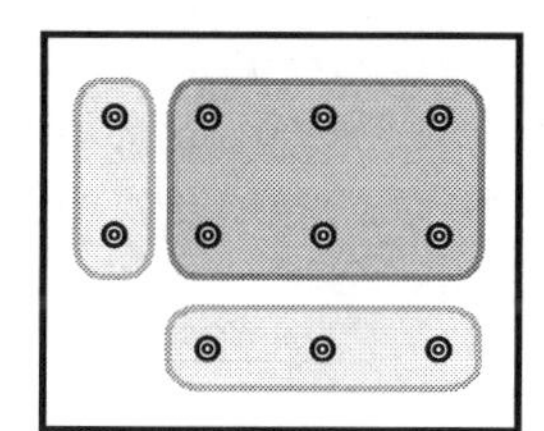
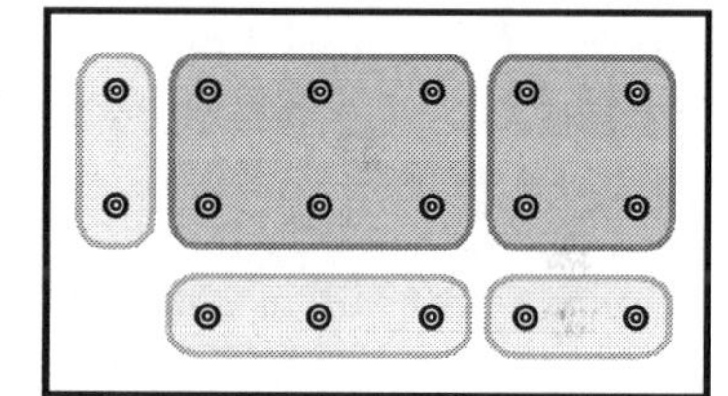

**Example 10.19 (Arithmetic, complex)** ◎

The categorification of $\mathbb{N}$ to FinSet is not optimal: it leaves unclear how to recover negative numbers, for example. Fortunately, there are multiple possible categorifications, some of which are more generalizable. Consider the subcategory FinVect of Vect consisting of finite-dimensional vector spaces over a field $\mathbb{F}$. Then one can lift $\mathbb{N}$ to this category by means of the decategorification dim: an $n$-dimensional vector space is the lift of $n \in \mathbb{N}$. Isomorphic objects in FinVect have the same dimension. In this categorification, the lift of addition is the direct sum, $\oplus$, and the lift of multiplication is to the tensor product (over $\mathbb{F}$), $\otimes$. There are (unique) identity objects for these operations: the 0-dimensional vector space, for $\oplus$, and the 1-dimensional vector space $\mathbb{F}$, for $\otimes$. By categorifying to a linear-algebraic setting, the morphisms between (even isomorphic) objects are richer and can store more data about relationships.

Other categorifications are more enlightening still. Consider FinChCo, the category of finite, finite-dimensional chain complexes (over a field $\mathbb{F}$), with chain maps as morphisms. The previous categorification to FinVect embeds in FinChCo as a sequence with one nonzero term. With complexes, one has a categorification of $\mathbb{Z}$ as follows. Given a two-term sequence (extended by zeros),

$$\cdots \longrightarrow 0 \longrightarrow 0 \longrightarrow V \xrightarrow{A} W \longrightarrow 0 \longrightarrow 0 \longrightarrow \cdots,$$

one can decategorify to $\mathbb{Z}$ by a difference of dimensions: $\dim V - \dim W$. The reader will recognize once again the appearance of the Euler characteristic $\chi\colon \mathsf{FinChCo} \to \mathbb{Z}$. By lifting $\mathbb{Z}$ to all of FinChCo, arbitrary (finite) sequences of vector spaces collapse via alternating sums of dimensions, providing a rich structure of sequences and chain maps comparing them. ⊚

**Example 10.20 (Co/Homology)** ⊚

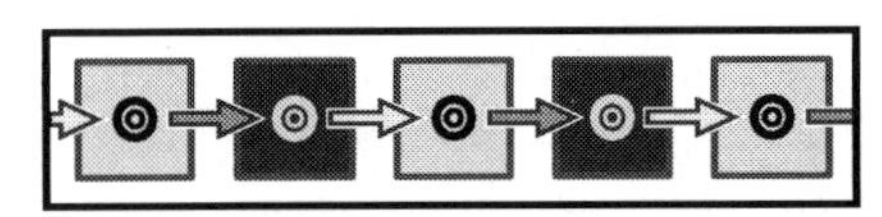

The two primal examples of decategorification – dimension and Euler characteristic – have appeared repeatedly in this text. By Theorem 3.7, they are together *complete* invariants of definable sets up to definable homeomorphism. Lifts of these two types of numerical invariants are central to algebraic topology. For a cell complex $X$, the combinatorics of adding together the number of cells, weighted by the parity of the cell dimension *seemed* in Chapter 3 to give a serendipitous topological invariant in $\chi$. Likewise, in Chapter 4, computing dimensions of simplicial homology groups to count cycles up to equivalence *seemed* to not depend on the simplicial structure. Mathematics knows no such generous deity, and coincidence hints at deeper reasons. The explanation given in Chapter 5 used the language of homology and exact sequences. More vocabulary is now available: *co/homology is a categorification.*

By converting the cells and assembly instructions into a chain complex $\mathcal{C} = (C_\bullet, \partial)$, one obtains a homology $H_\bullet$ that is functorial (maps between spaces yield chain maps that descend to homomorphisms on homology) and thus explains why topological invariance holds. Better still, natural equivalences between homology theories permit lifting $\chi$ and $H_\bullet$ to singular or non-cellular settings. The final ingredients are the corresponding decategorifications which send homology groups back to $\mathbb{N}$ (via dimension – a Betti number) or to $\mathbb{Z}$ (via the Euler characteristic). With the addition of cohomology, one picks up a multiplicative structure and a wealth of new dualities and relationships.

Nearly every tool in algebraic topology – long exact sequences, duality, naturality, excision, connecting homomorphisms, cup products, bundles, fibrations, sheaves – is built on the functoriality that categorification enables. Nearly every application in this text – to data, dynamics, games, networks, sensing, signals, clustering, coloring, motion planning, material defects, and more – rises from this functoriality. ⊚

**Example 10.21 (Co/Sheaves)** ⊚

Sheaves and sheaf cohomology (and their duals) provide categorifications of numerical data distributed over a space. One simple example from this text is a flow sheaf from

§9.4-9.5. Recall that for a given flow on a directed acyclic graph $X$ with capacity constraints, one cares about the net flow value at the source/target. The numerical capacities were lifted to a flow sheaf $\mathcal{F}$ by means of dimension – stalks over edges are vector spaces whose dimension equals the flow capacity at the edge. Restriction maps for the sheaf encode flow routing (or coding). The benefits of this categorification include (1) the ability to use sheaf cohomology to characterize information flows and flow values; (2) the ability to relate flow- and cut-values by means of long exact sequences; (3) the presence of duality in sheaf cohomology as the analogue of decomposition of flows into loops; and (4) the ability to use Euler characteristic to see obstructions to max-flow and min-cut values.

A better example of sheaves-as-categorification in this text is the Euler calculus of Chapter 3. As the Euler characteristic is a decategorification of co/homology, the Euler integral is a decategorification of compactly supported sheaf cohomology with coefficients in sheaves associated to constructible functions. As per §9.9, any constructible function $h \in \mathrm{CF}(X)$ on a tame set $X$ lifts to a complex of cellular sheaves $\mathcal{F}_h^\bullet$ compatible with the triangulation of $X$ induced by $h$. The decategorification of this complex by means of Euler characteristic *is* the Euler integral $\int_X \cdot d\chi \colon \mathrm{CF}(X) \to \mathbb{Z}$. The advantage of this perspective is an immediate access to functoriality, yielding the Fubini Theorem (Theorem 3.11), the integral transforms of §3.9, and more [95, 191, 272]. ⊚

## 10.11 "Bring out number"

With reflection, one suspects that many questions in applied mathematics are at heart functorial in nature and are profitably viewed through the lens of categorification. A few speculations appear below. Some of these may be realized as proper categorifications; others are a loose lifting that invoke the insubstantial ghost of functoriality. This text closes with a hope of provoking the reader into finding and using functoriality.

**Example 10.22 (Data analysis)** ⊚

Topological data analysis has generated a substantial body of evidence that topology facilitates the management and interpretation of large, unwieldy point clouds. How is this accomplished? One ingredient is the dextrous application of co/homology functors in order to capture features of a data set that are *global* (large-scale as opposed to fine detail), *computable* (thanks to Mayer-Vietoris and other exact sequences), and *robust* (topological invariants have good properties with respect to noise). The fact that homology is a functor from the category of topological spaces grants to topological data analysis a degree of nonchalance with respect to coordinates.

Data analysis begins with clustering, in the same way that homology begins with $H_0$. The clustering functors of §10.4 provide an immediate and satisfying categorification, but the problem of classifying and using novel clustering schemes based on functoriality is as yet incomplete [60]. The (persistent) homologies $H_k$ provide the higher-order terms in the Taylor series of shape for data. This is limited primarily in the restriction to a single parameter – multi-parameter persistence cannot access a suitably simple Structure Theorem as in 5.21. Perhaps a solution lies in the reformulation of persistent homology as a cosheaf over a 1-dimensional base space, as in §9.12: the classification of constructible cosheaves over the plane is much more complex that over the line, but not unimaginably so. Whether this, or more advanced representation-theoretic ideas, or deeper categorical methods arise to resolve the problem, it seems clear that the solution to multiparameter topological analysis of data lies in more abstraction, not less. ⊚

**Example 10.23 (Numerical analysis)** ⊚

When solving a partial differential equation [PDE] by means of a numerical method, one typically applies the method and examines the large-scale qualitative features of the solution. To validate that these results are not artifacts of the numerical scheme, it is common to run the method again on a refined grid or with a shorter time step, noting again the qualitative features of the solution. If there is a match (or, one might say, "equivalence") between these two solutions, then one infers that the solution is genuine. This is a physical instance of the maxim that morphisms and functors are more useful than objects. This example, though a bit cartoonish, illustrates the difficulty in building a careful categorification: what, exactly, is meant by the qualitative features of a numerical solution and an equivalence thereof?

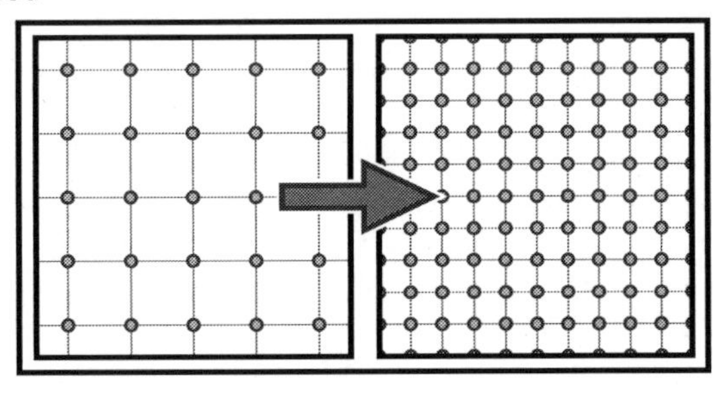

One way to proceed is to enrich a discretization with structure inherited from the PDE. Any numerical scheme works with functions on a discrete spatial domain with discrete time axis. At each such discretization point in space-time, one has a numerical value. The next time-value is given as a function of nearby space-time values, according to the relevant numerical scheme coming from the differential equation. The simplest such numerical schemes are agnostic as to the precise form of the differential equation. Recently, however, it has been shown that one can get better numerical results when the discretization is enriched and forced to preserve some structure inherited from the differential equation. For example, in the case where a PDE is conservative, such as Hamiltonian systems (*cf.* Examples 6.16 and 7.19), the dynamics must preserve an invariant differential form (say, a volume or symplectic form). Discretizing (decategorifying) a solution removes this structure. By modifying the discretization to retain a shadow of the appropriate invariant form, one can work to ensure that the numerical scheme preserves this structure, yielding more accurate simulations. This is precisely the motivating

idea behind the **discrete exterior calculus** [12, 13, 32, 94]. Though often presented as a structure-enriched variant of finite-element methods, it is perhaps better to think of it as a decategorification of differential forms and Hodge calculus that *forgets less* so as to retain enough functoriality to perform operations (wedge products, exterior derivatives, Hodge-stars, *etc.*). The specific decategorification is critical.

**Example 10.24 (Dynamics and index)** ⊚

Applications of algebraic topology in dynamics and differential equations rest on a foundation of index theory. In its first appearance in §3.4, the fixed point index $\mathcal{I}$ was described as an integer invariant for equilibria of planar vector fields, computable via a line integral and robust thanks to Green's Theorem. The Poincaré-Hopf Theorem (3.5) asserted that this index is *additive* and yields the Euler characteristic when *"integrated"* over the domain. In Example 4.23, $\mathcal{I}$ was revealed to be a degree, and therefore applicable to fixed points of self-maps as well as equilibria of vector fields, as per Example 7.20. This culminated in the Lefschetz index of §5.10, interpreted first as something like an Euler characteristic of degrees, then, in §7.7, revealed to be an integral with respect to Euler characteristic. In the language of Chapter 9, the fixed point index lifts to a complex of constructible sheaves, the Euler characteristic of which *is* the Lefschetz index.

This progression of indices exemplifies categorification nicely. It has been seen on the level of co/homology that the two simplest types of decategorification – dimension and Euler characteristic – work well with vector spaces and sequences of vector spaces respectively. When lifting to dynamics, these have analogues. The dimension of a vector space generalizes to the trace of a self-map; the Euler characteristic of a chain complex generalizes to the Lefschetz index of a chain map. Applying these to the identity map recovers dimension and Euler characteristic. Thus, trace and Lefschetz index are the appropriate numerical invariants of dynamics. To what do these categorify? In the same way that counting cells and faces lifts to co/homology and functoriality, counting fixed points and their indices lifts to the action of dynamics on co/homology.

| | |
|---|---|
| $V$ | dim |
| $A\colon V \to V$ | trace |
| $V_\bullet$ | $\chi$ |
| $A_\bullet\colon V_\bullet \to V_\bullet$ | $\tau$ |

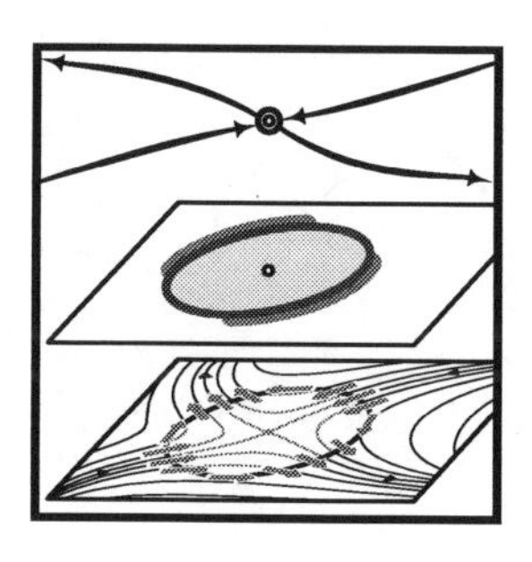

The story of categorification in dynamics is richer than it at first appears. Consider the gradient fields of Morse theory in Chapter 7. Here, the fixed point index is nearly mute, in that a nondegenerate critical point has fixed point index $\pm 1$. This lifts, however, to an $\mathbb{N}$-valued Morse index, $\mu$, which characterizes the local dynamics. Furthermore, with this richer index, one can stack the critical points into the Morse polynomial: the Morse inequalities of (7.1) are an instance of a richer algebra unveiled with slight added structure. This continues with the Morse homology of §7.3. By converting critical points into chains, the dynamics is shown to recover the homology of the underlying manifold (Theorem 7.3). Corollary 7.6 asserts that the Morse homology decategorifies to give the Euler characteristic.

As one ascends to higher types of structure, categorical language is essential. The flow category of Example 10.4 converts gradient dynamics into a category, with critical points as objects and morphisms as flowlines. This category, like Morse homology, remembers the underlying manifold, via the nerve construction of Example 10.7 and the geometric realization of simplicial sets as in Example 10.6: the flow category $C_h$ of a Morse function $h\colon M \to \mathbb{R}$ has nerve $\mathcal{N}(C_h)$ whose geometric realization is homotopic to $M$ [69].

One of the lessons of Chapter 7 is that the restrictions of classical Morse theory – manifolds, smoothness, nondegeneracy, gradients – are largely ignorable, given an appropriate ascension in technique. For example, the categorification of critical points in a nondegenerate smooth gradient field to Morse homology in §7.1 returns the homology of the base manifold, as in Theorem 7.3. This is mirrored in the degenerate setting of a discrete gradient field on a cell complex using discrete Morse homology, in Theorem 7.23. Better still is the ability to jettison the gradient assumption and work with dynamics, as occurs in the Conley index of §7.6. This, in its homotopic and homological variants, is the vanguard of efforts to categorify dynamics. The prototypical argument for existence of connecting orbits in Examples 7.15-7.16 hints at a functoriality of the Conley index which exceeds that of the index of Morse.

This story, however, is not completed. There are a multiplicity of extensions of the Conley index for different settings [128, 129, 266, 286], including, most notably, the setting of discrete dynamics based on the index pairs as in §7.7. Some recent work has focused on properties of the Conley index over a large parameter domain [11, 140], work that hints at sheaf-like properties of the index. This lies within the purview of the categorification of dynamics, but does not exhaust the possibilities. Space and time prevent an explanation of zeta-functions for counting periodic orbits [264], model-category structures for dynamical systems [184], the categorification of Floer homology to the **Fukaya category** [19, 275] and its applications [96], and the interaction of Lagrangian submanifolds with sheaves [170]. There is much more to be done. ⊙

**Example 10.25 (Optimization)** ⊙

There are numerous instances of *minimax theorems* in applied mathematics: in game theory, in optimization and linear programming, in differential equations, and more. The multiplicity of incarnations of a *minimax* nature lead to a suspicion of a deeper principle in action. Given the fact the minima and maxima lift to lattice notions of meet/join or categorical notions of limits/colimits, it may be conjectured that all minimax theorems are expressions of relationships between limits and colimits in an appropriate category.

One instance of this is the Max-Flow/Min-Cut Theorem, alluded to previously in Example 6.5 and §9.4. Recent results of Krishnan [201] provide a dramatic categorification. Recall that one begins with a directed (acyclic) graph $X$ from source to target nodes, $s \to t$, with edges having capacities in $\mathbb{R}^+$. The goal is to place a conservative flow on $X$ with maximal throughput at the source/target nodes. As a means of keeping flow conservation at the source and target, append to $X$ a feedback edge $e$ from target-to-source. Then, as noted in Example 6.5, a flow evokes a 1-cycle in homology, and a cut a dual cocycle. When framed this way, it seems clear that the max-flow min-cut theorem is a topological theorem. It is, and is best seen as such through the lens of categorification.

The first step is to categorify the capacity constraints as a cellular **capacity sheaf** over $X$. In the classical setting of numerical capacities $\mathrm{cap}(e) \in \mathbb{R}^+$ assigned to an edge $e$ of $X$, the capacity sheaf $\mathcal{F}$ over $e$ takes values in the interval $[0, \mathrm{cap}(e)]$ under addition, where the addition operation is (1) *non-invertible* – subtraction is forbidden; and (2) *partial* – not all pairs of numbers are allowed to be added. This partial addition encodes the constraint, since addition must not exceed the capacity. This kind of algebraic structure is a **partial commutative monoid**. It is a brilliant observation that one can categorify the constraints of the optimization problem as a sheaf, albeit over the algebraically intricate category of partial commutative monoids.

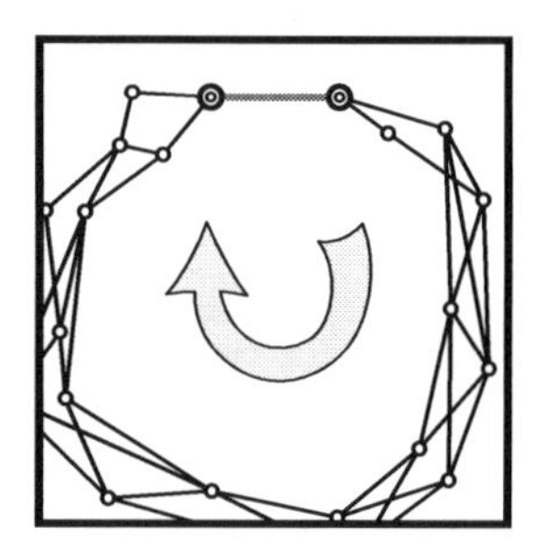

All other ingredients of the max-flow min-cut theorem lift likewise. Flow conservation mimics the homological cycle condition. However, to respect the directedness of the underlying graph, it is essential to build a homology theory that remembers and respects the directions of edges. This is encoded in an **orientation sheaf**, $\mathcal{O}$, taking values in $\mathbb{N}$-modules (in contrast to the un-oriented $\mathbb{Z}$-modules); this has the property that the stalks of $\mathcal{O}$ are copies of $\mathbb{N}$ summands, where each summand represents an independent directed local path. A directed homology theory $H_1$ taking values in sheaves of partial commutative monoids is constructed via an equalizer diagram (kernels being not well-defined), and the partial commutative monoid of flows on $X$ respecting the directions and constraints of $\mathcal{F}$ is, precisely,

$$H_1(X; \mathcal{F}) := H^0(X; \mathcal{F} \otimes \mathcal{O}) = \lim_{C \subset X} (\mathcal{F} \otimes \mathcal{O})(C). \tag{10.6}$$

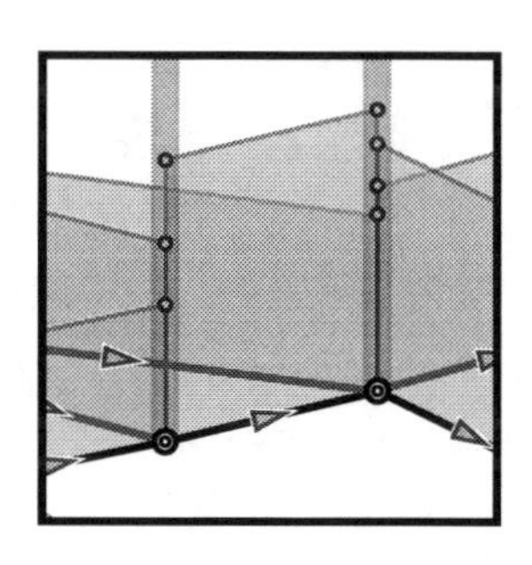

This definition/theorem has bound up within it a version of Poincaré duality for directed spaces proved by Krishnan [200] that both foretells and enables the sheaf-theoretic max-flow min-cut theorem. The left term of this equation becomes a categorification of global flows on $X$, and the right term becomes a categorification of local flow capacities. The subtle work is how to partially decategorify the left and right sides to flow-values and cut-values comparable in a common framework. This is done by mapping both sides above into a directed homology $H_0(X; \mathcal{F})$, expressible as something like a colimit. In the classical setting, this partial decategorification translates to the following equality of partial commutative monoids:

$$\bigcup_{\text{flow } \varphi} [0, \mathrm{val}(\varphi)] = \bigcap_{\text{cut } C} [0, \mathrm{val}(C)] \tag{10.7}$$

Thus, the maximal flow value equals the minimal cut value: but more than this, these coincident numbers descend from isomorphic structures in a categorification. The machinery required to complete the details is formidable and does not fit in this text, but the benefits are compelling. Because the duality holds for a very general category of capacity sheaves taking values in partial commutative monoids, other non-classical optimal flow problems (of commodities, signals, and logics) over networks can be solved via duality. This is a stellar example of categorification in applied mathematics. ⊚

# Notes

1. This text has worked with **small** categories whose objects and morphisms form sets. There are several unpleasant technicalities associated with categories that are not small.
2. There are relatively few computational tools for working with categories; one hopes that with increased awareness of applications that this will improve.
3. The interpretation of clustering as a functor in §10.4 is a very simple example to demonstrate the benefit of categorical thinking. Not all clustering methods used in practice are functors. One of the more popular methods, $k$-**means** clustering, takes as its input the point cloud $\mathcal{Q}$ along with a number $k \in \mathbb{N}$ and an initial $k$-tuple of points in $\mathcal{Q}$. It returns, via an iterative process, a partition of $\mathcal{Q}$ into $k$ clusters. It seems impossible to render such a method functorial, since the number of clusters is fixed *a priori*.
4. The interleaving approach to stability of persistent homology in §10.6 is a rapidly-developing subject. In the initial works on stability [70] a metric (called the **bottleneck distance**) on multi-sets of points in the birth-death plane was used. This has been shown to be isometric to the interleaving distance. Bubenik and Scott [51] use the language of natural transformations; Lesnick [207] uses the language of modules and structure theorems. The most recent work (at the time of publication) is a broad generalization of the categorical approach in §10.6 [50].
5. The terminology of limits and colimits in this text is not universal. Many sources use **direct limit**, **inductive limit**, or $\varinjlim$ to denote the *colimit*; the corresponding terms for a *limit* are **inverse limit**, **projective limit**, and $\varprojlim$. Yes, this *is* confusing.
6. Derived functors, of which sheaf cohomology is the precursor, are the true, unrealised, goal of this chapter. The interested reader should with all haste master the basics so as to ascend to this cornice.
7. Yes, dear reader, the progression from objects to morphism to functors does not end with natural transformations; nor, indeed, does it end at all. Mathematics is currently brimming with $A_\infty$ structures, $N$-categories, and other progressions which iterate the notions of ever-higher morphisms between ever-higher objects to dizzying heights. The use of $A_\infty$ structures seems to be particularly potent [31]– that this potency will descend to applications is to be hoped and remains to be seen.
8. Like the previous, this chapter is a shocking reduction of a vast intricate theory into a cartoonish sketch. The reader whose sense of adventure is aroused will find an unending country to explore. The book by Awodey [20] is particularly friendly to readers from Computer Science. Students of topology may wish to begin further exploration with adjunctions, exponentials, Kan extensions, homotopy co/limits, nerves, the Yoneda Lemma, and the other basic tools of category theory. The classic, complete reference

is [211]. For a serious high-level treatment with applications to sheaves, [192] is a valuable resource; for applications to logic, [160, 212] are the appropriate references. The most interesting new books on the subject are by Leinster [205] (extremely well-written) and Spivak [282] (extremely creative at making connections to data structures).

9. The author, in his youth, mocked category theory as so much unapplicable generalized nonsense. May the reader learn from his errors. *Miserere mihi peccatori.*

## Appendix A

# Background

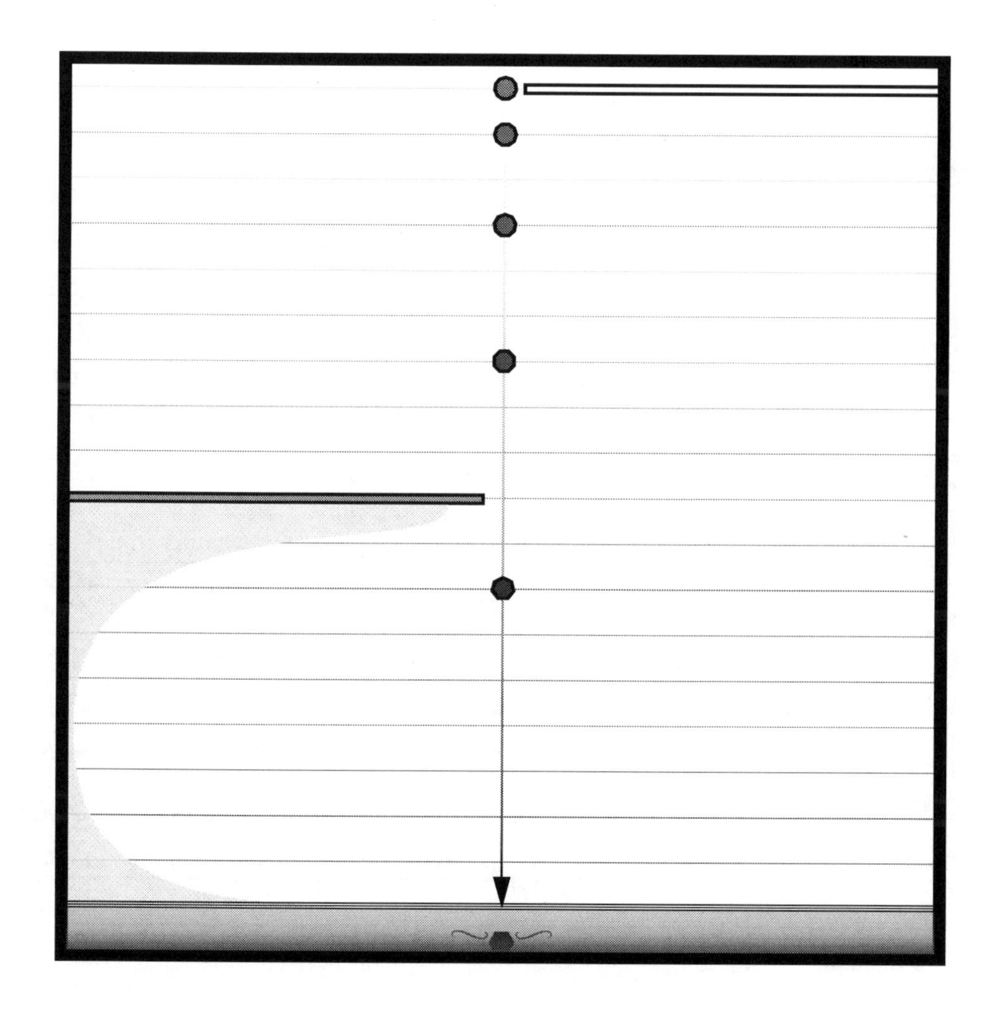

## A.1 On point-set topology

It is an unfortunate fact that most students' exposure to topology begins and ends with the set-theoretic foundations. To minimize such would be akin to minimizing grammar or arithmetic in primary school (which produces bad writers and no mathematicians). This text cannot cover all the foundations; however, a few definitions will perhaps provide appropriate pointers, with, *e.g.*, [238], as a source for more thorough coverage.

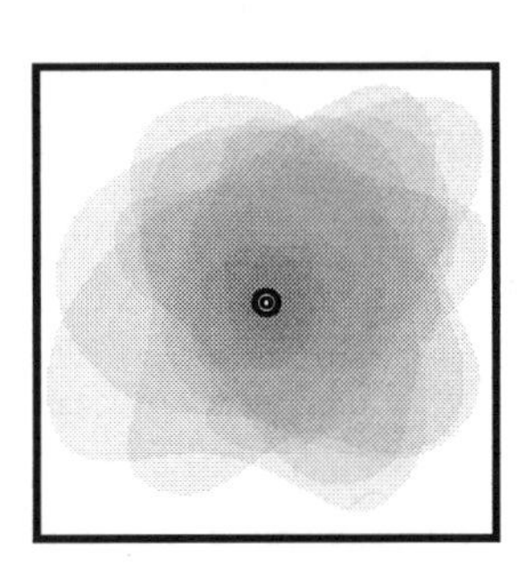

A **topology** on a set $X$ is a collection $\mathcal{T}_{op}$ of subsets of $X$ declared to be the open sets. The topology must satisfy three properties:

1. $\mathcal{T}_{op}$ is closed under arbitrary unions;
2. $\mathcal{T}_{op}$ is closed under finite intersections; and
3. $\mathcal{T}_{op}$ contains both $X$ and the empty set $\varnothing$.

From this spartan frame is the subject supported. A set $A \subset X$ is **open** if $A \in \mathcal{T}_{op}$ and **closed** if its complement $X - A \in \mathcal{T}_{op}$ is open. The **standard topology** on $\mathbb{R}^n$ is the topology whose elements consist of all possible unions of all open metric balls of all sufficiently small radii at all points. There are well-defined ways to generate topologies from such a **basis** for a topology.

The concept of a topology allows one to encode proximity without specifying a metric. This feature is increasingly relevant to applications in data, networks, and biology, where natural metrics may be obscured or nonexistent. The definition of a topology, though primal, is powerful and subtle.

**Example A.1 (Prime numbers)** ◎

The following brilliant proof is due to Furstenberg; it proves the infinitude of prime numbers using only the definition of primeness and basic properties of a topology. Consider the topology on $\mathbb{Z}$ generated by affine subsets $\{a\mathbb{Z} + b : a \neq 0\}$, meaning that nonempty open sets are generated by unions and finite intersections of these basis sets. Let the reader verify that: (1) this indeed forms a topology on $\mathbb{Z}$; (2) each basis affine set $\{a\mathbb{Z} + b : a \neq 0\}$ is both open and closed in this topology; and (3) no

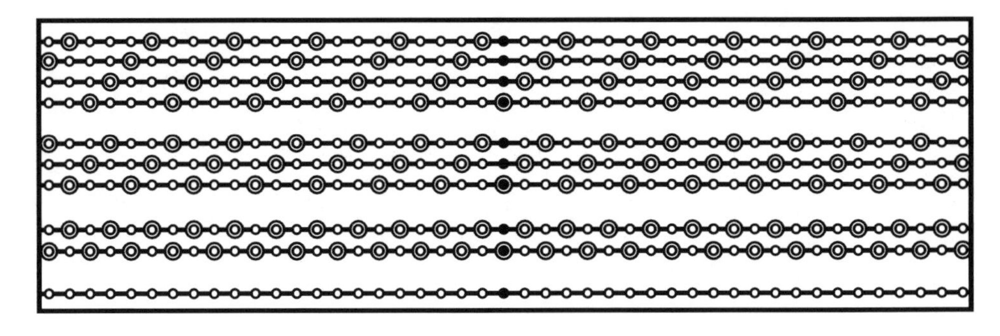

nonempty finite subset of $\mathbb{Z}$ is open. Then, consider the set:

$$S = \bigcup_p p\mathbb{Z} \quad p \text{ prime} .$$

As the union of open sets, $S$ is open. If the number of primes is finite, then $S$ is also closed, by the fact that a finite union of closed sets (being the complement of a finite intersection of open sets) is closed. However, the complement $\mathbb{Z}-S = \{-1, 1\}$ is finite: contradiction. Conclusion: there are infinitely many primes. Note that the only point at which arithmetic was used was in the definition of $S$ and the verification of the definition of a topology. ⊚

Other definitions common in point-set topology are:

1. The **interior** of a subset $A \subset X$ is the largest open subset in $A$, or, equivalently, the union of all open subsets of $A$.
2. The **closure** of $A \subset X$ is the smallest closed superset of $A$, or, equivalently, the intersection of all closed supersets of $A$.
3. The **boundary** of $A \subset X$, $\partial A$, is the complement of the interior in the closure.
4. A space $X$ is **compact** if every open cover of $X$ restricts to a finite subcover.
5. A function is **continuous** if the inverse image of open sets is open.
6. A **homeomorphism** is a continuous bijection with continuous inverse.

A topology on $X$ can induce topologies on spaces built from $X$. Given a pair of spaces $X$ and $Y$, the (cartesian) **product** $X \times Y$ is the set of ordered pairs $(x, y)$; the **product topology** is that with basis $U \times V$ for $U \in \mathcal{T}_{\mathrm{op}_X}$ and $V \in \mathcal{T}_{\mathrm{op}_Y}$. For $A$ any subset of a space $X$, the **subspace topology** on $A$ consists of the collection $\{U \cap A\}$ for all $U \in \mathcal{T}_{\mathrm{op}}$. This is the smallest or *weakest* topology on $A$ making the inclusion map $A \hookrightarrow X$ continuous. The subspace topology is assumed whenever one discusses a subset of a space as a space in its own right.

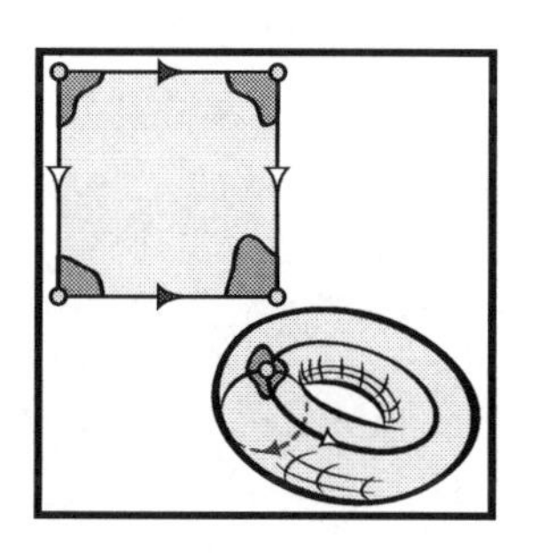

When collapsing a space $X$ to a quotient by identifying certain subsets, one imposes a topology on the image. Given a surjective function $q\colon X \to Y$, the **quotient** topology on $Y$ via $q$ is the collection of sets $V \subset Y$ such that $q^{-1}(V)$ is open in $X$. This is the largest or *strongest* topology on $Y$ making the quotient map continuous.

Most simple spaces have an obvious quotient or subspace topology, and the arcana of the subject emphasized in early texts is, for most practical purposes, inessential. There are, however, subtleties associated with topologies on infinite-dimensional spaces, particularly function spaces. The beginner should focus on the **compact-open** topology on the space $C(X, Y)$ of maps $f\colon X \to Y$. This topology is the smallest generated by sets of the form $C(K, U)$, where $K \subset X$ is compact and $U \subset Y$ is open.

## A.2 On linear and abstract algebra

Deep knowledge of abstract algebra is not a prerequisite for reading this book, nor for learning many basic aspects of algebraic topology. The reader who *does* know algebra well will find many of the important tools not in this book (Hom, Ext, Tor, ⊗, *etc.*) natural and implicit. A few basic concepts suffice for those coming from a minimal background. At the very least, the reader needs proper training in linear algebra. From this subject, little need be said, with two crucial exceptions.

First: quotient spaces. Given a vector space $V$ and subspace $W$, the quotient space $V/W$ consists of equivalence classes $[v]$ of vectors in $V$ modulo vectors in $W$. Specifically, $[v] = [v']$ if and only if $v' - v \in W$. These **cosets** are often illustrated in terms of the orthogonal complement $W^\perp$ of $W$ in $V$. This is erroneous, as the orthogonal complement requires the existence of a well-defined inner product on $V$ and *such is not required* to define $V/W$. Another common error is, via conflation of $V/W$ and $W^\perp$, to assume that $V/W$ is a subspace of $V$. *It is not.*

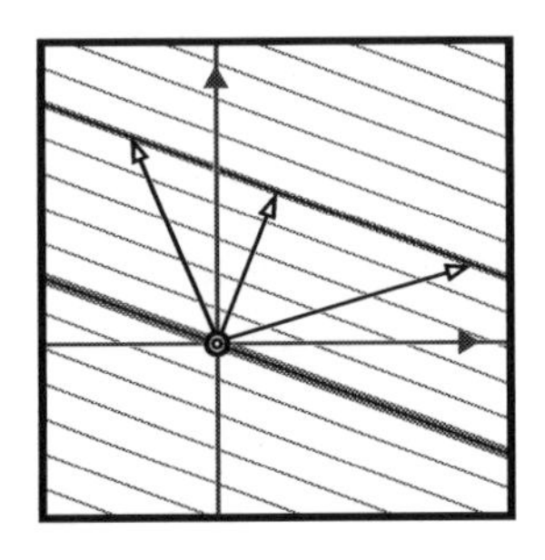

Second: transformations. More important that vector spaces themselves are linear transformations between them. One characterizes $A\colon V \to W$ in terms of auxiliary vector spaces and simple transformations. The **image** of $A$ is a subspace im $A$ of the codomain $W$; the **kernel** of $A$ is $A^{-1}(0)$, a subspace ker $A$ of the domain $V$. Less familiar to students is the equally-important **cokernel**, coker $A$, the quotient space $W/\text{im } A$ of the codomain by the image.

While many of the constructs in this text can be accomplished using only the language of linear algebra, it quickly becomes important to grasp the more general algebraic structures available. A **group** is a set $\mathbf{G}$ together with a binary operation $\bullet\colon \mathbf{G} \times \mathbf{G} \to \mathbf{G}$ (often called *multiplication*) that satisfies the following:

1. The operation is associative.
2. There is an identity element $e \in \mathbf{G}$, with $e \bullet g = g \bullet e = g$ for all $g \in \mathbf{G}$.
3. There is an inverse operation $g \mapsto \overline{g}$ so that $\overline{g} \bullet g = e = g \bullet \overline{g}$ for all $g \in \mathbf{G}$.

Examples include number systems ($\mathbb{Z}$, $\mathbb{Q}$, $\mathbb{R}$, and $\mathbb{C}$, but not $\mathbb{N}$) under addition (but not multiplication); vector spaces under vector addition; square matrix groups ($\mathrm{SL}_n$, $\mathrm{SO}_n$, $\mathrm{U}_n$) under matrix multiplication; and polynomials $\mathbb{Z}[x]$ under addition.

Groups are broader than vector spaces, yet have some structural similarities. A **subgroup** $\mathbf{H} < \mathbf{G}$ is a subset of a group which is itself a group under the group operation, meaning, in particular, that it is closed under the group operation. A **homomorphism** is a function $\phi\colon \mathbf{G} \to \mathbf{K}$ between groups which is linear in the sense that $\phi(g \bullet g') = \phi(g) \circ \phi(g')$, where $\circ$ denotes the group operation in $\mathbf{K}$. The **kernel** of a homomorphism is the subgroup $\ker \phi = \phi^{-1}(e)$, where $e$ denotes the identity element of $\mathbf{K}$.

Groups are often specified using a **presentation**, in which a collection of **generators** $\{g_\alpha\}$ and their inverses form finite words with the usual associativity, identity, and inverse rules applying. To these words are applied a set of **relations**, thought of as replacement rules of the form $r_\beta = e$, for some collection of finite words $r_\beta$. For example, the group $\mathbb{Z}^2$ under addition has the presentation

$$\mathbb{Z}^2 \cong \langle x, y \ :\ x \bullet y \bullet x^{-1} \bullet y^{-1} = e \rangle$$

Rewriting the (sole) relation yields the usual commutativity rule for multiplication of $x$ and $y$. The group presented with $N$ generators and *no* relations is the **free**

**group** $F_N$; in general, the free product $G * H$ of two finitely presented groups has presentation given by combining the generators and relations of each factor, with no additional relations. Presentations can be very complicated, implicating infinitely many generators and/or relations. Determining when two presentations yield isomorphic groups is uncomputable in general.

For purposes of doing homology and cohomology, it is convenient to work with **abelian** groups – groups for which the operation is commutative. An abelian group operation is almost always written in additive notation: $+\colon \mathbf{G} \times \mathbf{G} \to \mathbf{G}$, the identity is written as "0", and the inverse of $g \in \mathbf{G}$ is written "$-g$." The **quotient** of an abelian group $\mathbf{G}$ by a subgroup $\mathbf{H} < \mathbf{G}$ consists of equivalence classes $[g]$ for $g \in \mathbf{G}$, where $[g] = [g']$ if and only if $g' - g \in \mathbf{H}$.

Abelian groups generalize to a hierarchy of algebraic structures ascending to vector spaces. A **ring** is an abelian group $\mathbf{R}$ whose group operation $+$ is paired with a multiplication $\bullet\colon \mathbf{R} \times \mathbf{R} \to \mathbf{R}$ that is associative and distributive with a multiplicative identity. The canonical examples are $\mathbb{Z}$ and $\mathbb{Z}[x]$ with 1 as multiplicative inverse. When multiplication is commutative and a multiplicative inverse exists (on the complement of the additive identity 0), a ring ascends to a **field**. The familiar $\mathbb{R}$-vector space generalizes to have scalars in an arbitrary field $\mathbb{F}$. One can relax the definition of a vector space yet further, allowing scalars to reside not in a field but rather a ring $\mathbf{R}$. The resulting structure $\mathbf{M}$ is called an **R-module**. Other relaxations of algebraic objects can be beneficial, even those which yield structures more primitive than groups: for example, dropping from the definition of a group the existence of inverses leads to a **monoid**; further dropping the existence of an identity yields a **semigroup**.

Other structures, though simpler, can capture aspects of algebraic operations that are crucial in applications. For example, a **poset**, or **partially-ordered set**, is a set $P$ together with a binary relation $\trianglelefteq$ that is reflexive, antisymmetric, and transitive. Such a structure encodes, *e.g.*, inclusion in a topology, or the face relation of cells.

As this text demonstrates at several points, the rewards of even moderately increased algebraic generality are substantial. Coefficients in finite fields can provide computational accuracy as compared to real coefficients. Homology and cohomology of chain complexes as $\mathbb{Z}$-modules are critical for defining winding numbers, degrees, and more. Ring structures enable cup products in cohomology. Monoids are the critical structures for encoding constraints in network flow problems. Deeper truths about homology and cohomology – in particular, the understanding and management of torsional elements – require yet deeper tools from homological algebra (Ext, Tor) that linear algebra does not immediately presage.

# Notes on Figures

The figures in this book were intentionally drawn without captions or textual markings, since much of Mathematics consists of solving riddles in pictures. The chapter heading illustrations are amalgamations of mathematical, scientific, and literary allusions. The latter are inspired by the following myths: *Genesis*, the *Theogony*, the *Odyssey*, the *Metamorphoses*, the *Commedia*, the *Decameron*, *Paradise Lost*, *Ulysses*, and, perhaps, more. Not only the title, but the plan and a good deal of the incidental symbolism of this text were suggested by these works. Together, like the myths, the figures help reveal truth.

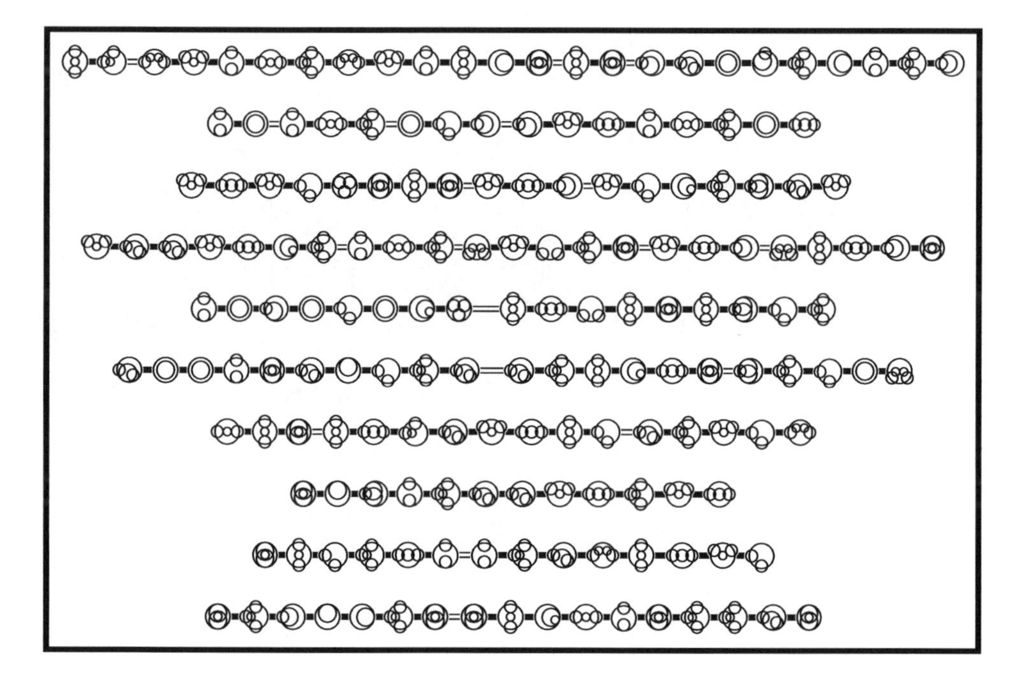

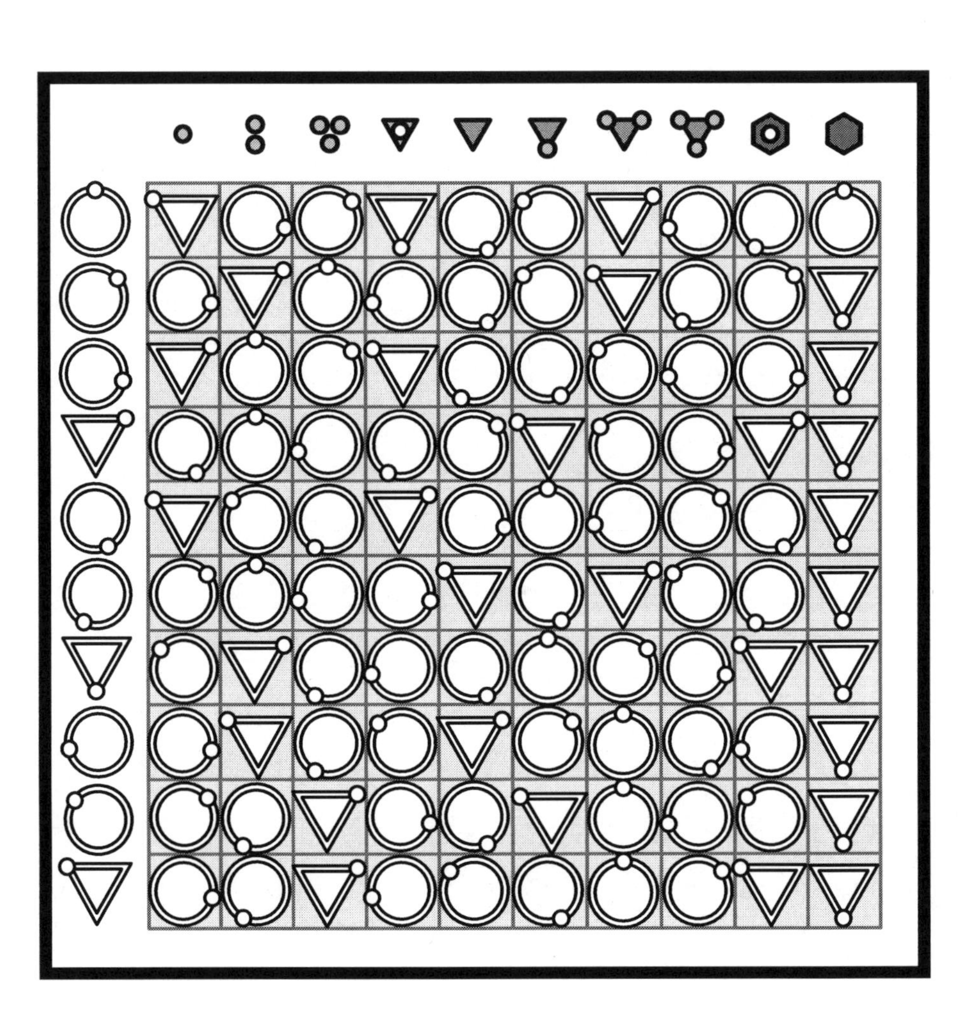

# Bibliography

[1] Abraham R, Marsden JE, & Ratiu T, *Manifolds, Tensor Analysis, and Applications*, volume 75 of *Applied Mathematical Sciences* (Springer-Verlag, New York), second edition (1988).

[2] Abrams A, "Configuration spaces of colored graphs". *Geom. Dedicata*, 92:185–194 (2002), dedicated to John Stallings on the occasion of his 65th birthday.

[3] Adler RJ, *The Geometry of Random Fields* (Society for Industrial and Applied Mathematics) (1981).

[4] Adler RJ, Bobrowski O, *et al.*, "Persistent homology for random fields and complexes", in *Borrowing Strength: Theory Powering Applications*, 124–143 (IMS Collections) (2010).

[5] Adler RJ & Taylor JE, *Random Fields and Geometry*, Springer Monographs in Mathematics (Springer, New York) (2007).

[6] Alexander GP, Chen BG, *et al.*, "Colloquium: Disclination loops, point defects, and all that in nematic liquid crystals". *Rev. Mod. Phys.*, 84:497–514 (2012).

[7] Alexander JW, "A lemma on systems of knotted curves". *Proc. Nat. Acad. Sci.*, 9:93–95 (1923).

[8] Alexandrov P, "Über den allgemeinen Dimensionsbegriff und seine Beziehungen zur elementaren geometrischen Anschauung". *Mathematische Annalen*, 98:617–635 (1928).

[9] Amenta N, "Helly theorems and generalized linear programming", in *Proceedings of the Ninth Annual Symposium on Computational Geometry*, SCG '93, 63–72 (ACM, New York, NY, USA) (1993).

[10] Angenent S & van der Vorst R, "A superquadratic indefinite elliptic system and its Morse-Conley-Floer homology". *Math. Z.*, 231 (2):203–248 (1999).

[11] Arai Z, Kalies W, *et al.*, "A database schema for the analysis of global dynamics of multiparameter systems". *SIAM J. Appl. Dyn. Syst.*, 8 (3):757–789 (2009).

[12] Arnold DN, "Differential complexes and numerical stability", in *Proceedings of the International Congress of Mathematicians, Vol. I (Beijing, 2002)*, 137–157 (Higher Ed. Press, Beijing) (2002).

[13] Arnold DN, Falk RS, & Winther R, "Finite element exterior calculus, homological techniques, and applications". *Acta Numer.*, 15:1–155 (2006).

[14] Arnol'd VI, "The asymptotic Hopf invariant and its applications". *Selecta Math. Soviet.*, 5 (4):327–345 (1986).

[15] ———, *Mathematical Methods of Classical Mechanics*, volume 60 of *Graduate Texts in Mathematics* (Springer-Verlag, New York) (1989), translated from the 1974 Russian original by K. Vogtmann and A. Weinstein.

[16] Arnol'd VI & Khesin BA, *Topological Methods in Hydrodynamics*, volume 125 of *Applied Mathematical Sciences* (Springer-Verlag, New York) (1998).

[17] Arone G, "A note on the homology of $\Sigma_n$, the Schwartz genus, and solving polynomial equations", in *An Alpine Anthology of Homotopy Theory*, volume 399 of *Contemp. Math.*, 1–10 (Amer. Math. Soc., Providence, RI) (2006).

[18] Arrow KJ & Debreu G, "Existence of an equilibrium for a competitive economy". *Econometrica*, 22:265–290 (1954).

[19] Auroux D, "A beginner's introduction to Fukaya categories" (2013), arXiv:1301.7056.

[20] Awodey S, *Category Theory* (Oxford University Press), 2nd edition (2010).

[21] Banyaga A & Hurtubise D, *Morse Homology* (Springer) (2004).

[22] Baryshnikov Y, "Unifying impossibility theorems: a topological approach". *Adv. in Appl. Math.*, 14 (4):404–415 (1993).

[23] Baryshnikov Y, Bubenik P, & Kahle M, "Min-type morse theory for configuration spaces of hard spheres". *International Mathematics Research Notices* (2013).

[24] Baryshnikov Y & Ghrist R, "Target enumeration via Euler characteristic integrals". *SIAM J. Appl. Math.*, 70 (3):825–844 (2009).

[25] ———, "Euler integration over definable functions". *Proc. Natl. Acad. Sci. USA*, 107 (21):9525–9530 (2010).

[26] ———, "Unimodal category and topological statistics", in *Proc. NOLTA: Nonlinear Theory & Applications*, 196–199 (2011).

[27] Baryshnikov Y, Ghrist R, & Lipsky D, "Inversion of Euler integral transforms with applications to sensor data". *Inverse Problems*, 27 (12) (2011).

[28] Baryshnikov Y, Ghrist R, & Wright M, "Hadwiger's Theorem for definable functions". *Adv. Math.*, 245:573–586 (2013).

[29] Basener WF, *Topology and its Applications*, Pure and Applied Mathematics (Wiley-Interscience, Hoboken, NJ) (2006).

[30] Bauer U, Kerber M, & Reininghaus J, "Clear and compress: computing persistent homology in chunks", in *Topological Methods in Data Analysis and Visualization III*, Mathematics and Visualization, 103–117 (2014).

[31] Belchí F & Murillo A, "A-infinity persistence" (2014), arXiv:1403.2395.

[32] Bell N & Hirani AN, "PyDEC: software and algorithms for discretization of exterior calculus". *ACM Trans. Math. Software*, 39 (1):41 (2012).

[33] Ben-El-Mechaiekh H, Bich P, & Florenzano M, "General equilibrium and fixed point theory: a partial survey". *J. Fixed Point Theory Appl.*, 6 (2):207–226 (2009).

[34] Bendich P & Harer J, "Persistent intersection homology". *Found. Comput. Math.*, 11 (3):305–336 (2010).

[35] Bhattacharya S, Lipsky D, *et al.*, "Invariants for homology classes with application to optimal search and planning problem in robotics". *Ann. Math. Artif. Intell.*, 67 (3-4):251–281 (2013).

[36] Billera LJ, "Homology of smooth splines: generic triangulations and a conjecture of Strang". *Trans. Amer. Math. Soc.*, 310 (1):325–340 (1988).

[37] Billera LJ, Holmes SP, & Vogtmann K, "Geometry of the space of phylogenetic trees". *Adv. in Appl. Math.*, 27 (4):733–767 (2001).

[38] Birman JS, *Braids, Links, and Mapping Class Groups*, number 82 in Annals of Mathematics Studies (Princeton University Press, Princeton, NJ.) (1974).

[39] Birman JS & Williams RF, "Knotted periodic orbits in dynamical systems. I. Lorenz's equations". *Topology*, 22 (1):47–82 (1983).

[40] ———, "Knotted periodic orbits in dynamical systems. II. Knot holders for fibered knots", in *Low-dimensional topology (San Francisco, Calif., 1981)*, volume 20 of *Contemp. Math.*, 1–60 (Amer. Math. Soc., Providence, RI) (1983).

[41] Blagojević PVM, Matschke B, & Ziegler GM, "A tight colored Tverberg theorem for maps to manifolds". *Topology Appl.*, 158 (12):1445–1452 (2011).

[42] Blum L, Shub M, & Smale S, "On a theory of computation and complexity over the real numbers: NP-completeness, recursive functions and universal machines". *Bull. Amer. Math. Soc. (N.S.)*, 21 (1):1–46 (1989).

[43] Blumberg AJ, Gal I, *et al.*, "Robust statistics, hypothesis testing, and confidence intervals for persistent homology on metric measure spaces". *Found. Comput. Math.*, 14 (4):745–789 (2014).

[44] Boczko E, Gedeon T, & Mischaikow K, "Dynamics of a simple regulatory switch". *J. Math. Biol.*, 55 (5-6):679–719 (2007).

[45] Bolsinov AV & Fomenko AT, *Integrable Hamiltonian systems: Geometry, Topology, Classification* (Chapman & Hall/CRC, Boca Raton, FL) (2004), translated from the 1999 Russian original.

[46] Bott R & Tu L, *Differential Forms in Algebraic Topology* (Springer) (1982).

[47] Bredon G, *Sheaf Theory* (Springer) (1997).

[48] Bröcker L & Kuppe M, "Integral geometry of tame sets". *Geom. Dedicata*, 82 (1-3):285–323 (2000).

[49] Brown RF, *The Lefschetz Fixed Point Theorem* (Scott, Foresman and Co., Glenview, Ill.-London) (1971).

[50] Bubenik P, de Silva V, & Scott J, "Metrics for generalized persistence modules" (2014), arXiv:1312.3829 [math.AT].

[51] Bubenik P & Scott JA, "Categorification of persistent homology". *Discrete Comput. Geom.*, 51 (3):600–627 (2014).

[52] Burghelea D & Dey TK, "Topological persistence for circle-valued maps". *Discrete and Computational Geometry*, 50 (1):1–30 (2011).

[53] Cagliari F, Ferri M, & Pozzi P, "Size functions from a categorical viewpoint". *Acta Applicandae Mathematicae*, 67 (3):225–235 (2001).

[54] Carbinatto MC, Kwapisz J, & Mischaikow K, "Horseshoes and the Conley index spectrum". *Ergodic Theory Dynam. Systems*, 20 (2):365–377 (2000).

[55] Carlsson G, "Topology and data". *Bull. Amer. Math. Soc. (N.S.)*, 46 (2):255–308 (2009).

[56] Carlsson G & de Silva V, "Zigzag persistence". *Found. Comput. Math.*, 10 (4):367–405 (2010).

[57] Carlsson G, de Silva V, & Morozov D, "Zigzag persistent homology and real-valued functions", in *Proceedings 25th ACM Symposium on Computational Geometry (SoCG)*, 247–256 (2009).

[58] Carlsson G, Gorham J, *et al.*, "Computational topology for configuration spaces of hard disks". *Phys. Rev. E*, 85 (2012).

[59] Carlsson G, Ishkhanov T, *et al.*, "On the local behavior of spaces of natural images". *International Journal of Computer Vision*, 76 (1):1–12 (2008).

[60] Carlsson G & Mémoli F, "Classifying clustering schemes". *Found. Comput. Math.*, 13 (2):221–252 (2013).

[61] Carlsson G, Singh G, & Zomorodian A, "Computing multidimensional persistence". *J. Comput. Geom.*, 1 (1):72–100 (2010).

[62] Carlsson G & Zomorodian A, "The theory of multidimensional persistence". *Discrete Comput. Geom.*, 42 (1):71–93 (2009).

[63] Čech E, "Théorie générale de l'homologie dans un espace quelconque". *Fund. Math.*, 19:149–183 (1932).

[64] Chambers EW, de Silva V, *et al.*, "Vietoris-Rips complexes of planar point sets". *Discrete Comput. Geom.*, 44 (1):75–90 (2010).

[65] Chen B, "Minkowski algebra. I. A convolution theory of closed convex sets and relatively open convex sets". *Asian J. Math.*, 3 (3):609–634 (1999).

[66] Chichilnisky G & Heal G, "Necessary and sufficient conditions for a resolution of the social choice paradox". *J. Econom. Theory*, 31 (1):68–87 (1983).

[67] Cluckers R & Edmundo M, "Integration of positive constructible functions against Euler characteristic and dimension". *J. Pure Appl. Algebra*, 208 (2):691–698 (2007).

[68] Cluckers R & Loeser F, "Constructible motivic functions and motivic integration". *Invent. Math.*, 173 (1):23–121 (2008).

[69] Cohen RL, Jones JDS, & Segal GB, "Morse theory and classifying spaces" (1995), preprint.

[70] Cohen-Steiner D, Edelsbrunner H, & Harer J, "Stability of persistence diagrams". *Discrete Comput. Geom.*, 37 (1):103–120 (2007).

[71] Coifman R, Shkolnisky Y, *et al.*, "Graph Laplacian tomography from unknown random projections". *Image Processing, IEEE Transactions on*, 17 (10):1891–1899 (2008).

[72] Conley C, *Isolated Invariant Sets and the Morse Index*, Regional conference series in mathematics (American Mathematical Society) (1978).

[73] Conley C & Gardner R, "An application of the generalized Morse index to travelling wave solutions of a competitive reaction-diffusion model". *Indiana Univ. Math. J.*, 33 (3):319–343 (1984).

[74] Cornea O, Lupton G, *et al.*, *Lusternik-Schnirelmann Category*, volume 103 of *Mathematical Surveys and Monographs* (American Mathematical Society, Providence, RI) (2003).

[75] Cowen N, Weingerten J, & Koditschek DE, "Visual servoing via navigation functions". *IEEE Transactions on Robotics and Automation*, 18 (4):521–533 (2002).

[76] Călugăreanu G, "Sur les classes d'isotopie des noeuds tridimensionels et leurs invariants". *Czech. Math. J.*, 11:588–625 (1961).

[77] Curry J, *Sheaves, Cosheaves and Applications*, Ph.D. thesis, University of Pennsylvania (2014).

[78] Curry J, Ghrist R, & Nanda V, "Discrete Morse theory for computing cellular sheaf cohomology". *ArXiv e-prints* (2013).

[79] Curry J, Ghrist R, & Robinson M, "Euler calculus with applications to signals and sensing", in *Advances in Applied and Computational Topology*, volume 70 of *Proc. Sympos. Appl. Math.*, 75–145 (Amer. Math. Soc., Providence, RI) (2012).

[80] Curto C & Itskov V, "Cell groups reveal structure of stimulus space". *PLoS Comput. Biol.*, 4 (10):e1000205, 13 (2008).

[81] Day S, Junge O, & Mischaikow K, "A rigorous numerical method for the global analysis of infinite-dimensional discrete dynamical systems". *SIAM J. Appl. Dyn. Syst.*, 3 (2):117–160 (2004).

[82] Day S, Kalies WD, & Wanner T, "Verified homology computations for nodal domains". *Multiscale Model. Simul.*, 7 (4):1695–1726 (2009).

[83] Day S, Kalies WD, *et al.*, "Probabilistic and numerical validation of homology computations for nodal domains". *Electron. Res. Announc. Amer. Math. Soc.*, 13:60–73 (electronic) (2007).

[84] Day S, Lessard JP, & Mischaikow K, "Validated continuation for equilibria of PDEs". *SIAM J. Numer. Anal.*, 45 (4):1398–1424 (2007).

[85] De Concini C, Procesi C, & Salvetti M, "Arithmetic properties of the cohomology of braid groups". *Topology*, 40 (4):739–751 (2001).

[86] de Silva V & Carlsson G, "Topological estimation using witness complexes", in *Eurographics Symposium on Point-based Graphics*, M Alexa & S Rusinkiewicz, eds. (2004).

[87] de Silva V & Ghrist R, "Coordinate-free coverage in sensor networks with controlled boundaries via homology". *International Journal of Robotics Research*, 25 (12):1205–1222 (2006).

[88] ———, "Coverage in sensor networks via persistent homology". *Algebraic & Geometric Topology*, 7:339–358 (2007).

[89] de Silva V, Morozov D, & Vejdemo-Johansson M, "Persistent cohomology and circular coordinates". *Discrete Comput. Geom.*, 45 (4):737–759 (2011).

[90] de Silva V, Robbin JW, & Salamon DA, "Combinatorial Floer homology". *Mem. Amer. Math. Soc.*, 230 (1080):1–114 (2014).

[91] Delfinado CJA & Edelsbrunner H, "An incremental algorithm for Betti numbers of simplicial complexes on the 3-sphere". *Computer Aided Geometric Design*, 12 (7):771–784 (1995).

[92] Denef J & Loeser F, "Motivic integration and the Grothendieck group of pseudo-finite fields", in *Proceedings of the International Congress of Mathematicians, Vol. II (Beijing, 2002)*, 13–23 (Higher Ed. Press, Beijing) (2002).

[93] Dequeant M, Ahnert S, *et al.*, "Comparison of pattern detection methods in microarray time series of the segmentation clock". *PLOS ONE*, 3 (8) (2008).

[94] Desbrun M, Leok M, & Marsden JE, "Discrete Poincaré lemma". *Appl. Numer. Math.*, 53 (2-4):231–248 (2005).

[95] Dimca A, *Sheaves in Topology* (Springer) (2004).

[96] Dimitrov G, Fabian H, *et al.*, "Dynamical systems and categories" (2013), arXiv:1307.8418.

[97] Dłotko P, Ghrist R, *et al.*, "Distributed computation of coverage in sensor networks by homological methods". *Appl. Algebra Engrg. Comm. Comput.*, 23 (1-2):29–58 (2012).

[98] Dold A, *Lectures on Algebraic Topology*, Classics in Mathematics (Springer-Verlag, Berlin) (1995), reprint of the 1972 edition.

[99] Dowden R, "World-wide lightning localization using VLF propagation in the earth-ionosphere waveguide". *Antennas and Propagation Magazine*, 50 (5):40–60 (2008).

[100] Dowker C, "Homology groups of relations". *Annals of Mathematics*, 84–95 (1952).

[101] Dwyer WG & Spaliński J, "Homotopy theories and model categories", in *Handbook of Algebraic Topology*, 73–126 (North-Holland, Amsterdam) (1995).

[102] Eckmann B, "Räume mit Mittelbildungen". *Comment. Math. Helv.*, 28:329–340 (1954).

[103] Edelsbrunner H & Harer J, "Persistent homology — a survey", in *Surveys on Discrete and Computational Geometry: Twenty Years Later.*, JE Goodman, J Pach, & R Pollack, eds., volume 453 of *Contemporary Mathematics*, 257–282 (American Mathematical Society) (2008).

[104] ———, *Computational Topology: an Introduction* (American Mathematical Society, Providence, RI) (2010).

[105] Edelsbrunner H, Letscher D, & Zomorodian A, "Topological persistence and simplification". *Discrete and Computational Geometry*, 28:511–533 (2002).

[106] Eilenberg S & Montgomery D, "Fixed point theorems for multi-valued transformations". *Amer. J. Math.*, 68:214–222 (1946).

[107] Emrani S, Gentimis T, & Krim H, "Persistent homology of delay embeddings and its application to wheeze detection". *IEEE Signal Process. Lett.*, 21 (4):459–463 (2014).

[108] Erdmann M, "On the topology of discrete strategies". *Int. J. Rob. Res.*, 29 (7):855–896 (2010).

[109] ———, "On the topology of discrete planning with uncertainty", in *Advances in Applied and Computational Topology*, volume 70 of *Proc. Sympos. Appl. Math.*, 147–194 (Amer. Math. Soc., Providence, RI) (2012).

[110] Escolar E & Hiraoka Y, "Persistence modules on commutative ladders of finite type" (2014), preprint.

[111] Fair R, "Digital microfluidics: Is a true lab-on-a-chip possible?" *Microfluidics and Nanofluidics*, 3 (3):245–281 (2007).

[112] Farber M, "Topological complexity of motion planning". *Discrete Comput. Geom.*, 29 (2):211–221 (2003).

[113] ———, *Invitation to Topological Robotics*, Zurich Lectures in Advanced Mathematics (European Mathematical Society (EMS), Zürich) (2008).

[114] Farber M & Grant M, "Robot motion planning, weights of cohomology classes, and cohomology operations". *Proc. Amer. Math. Soc.*, 136 (9):3339–3349 (2008).

[115] ———, "Topological complexity of configuration spaces". *Proc. Amer. Math. Soc.*, 137 (5):1841–1847 (2009).

[116] Farber M, Tabachnikov S, & Yuzvinsky S, "Topological robotics: motion planning in projective spaces". *Int. Math. Res. Not.*, 34:1853–1870 (2003).

[117] Farley D & Sabalka L, "On the cohomology rings of tree braid groups". *J. Pure Appl. Algebra*, 212 (1):53–71 (2008).

[118] ———, "Presentations of graph braid groups". *Forum Math.*, 24 (4):827–859 (2012).

[119] Federer H, *Geometric Measure Theory*, Die Grundlehren der mathematischen Wissenschaften, Band 153 (Springer-Verlag) (1969).

[120] Ferri M, Frosini P, & Landi C, "Stable shape comparison by persistent homology". *Atti Semin. Mat. Fis. Univ. Modena Reggio Emilia*, 58:143–162 (2011).

[121] Flapan E, *When Topology Meets Chemistry* (Cambridge University Press) (2000).

[122] Floer A, "Morse theory for Lagrangian intersections". *J. Differential Geom.*, 28 (3):513–547 (1988).

[123] Forman R, "Morse theory for cell complexes". *Adv. Math.*, 134 (1):90–145 (1998).

[124] ———, "Morse theory and evasiveness". *Combinatorica*, 20 (4):489–504 (2000).

[125] ———, "Combinatorial Novikov-Morse theory". *Internat. J. Math.*, 13 (4):333–368 (2002).

[126] ———, "A user's guide to discrete Morse theory". *Sém. Lothar. Combin.*, 48 (2002).

[127] Fragouli C & Soljanin E, "Network coding fundamentals". *Foundations and Trends in Networking*, 2 (1):1–133 (2007).

[128] Franks J & Richeson D, "Shift equivalence and the Conley index". *Trans. Amer. Math. Soc.*, 352 (7):3305–3322 (2000).

[129] Franzosa RD, "The connection matrix theory for Morse decompositions". *Trans. Amer. Math. Soc.*, 311 (2):561–592 (1989).

[130] Freedman M, Gompf R, *et al.*, "Man and machine thinking about the smooth 4-dimensional Poincaré conjecture". *Quantum Topology*, 1:171–208 (2010).

[131] Freedman MH & He ZX, "Divergence-free fields: energy and asymptotic crossing number". *Ann. of Math. (2)*, 134 (1):189–229 (1991).

[132] Frosini P, "Measuring shape by size functions", in *Proceedings of SPIE on Intelligent Robotic Systems*, volume 1607, 122–133 (1991).

[133] ———, "Discrete computation of size functions". *Journal of Combinatorics, Information & System Sciences*, 17 (3–4):232–250 (1992).

[134] Fu JHG, "Curvature measures of subanalytic sets". *Amer. J. Math.*, 116 (4):819–880 (1994).

[135] Fulton W, *Algebraic Topology: A First Course*, volume 153 of *Graduate Texts in Mathematics* (Springer-Verlag) (1991).

[136] Gabriel P, "Unzerlegbare Darstellungen I". *Manuscripta Mathematica*, 6:71–103 (1972).

[137] Gal ŚR, "Euler characteristic of the configuration space of a complex". *Colloq. Math.*, 89 (1):61–67 (2001).

[138] Gale D, "The game of Hex and the Brouwer fixed-point theorem". *Amer. Math. Monthly*, 86 (10):818–827 (1979).

[139] Gameiro M, Hiraoka Y, *et al.*, "Topological measurement of protein compressibility via persistent diagrams". *Japan J. Industrial & Applied Mathematics* (2014), to appear.

[140] Gameiro M, Lessard JP, & Mischaikow K, "Validated continuation over large parameter ranges for equilibria of PDEs". *Math. Comput. Simulation*, 79 (4):1368–1382 (2008).

[141] Gameiro M, Mischaikow K, & Kalies W, "Topological characterization of spatial-temporal chaos". *Phys. Rev. E (3)*, 70 (3):035203, 4 (2004).

[142] Gelfand SI & Manin YI, *Methods of Homological Algebra*, Springer Monographs in Mathematics (Springer-Verlag, Berlin), second edition (2003).

[143] Ghrist R, "Branched two-manifolds supporting all links". *Topology*, 36 (2):423–448 (1997).

[144] ———, "Barcodes: the persistent topology of data". *Bull. Amer. Math. Soc. (N.S.)*, 45 (1):61–75 (2008).

[145] ———, "Configuration spaces, braids, and robotics", in *Braids*, volume 19 of *Lect. Notes Ser. Inst. Math. Sci. Natl. Univ. Singap.*, 263–304 (World Sci. Publ., Hackensack, NJ) (2010).

[146] Ghrist R & Hiraoka Y, "Sheaves for network coding", in *Proc. NOLTA: Nonlinear Theory and Applications*, 266–269 (2011).

[147] Ghrist R & Holmes PJ, "An ODE whose solutions contain all knots and links". *Internat. J. Bifur. Chaos Appl. Sci. Engrg.*, 6 (5):779–800 (1996).

[148] Ghrist R & Kin E, "Flowlines transverse to knot and link fibrations". *Pacific J. Math.*, 217 (1):61–86 (2004).

[149] Ghrist R & Koditschek DE, "Safe cooperative robot dynamics on graphs". *SIAM J. Control Optim.*, 40 (5):1556–1575 (2002).

[150] Ghrist R & Krishnan S, "A topological max-flow-min-cut theorem", in *Proc. Global Sig. Inf. Proc.* (2013).

[151] Ghrist R, Lipsky D, *et al.*, "Surrounding nodes in coordinate-free networks", in *Algorithmic foundation of robotics VII*, volume 47 of *Springer Tracts Adv. Robot.*, 409–424 (Springer, Berlin) (2008).

[152] ———, "Topological landmark-based navigation and mapping" (2012), preprint.

[153] Ghrist R & Peterson V, "The geometry and topology of reconfiguration". *Adv. in Appl. Math.*, 38 (3):302–323 (2007).

[154] Ghrist R & Robinson M, "Euler-Bessel and Euler-Fourier transforms". *Inverse Problems*, 27 (12) (2011).

[155] Ghrist R, Van den Berg JB, & Vandervorst RC, "Morse theory on spaces of braids and Lagrangian dynamics". *Invent. Math.*, 152 (2):369–432 (2003).

[156] Ghrist R & Vandervorst RC, "Scalar parabolic PDEs and braids". *Trans. Amer. Math. Soc.*, 361 (5):2755–2788 (2009).

[157] Godement R, *Topologie Algebrique et Théorie des Faisceaux* (Herman, Paris) (1958).

[158] Goerss PG & Jardine JF, *Simplicial Homotopy Theory*, Modern Birkhäuser Classics (Birkhäuser Verlag, Basel) (2009), reprint of the 1999 edition.

[159] Goguen JA, "Sheaf semantics for concurrent interacting objects". *Mathematical Structures in Computer Science*, 2 (02):159–191 (1992).

[160] Goldblatt R, *Topoi: the Categorial Analysis of Logic*, volume 98 of *Studies in Logic and the Foundations of Mathematics* (North-Holland Publishing Co., Amsterdam), second edition (1984).

[161] Golubitsky M & Guillemin V, *Stable Mappings and Their Singularities* (Springer-Verlag, New York) (1973).

[162] Gompf RE, "An exotic menagerie". *J. Differential Geom.*, 37 (1):199–223 (1993).

[163] Goresky M & MacPherson R, *Stratified Morse Theory*, volume 14 of *Ergebnisse Der Mathematik Und Ihrer Grenzgebiete.* (Springer-Verlag) (1988).

[164] ———, "Local contribution to the Lefschetz fixed point formula". *Invent. Math.*, 111 (1):1–33 (1993).

[165] Gottlieb DH, "Topology and the robot arm". *Acta Appl. Math.*, 11 (2):117–121 (1988).

[166] Gromov M, "Pseudoholomorphic curves in symplectic manifolds". *Invent. Math.*, 82 (2):307–347 (1985).

[167] Guckenheimer J & Holmes P, *Nonlinear Oscillations, Dynamical Systems, and Bifurcations of Vector Fields*, volume 42 of *Applied Mathematical Sciences* (Springer-Verlag, New York) (1983).

[168] Guibas L, Ramshaw L, & Stolfi J, "A kinetic framework for computational geometry", in *Proc. IEEE Sympos. Found. Comput. Sci.* (1983).

[169] Guillemin V & Pollack A, *Differential Topology* (AMS Chelsea Publishing, Providence, RI) (2010), reprint of the 1974 original.

[170] Guillermou S, Kashiwara M, & Schapira P, "Sheaf quantization of Hamiltonian isotopies and applications to nondisplaceability problems". *Duke Math. J.*, 161 (2):201–245 (2012).

[171] Gundert A & Wagner U, "On Laplacians of random complexes", in *Proceedings of the Twenty-eighth Annual Symposium on Computational Geometry*, SoCG '12, 151–160 (ACM, New York, NY, USA) (2012).

[172] Guseĭn-Zade SM, "Integration with respect to the Euler characteristic and its applications". *Uspekhi Mat. Nauk*, 65 (3):5–42 (2010).

[173] Gutierrez A, Monaghan D, *et al.*, "Persistent homology for 3D reconstruction evaluation", in *Computational topology in image context*, volume 7309 of *Lecture Notes in Comput. Sci.*, 139–147 (Springer, Heidelberg) (2012).

[174] Hadwiger H, "Integralsätze im Konvexring". *Abh. Math. Sem. Univ. Hamburg*, 20:136–154 (1956).

[175] Hales T, "What is motivic measure?" *Bull. Amer. Math. Soc.*, 42 (2):119–135 (2005).

[176] Hatcher A, *Algebraic Topology* (Cambridge University Press) (2002).

[177] Haynes GC, Cohen FR, & Koditschek DE, "Gait transitions for quasi-static hexapedal locomotion on level ground", in *Robotics Research*, C Pradalier, R Siegwart, & G Hirzinger, eds., volume 70 of *Springer Tracts in Advanced Robotics*, 105–121 (Springer Berlin Heidelberg) (2011).

[178] Herlihy M & Shavit N, "The topological structure of asynchronous computability". *J. ACM*, 46 (6):858–923 (1999).

[179] Hirsch MW, *Differential Topology*, volume 33 of *Graduate Texts in Mathematics* (Springer-Verlag, New York) (1994), corrected reprint of the 1976 original.

[180] Husseini SY, Lasry JM, & Magill MJP, "Existence of equilibrium with incomplete markets". *J. Math. Econom.*, 19 (1-2):39–67 (1990).

[181] Hutson V, Mischaikow K, & Poláčik P, "The evolution of dispersal rates in a heterogeneous time-periodic environment". *J. Math. Biol.*, 43 (6):501–533 (2001).

[182] Hutson V, Mischaikow K, & Vickers GT, "Multiple travelling waves in evolutionary game dynamics". *Japan J. Indust. Appl. Math.*, 17 (3):341–356 (2000).

[183] Iversen B, *Cohomology of Sheaves* (Springer) (1986).

[184] Jardine JF, "Homotopy theories of diagrams". *Theory Appl. Categ.*, 28 (11):269–303 (2013).

[185] Jiang X, Lim LH, *et al.*, "Statistical ranking and combinatorial Hodge theory". *Math. Program.*, 127 (1, Ser. B):203–244 (2011).

[186] Kaczynski T, Mischaikow K, & Mrozek M, *Computational Homology*, volume 157 of *Applied Mathematical Sciences* (Springer-Verlag, New York) (2004).

[187] Kakutani S, "A generalization of Brouwer's fixed point theorem". *Duke Math. J.*, 8:457–459 (1941).

[188] Kalies W, Mischaikow M, & Vandervorst RC, "Conley theory" (2014), in preparation.

[189] Kapovich M & Millson JJ, "Universality theorems for configuration spaces of planar linkages". *Topology*, 41 (6):1051–1107 (2002).

[190] Kashiwara M, "On the maximally overdetermined system of linear differential equations. I". *Publ. Res. Inst. Math. Sci.*, 10:563–579 (1974/75).

[191] Kashiwara M & Schapira P, *Sheaves on Manifolds* (Springer) (1990).

[192] ———, *Categories and Sheaves*, volume 332 of *Grundlehren der Mathematischen Wissenschaften* (Springer-Verlag) (2006).

[193] Kato G, *The Heart of Cohomology* (Springer, Dordrecht) (2006).

[194] Kleinberg J, "An impossibility theorem for clustering", in *Proc. NIPS*, 446–453 (MIT Press) (2002).

[195] Koditschek D & Buehler M, "Analysis of a simplified hopping robot". *International Journal of Robotics Research*, 10 (6):587–605 (1991).

[196] Koetter R & Medard M, "An algebraic approach to network coding". *IEEE/ACM Transactions on Networking*, 11 (5):782–795 (2003).

[197] Kozlov D, "Discrete Morse theory for free chain complexes". *Comptes Rendus Mathematique*, 340:867–872 (2005).

[198] ———, *Combinatorial Algebraic Topology*, volume 21 of *Algorithms and Computation in Mathematics* (Springer) (2008).

[199] Kramár M, Goullet A, *et al.*, "Quantifying force networks in particulate systems". *Phys. D*, 283:37–55 (2014).

[200] Krishnan S, "Directed Poincaré duality" (2013), preprint.

[201] ———, "Flow-cut dualities for sheaves" (2014), arXiv:1409.6712.

[202] Kuperberg K, "Counterexamples to the Seifert conjecture", in *Proceedings of the International Congress of Mathematicians, Vol. II (Berlin, 1998)*, Extra Vol. II, 831–840 (electronic) (1998).

[203] Kurland HL, "The Morse index of an isolated invariant set is a connected simple system". *J. Differential Equations*, 42 (2):234–259 (1981).

[204] Latschev J, "Vietoris-Rips complexes of metric spaces near a closed Riemannian manifold". *Archiv der Mathematik*, 77:522–528 (2001).

[205] Leinster T, *Basic Category Theory* (Cambridge University Press) (2014).

[206] Leray J, "Sur la forme des espaces topologiques et sur les points fixes des representations". *Journal de Math.*, 24:95–167 (1945).

[207] Lesnick M, "The theory of the interleaving distance on multidimensional persistence modules" (2011), arXiv:1106.5305.

[208] Lewiner T, Lopes H, & Tavares G, "Applications of Forman's discrete Morse theory to topology visualization and mesh compression". *IEEE Trans. Visualization & Comput. Graphics*, 10 (5):499–508 (2004).

[209] Lewis R & Zomorodian A, "Multicore homology via Mayer Vietoris" (2014), arXiv:1407.2275.

[210] Lipsky D, Skraba P, & Vejdemo-Johansson M, "A spectral sequence for parallelized persistence" (2011), arXiv:1112.1245.

[211] Mac Lane S, *Categories for the Working Mathematician*, volume 5 of *Graduate Texts in Mathematics* (Springer-Verlag, New York), second edition (1998).

[212] Mac Lane S & Moerdijk I, *Sheaves in Geometry and Logic: A first introduction to Topos Theory*, Universitext (Springer-Verlag, New York) (1994), corrected reprint of the 1992 edition.

[213] MacPherson R & Schweinhart B, "Measuring shape with topology". *J. Math. Phys.*, 53 (7):073516, 13 (2012).

[214] MacPherson R & Srolovitz D, "The von Neumann relation generalized to coarsening of three-dimensional microstructures". *Nature*, 446:1053–1055 (2007).

[215] MacPherson RD, "Chern classes for singular algebraic varieties". *Ann. of Math. (2)*, 100:423–432 (1974).

[216] Margalef-Roig J & Outerelo Dominguez E, *Differential Topology* (North-Holland), first edition (1992).

[217] Marsden JE & Ratiu TS, *Introduction to Mechanics and Symmetry*, volume 17 of *Texts in Applied Mathematics* (Springer-Verlag, New York), second edition (1999).

[218] Massey WS, *A Basic Course in Algebraic Topology*, volume 127 of *Graduate Texts in Mathematics* (Springer-Verlag, New York) (1991).

[219] Matoušek J, *Using the Borsuk-Ulam Theorem: Lectures on Topological Methods in Combinatorics and Geometry* (Springer) (2003).

[220] May JP, *Simplicial Objects in Algebraic Topology*, Chicago Lectures in Mathematics (University of Chicago Press, Chicago, IL) (1992), reprint of the 1967 original.

[221] ———, *A Concise Course in Algebraic Topology*, Chicago Lectures in Mathematics (University of Chicago Press, Chicago, IL, University of Chicago Press) (1999).

[222] McCord C & Mischaikow K, "Connected simple systems, transition matrices, and heteroclinic bifurcations". *Trans. Amer. Math. Soc.*, 333 (1):397–422 (1992).

[223] McDonald T & Schenck H, "Piecewise polynomials on polyhedral complexes". *Adv. in Appl. Math.*, 42 (1):82–93 (2009).

[224] McDuff D & Salamon D, *Introduction to Symplectic Topology*, Oxford Mathematical Monographs (Oxford University Press), second edition (1998).

[225] Milnor J, "Differential topology forty-six years later". *Notices Amer. Math. Soc.*, 58 (6):804–809 (2011).

[226] Milnor JW & Stasheff JD, *Characteristic Classes*, number 76 in Annals of Mathematics Studies (Princeton University Press) (1974).

[227] Mirollo RE & Strogatz SH, "Synchronization of pulse-coupled biological oscillators". *SIAM Journal on Applied Mathematics*, 50 (6):1645–1662 (1990).

[228] Mischaikow K, "Topological techniques for efficient rigorous computation in dynamics". *Acta Numer.*, 11:435–477 (2002).

[229] Mischaikow K, Mrozek M, *et al.*, "Construction of symbolic dynamics from experimental time series". *Physical Review Letters*, 82 (6):1144 (1999).

[230] Mischaikow K & Nanda V, "Morse theory for filtrations and efficient computation of persistent homology". *Discrete Comput. Geom.*, 50 (2):330–353 (2013).

[231] Misner CW, Thorne KS, & Wheeler JA, *Gravitation* (W. H. Freeman and Co., San Francisco, Calif.) (1973).

[232] Moffatt HK, "The degree of knottedness of tangled vortex lines". *Journal of Fluid Mechanics*, 35 (01):117–129 (1969).

[233] Morgan J & Tian G, *Ricci Flow and the Poincaré Conjecture*, volume 3 of *Clay Mathematics Monographs* (2007).

[234] Morvan JM, *Generalized Curvatures*, volume 2 of *Geometry and Computing* (Springer-Verlag, Berlin) (2008).

[235] Muhammad A & Jadbabaie A, "Distributed computation of homology groups by gossip.", in *Proceedings of American Control Conference (ACC)* (2007).

[236] Mumford D, "Pattern theory: the mathematics of perception". *Proc. Intl. Congress of Mathematicians, Vol. III*, 1–21 (2002).

[237] Mumford D, Lee A, & Pedersen K, "The nonlinear statistics of high-contrast patches in natural images". *Intl. J. Computer Vision*, 54:83–103 (2003).

[238] Munkres J, *Topology* (Prentice Hall) (2000).

[239] Nash JF Jr, "Equilibrium points in *n*-person games". *Proc. Nat. Acad. Sci. U. S. A.*, 36:48–49 (1950).

[240] Nicolaescu LI, "Tame flows". *Mem. Amer. Math. Soc.*, 208 (980) (2010).

[241] Niyogi P, Smale S, & Weinberger S, "Finding the homology of submanifolds with high confidence from random samples". *Discrete & Computational Geometry*, 39:419–441 (2008).

[242] Pachter L & Sturmfels B, "The mathematics of phylogenomics". *SIAM Rev.*, 49 (1):3–31 (2007).

[243] Penrose R, "La cohomologie des figures impossibles". *Structural Topology*, 17:11–16 (1991).

[244] Pratt V, "Modelling concurrency with geometry", in *Proc. 18th Symp. on Principles of Programming Languages* (1991).

[245] Prue P & Scrimshaw T, "Abrams's stable equivalence for graph braid groups". *Topology and Its Applications*, 178:136–145 (2014).

[246] Quillen DG, *Homotopical Algebra*, Lecture Notes in Mathematics, No. 43 (Springer-Verlag, Berlin-New York) (1967).

[247] Rice SO, "Mathematical analysis of random noise". *Bell System Tech. J.*, 24:46–156 (1945).

[248] Rimon E & Koditschek DE, "Exact robot navigation using artificial potential functions". *IEEE Transactions on Robotics and Automation*, 8 (5):501–518 (1992).

[249] Rimon E, Mason R, *et al.*, "A general stance stability test based on stratified Morse theory with application to quasi-static locomotion planning". *IEEE Transactions on Robotics*, 626–641 (2008).

[250] Rizzi AA & Koditschek DE, "The control of a robot juggler", in *Proceedings Third International Symposium on Experimental Robotics* (Kyoto, Japan) (1993).

[251] Robins V, "Towards computing homology from finite approximations", in *Proceedings of the 14th Summer Conference on General Topology and its Applications (Brookville, NY, 1999)*, volume 24, 503–532 (1999).

[252] Robins V, Wood P, & Sheppard A, "Theory and algorithms for constructing discrete Morse complexes from grayscale digital images". *IEEE Transactions on Pattern Analysis and Machine Intelligence*, 33 (8):1646–1658 (2011).

[253] Robinson M, "Inverse problems in geometric graphs using internal measurements, `arxiv:1008.2933`" (2010).

[254] ———, "Asynchronous logic circuits and sheaf obstructions". *Electronic Notes in Theoretical Computer Science*, 159–177 (2012).

[255] ———, "The Nyquist theorem for cellular sheaves", in *Proc. Global Sig. Inf. Proc.* (2013).

[256] ———, *Topological Signal Processing* (Springer, Heidelberg) (2014).

[257] Robinson M & Ghrist R, "Topological localization via signals of opportunity". *IEEE Trans. Signal Process.*, 60 (5):2362–2373 (2012).

[258] Robinson RC, *An Introduction to Dynamical Systems — Continuous and Discrete*, volume 19 of *Pure and Applied Undergraduate Texts* (American Mathematical Society, Providence, RI), second edition (2012).

[259] Rodriguez A, Mason M, & Ferry S, "From caging to grasping". *International Journal of Robotics Research*, 31 (7):886–900 (2012).

[260] Rolfsen D, *Knots and Links*, volume 7 of *Mathematics Lecture Series* (Publish or Perish, Inc., Houston, TX) (1990), corrected reprint of the 1976 original.

[261] Rot TO & Vandervorst RCAM, "Morse-Conley-Floer homology". *J. Topol. Anal.*, 6 (3):305–338 (2014).

[262] Rota GC, "On the combinatorics of the Euler characteristic", in *Studies in Pure Mathematics (Presented to Richard Rado)*, 221–233 (Academic Press, London) (1971).

[263] Roweis S & Saul L, "Nonlinear dimensionality reduction by locally linear embedding". *Science*, 290:2323–2326 (2000).

[264] Ruelle D, "Dynamical zeta functions and transfer operators". *Notices Amer. Math. Soc.*, 49 (8):887–895 (2002).

[265] Sahai T, Speranzon A, & Banaszuk A, "Wave equation based algorithm for distributed eigenvector computation", in *Proceedings of IEEE Conference on Decision and Control (CDC)* (2010).

[266] Salamon D, "Connected simple systems and the Conley index of isolated invariant sets". *Trans. Amer. Math. Soc.*, 291 (1):1–41 (1985).

[267] Salehi AT & Jadbabaie A, "Distributed coverage verification in sensor networks without location". *IEEE Transactions on Automatic Control*, 55 (8):1837–1849 (2010).

[268] Schanuel SH, "Negative sets have Euler characteristic and dimension", in *Category Theory (Como, 1990)*, volume 1488 of *Lecture Notes in Math.*, 379–385 (Springer, Berlin) (1991).

[269] Schapira P, "Operations on constructible functions". *J. Pure Appl. Algebra*, 72 (1):83–93 (1991).

[270] ———, "Tomography of constructible functions", in *Applied Algebra, Algebraic Algorithms and Error-Correcting Codes*, 427–435 (Springer) (1995).

[271] Schenck H, "A spectral sequence for splines". *Adv. in Appl. Math.*, 19 (2):183–199 (1997).

[272] Schürmann J, *Topology of Singular Spaces and Constructible Sheaves*, volume 63 of *Mathematics Institute of the Polish Academy of Sciences. Mathematical Monographs (New Series)* (Birkhäuser Verlag, Basel) (2003).

[273] Schwartzman S, "Asymptotic cycles". *Ann. of Math. (2)*, 66:270–284 (1957).

[274] Schwarz M, *Morse Homology*, volume 111 of *Progress in Mathematics* (Birkhäuser Verlag, Basel) (1993).

[275] Seidel P, "Fukaya categories and deformations", in *Proceedings of the International Congress of Mathematicians, Vol. II (Beijing, 2002)*, 351–360 (Higher Ed. Press, Beijing) (2002).

[276] Shepard A, *A Cellular Description of the Derived Category of a Stratified Space*, Ph.D. thesis, Brown University (1985).

[277] Shiota M, *Geometry of Subanalytic and Semialgebraic Sets*, Progress in Mathematics (Birkhäuser) (1997).

[278] Smale S, "Algorithms for solving equations", in *Proceedings of the International Congress of Mathematicians, Vol. 1, 2 (Berkeley, Calif., 1986)*, 172–195 (Amer. Math. Soc., Providence, RI) (1987).

[279] Smoller J, *Shock Waves and Reaction-Diffusion Equations*, volume 258 of *Grundlehren der Mathematischen Wissenschaften* (Springer-Verlag, New York-Berlin) (1983).

[280] Spanier EH, *Algebraic Topology* (McGraw-Hill Book Co., New York) (1966).

[281] Spielman D, "Algorithms, graph theory, and linear equations in Laplacian matrices", in *Proc. Intl. Congress of Mathematicians* (2010).

[282] Spivak D, "Category theory for scientists" (2013), arXiv:1302.6946.

[283] Su FE, "Rental harmony: Sperner's lemma in fair division". *Amer. Math. Monthly*, 106 (10):930–942 (1999).

[284] Sullivan D, "Cycles for the dynamical study of foliated manifolds and complex manifolds". *Invent. Math.*, 36:225–255 (1976).

[285] Sumners DW, "Lifting the curtain: using topology to probe the hidden action of enzymes". *Notices Amer. Math. Soc.*, 42 (5):528–537 (1995).

[286] Szymczak A, "The Conley index for discrete semidynamical systems". *Topology Appl.*, 66 (3):215–240 (1995).

[287] Tanner HG, Jadbabaie A, & Pappas GJ, "Flocking in fixed and switching networks". *Automatic Control, IEEE Transactions on*, 52 (5):863–868 (2007).

[288] Tenenbaum JB, de Silva V, & Langford JC, "A global geometric framework for nonlinear dimensionality reduction". *Science*, 290:2319–2323 (2000).

[289] Tesfatsion L, "Pure strategy Nash equilibrium points and the Lefschetz fixed point theorem". *Internat. J. Game Theory*, 12 (3):181–191 (1983).

[290] Thurston WP, *Three-Dimensional Geometry and Topology. Vol. 1*, volume 35 of *Princeton Mathematical Series* (Princeton University Press, Princeton, NJ) (1997), edited by Silvio Levy.

[291] Univalent Foundations Program, *Homotopy Type Theory: Univalent Foundations of Mathematics* (`http://homotopytypetheory.org/book`, Institute for Advanced Study) (2013).

[292] van den Berg JB, Mireles-James JD, *et al.*, "Rigorous numerics for symmetric connecting orbits: even homoclinics of the Gray-Scott equation". *SIAM J. Math. Anal.*, 43 (4):1557–1594 (2011).

[293] van den Dries L, *Tame Topology and O-minimal Structures*, London Mathematical Society Lecture Note Series (Cambridge University Press) (1998).

[294] van Hateren J & van der Schaff A, "Independent component filters of natural images compared with simple cells in primary visual cortex". *Proc. R. Soc. of London, vol B*, 265:359–366 (1998).

[295] Vasil'ev VA, "Cohomology of braid groups and the complexity of algorithms". *Funktsional. Anal. i Prilozhen.*, 22 (3):15–24, 96 (1988).

[296] Vietoris L, "Über den höheren Zusammenhang kompakter Räume und eine Klasse von zusammenhangstreuen Abbildungen". *Math. Ann.*, 97 (1):454–472 (1927).

[297] Viro O, "Some integral calculus based on Euler characteristic". *Lecture Notes in Mathematics*, 1346:127–138 (1988).

[298] Vybornov M, "Constructible sheaves on simplicial complexes and Koszul duality". *Mathematical Research Letters*, 5 (675):675–683 (1998).

[299] Walker K, *Configuration spaces of linkges*, Undergraduate thesis, Princeton University (1985).

[300] Weibel CA, *An Introduction to Homological Algebra*, volume 38 of *Cambridge Studies in Advanced Mathematics* (Cambridge University Press, Cambridge) (1994).

[301] Weinberger S, "On the topological social choice model". *J. Econom. Theory*, 115 (2):377–384 (2004).

[302] Worsley KJ, "Local maxima and the expected Euler characteristic of excursion sets of $\chi^2$, $F$ and $t$ fields". *Adv. in Appl. Probab.*, 26 (1):13–42 (1994).

[303] ———, "Estimating the number of peaks in a random field using the Hadwiger characteristic of excursion sets, with applications to medical images". *Ann. Statist.*, 23 (2):640–669 (1995).

[304] Yuzvinsky S, "Modules of splines on polyhedral complexes". *Math. Z.*, 210 (2):245–254 (1992).

[305] Zomorodian A, *Topology for Computing* (Cambridge Univ Press) (2005).

[306] Zomorodian A & Carlsson G, "Computing persistent homology". *Discrete Comput. Geom.*, 33 (2):249–274 (2005).

# Index


A
abelian . . . 241
additivity
  of Conley index . . . 145
  of Euler characteristic . . . 50
  of homology . . . 88
  of intrinsic volumes . . . 128
Alexander
  -Spanier cohomology . . . 132
  duality . . . 118–119
algebra . . . 239
  Boolean . . . 210
alpha complex . . . 42
ambient isotopy . . . 163
ambiguity sheaf . . . 196
antipodal map . . . 11
arbitrage . . . 115
Arnol'd conjecture . . . 147
asymptotic linking number . . . 127
atlas . . . 10
attractor . . . 146
  Lorenz . . . 165

B
bandlimited . . . 195
barcode . . . 104–106, 202
base space . . . 171
basepoint . . . 158
basis . . . 238
Betti numbers, $\beta_k$ . . . 65
Bézier curve . . . 200
Boolean algebra . . . 210
Borel-Moore homology . . . 132
Borsuk-Ulam theorem . . . 97–98
bottleneck distance . . . 234
boundary
  cycle . . . 63
  map . . . 62
  of manifold . . . 22
  of set . . . 239
braid . . . 16
  closure of . . . 164
  group . . . 164
  index . . . 164
Brouwer theorem . . . 80
bundle
  cotangent . . . 112
  fiber . . . 171
  jet . . . 24
  tangent . . . 14
  trivial . . . 171
  vector . . . 171
butterfly network . . . 188

C
$\check{C}_\epsilon$ . . . 30
$\mathcal{C}^n$ . . . 16
cap product, $\frown$ . . . 132
capacity . . . 114, 186, 233
categorification . . . 226–229
category . . . 208
  equivalent/isomorphic . . . 219
  geometric . . . 152
  index . . . 220
  LS . . . 151
  model . . . 175
  opposite . . . 213
  sectional . . . 173
  small . . . 234
  unimodal . . . 153
Cauchy-Riemann equations . . . 192
Čech
  complex . . . 30
  homology . . . 70

sheaf cohomology ............. 190
cellular
approximation theorem ......... 109
complex ........................ 28
homology .................. 64, 90
sheaf................180–183, 224
sheaf cohomology ............. 184
CF ................................. 50
chain
complex....................62, 68
homotopy.......................84
map ....................... 74, 84
characteristic class ................. 177
chart...............................10
circular coordinates.................131
closed set .......................... 238
closure ............................ 239
clustering...........................215
cochain complex .................... 113
cocomplete .........................224
cocone .............................223
cocycle.............................113
codensity ..........................106
coding map..........................189
coface ............................ 113
cofibration..........................175
cohomology
Alexander-Spanier ............. 132
local ......................... 182
of cellular sheaf .............. 184
of cochain complex ............ 113
of topological sheaf............192
de Rham ..................... 124
-ring ......................... 125
with compact supports.... 118, 124
colimit........................223–224
coloring ........................... 96
compact ........................... 239
complete ...........................224
complex
alpha .......................... 42
Čech ........................... 30
chain ..................... 62, 68
cochain ........................113
CW ............................ 28
double.........................109
Dowker .........................73
flag ........................... 30
independence...................26
input/output ...................36
link.............................33
Morse ........................ 138
nerve ...........................31
of sheaves .................... 198
simplicial.......................26
singular ........................69
state .......................... 39
strategy ....................... 34
Vietoris-Rips .................. 28
witness ........................ 29
Condorcet paradox ................. 117
cone ............................. 221
configuration space
discretized .................... 37
of linkage ..................... 12
of points........................16
Conley index ................ 143–146
connecting
homomorphism ............87, 218
orbit ......................... 138
conormal cycle ..................... 128
constant sheaf ............... 181, 192
constructible
function ....................... 50
sheaf..........................197
continuation ...................... 144
continuous ........................ 239
functor ...................... 224
contractible
loop ..........................158
space ...........................4
contravariant ...................... 213
control point ..................... 200
convolution ........................54
coordination space ................. 41
coproduct.....................12, 223
coset ............................. 240
cosheaf
cellular ...................... 198
homology ......................199
homology fiber ................200
costalk ............................198
cotangent space, bundle ............ 112
covariant ..........................213
cover
acyclic ...................70, 119
by charts ..................... 10
classification of .............. 162
convex ....................... 119
equivalence of.................161

of a space . . . 160
open . . . 70
sensor . . . 92–94
universal . . . 161
critical
cell . . . 149
point . . . 136
cup
length . . . 152
product, $\smile$ . . . 125
curl . . . 122
current . . . 126
cut, cut value . . . 114
CW complex . . . 28
cycle . . . 63
normal, conormal . . . 128

D
$\delta$ . . . 87
$\mathcal{D}^n$ . . . 37
$\partial$ . . . 62, 67
$d$ . . . 113, 122
$d\chi$ . . . 50
$\lfloor d\chi \rfloor$, $\lceil d\chi \rceil$ . . . 140
decategorification . . . 226–229
definable
Euler integration . . . 140
functional . . . 140
homeomorphism . . . 49
space . . . 49
deformation retraction . . . 4
degeneracy map . . . 214
degree
local . . . 94
of chain homotopy . . . 84
of map . . . 78
dendrogram . . . 216
diagram . . . 220
diffeomorphism . . . 10
differential forms . . . 121
direct image . . . 193
director field . . . 79
disclination . . . 79
discrete
exterior calculus . . . 231
flow . . . 150
Laplacian . . . 130
Morse theory . . . 149–151
vector field . . . 149
discretized configuration space . . . 37
disjoint union . . . 12
divergence . . . 122
DNA . . . 164
Dold-Thom theorem . . . 177
double complex . . . 109
Dowker complex . . . 73
dual
cell structure . . . 117
space . . . 112
duality
Alexander . . . 118–119
Poincaré . . . 117–118, 139
Verdier . . . 132

E
Eilenberg-MacLane space . . . 169
electric field . . . 122, 172
epic . . . 211
equalizer . . . 212, 222
equilibrium
Nash . . . 100
of vector field . . . 14
price . . . 80
étale space . . . 204
Euclidean neighborhood retract . . . 100
Euler
-Bessel tranform . . . 141
-Fourier transform . . . 141
-Morse index . . . 142
angles . . . 163
characteristic, $\chi$
additivity of . . . 50, 99
homological . . . 99
of sequence . . . 98
of sheaf . . . 195
of tame/cellular space . . . 44
class . . . 172
integral . . . 50
evasive . . . 151
exact sequence . . . 85
exchange rates . . . 115
excision theorem . . . 71
excursion set . . . 59, 136
exit set . . . 144
extension map . . . 198
exterior derivative . . . 122

F
face map . . . 63, 214
fiber . . . 51, 52, 161

bundle ... 171
homology sheaf ... 183
fibered knot ... 172
fibration ... 175
Hopf ... 171, 176
field
algebraic ... 241
director ... 79
electric/magnetic ... 122, 172
navigation ... 17
random ... 58
ruler ... 112
vector ... 14
5-lemma ... 89
fixed point ... 14, 79
index ... 47, 148
theorem
Brouwer ... 80
Kakutani ... 101
Lefschetz ... 100
flag complex ... 30
Floer homology ... 146
flow
cateogry ... 210
discrete- ... 150
information- ... 189
lines ... 165
network- ... 114, 186
of vector field ... 14
semi- ... 165
sheaf ... 186–188
value ... 114, 186
fluid dynamics ... 123
forcing ... 145
form
$p$- ... 121
1- ... 112
field ... 121
harmonic ... 129
symplectic ... 147
volume ... 129
Fourier-Sato transform ... 56, 198
Fredholm index ... 146
Fubini theorem ... 53, 198, 229
functor ... 213
functorial ... 213
functoriality ... 74–75, 159
fundamental group, $\pi_1$ ... 158

G
$\mathbb{G}_k^n$ ... 11
game of Hex ... 102
Ganea conjecture ... 152
gates ... 183
Gauss-Bonnet theorem ... 45, 56
Gaussian random field ... 58
generator
of group ... 240
of reconfiguration ... 39
genus
of knot ... 163
of surface ... 10
geometric
category ... 152
realization ... 27, 214
gluing axiom ... 191, 225
gradient ... 112, 122
discrete ... 150
graph ... 26
Laplacian ... 130
Grassmannian ... 11
group ... 240
braid ... 164
fundamental ... 158
homotopy, $\pi_n$ ... 166
surface ... 159
groupoid ... 209

H
$H_\bullet$ ... 64
$H^\bullet$ ... 113
$H_c^\bullet$ ... 118
Hadwiger theorem ... 57
hairy ball theorem ... 46
Hardt theorem ... 53
harmonic forms ... 129
helicity ... 127
Helly theorem ... 119
Hex game, theorem ... 102
Hodge
star, $\star$ ... 129
theorem ... 129
homeomorphism ... 4, 239
local ... 161
homology
barcode ... 202
Borel-Moore ... 132
Čech ... 70
cellular ... 64, 90

class, of cycle . . . 64
fiber cosheaf . . . 200
Floer . . . 146
intersection . . . 142
local . . . 72
Morse . . . 138–140
of cosheaf . . . 199
persistent . . . 104–106
reduced . . . 69
relative . . . 70–72
simplicial . . . 62–64
singular . . . 68–69

homotopy
group, $\pi_n$ . . . 166
lifting . . . 162, 175
of chain maps . . . 84
of maps . . . 4
type theory . . . 176

Hopf
fibration . . . 171, 176
theorem . . . 78

Hurewicz theorem . . . 169

I

$\mathfrak{I}$ . . . 47
incidence number . . . 72
independence complex . . . 26, 151
index
braid . . . 164
category . . . 220
Conley . . . 143–146
Euler-Morse . . . 142
fixed point . . . 47, 78, 148
Fredholm . . . 146
Lefschetz . . . 99, 101, 147
Morse . . . 136
pair . . . 142, 144, 148, 154
triple . . . 146

indicence number . . . 91
information flow . . . 189
initial . . . 212
injectivity radius . . . 31
input complex . . . 36
integration
Euler
constructible . . . 50–51, 197
definable . . . 140
numerical . . . 120
of cochains . . . 113
of forms . . . 123–125

interior . . . 239
interleaving . . . 110, 137, 219–220
intersection homology . . . 142
interval indecomposable . . . 105, 203
intrinsic volume . . . 56, 128
invariant set . . . 143
inverse image . . . 193
iso . . . 212
isolating block . . . 143

J

jet bundle . . . 24
Jordan curve theorem . . . 82

K

$K^2$ . . . 10
$K(\mathbf{G}, n)$ . . . 169
Kakutani theorem . . . 101
kernel . . . 240
kinematic map . . . 76
Kirchhoff
current rule . . . 66
voltage rule . . . 114

Klein bottle . . . 10
knot . . . 78, 163
fibered . . . 172

Künneth formula . . . 65

L

$\mathbf{\Lambda}^p$ . . . 121
Laplacian, $\Delta$ . . . 129–130
Lefschetz
-Hopf theorem . . . 148
fixed point theorem . . . 100
index . . . 99, 101, 147

lens space . . . 161
lift . . . 161
lifting
criterion . . . 162
homotopy . . . 175

limit . . . 220–222
link
of periodic orbits . . . 127, 165
simplicial . . . 33

linkage . . . 12
linking
asymptotic . . . 127
number, knots . . . 78

liquid crystals . . . 78, 168
LMD . . . 142

local
- cohomology . . . . 182
- degree . . . . 94
- homeomorphism . . . . 161
- homology . . . . 72
- Morse data . . . . 142
- orientation . . . . 72

long exact sequence . . . . 86
- of fibration . . . . 175
- of pair . . . . 87, 186
- of triple . . . . 89
- sheaf cohomology . . . . 185

loop . . . . 158
Lorenz attractor . . . . 165
Lusternik-Schnirelman
- category . . . . 151
- theorem . . . . 98

M

$\mu(\cdot)$ . . . . 136
$\mu_k$ . . . . 56
MacPherson-Srolovitz formula . . . . 58
magnetic field . . . . 122, 172
manifold . . . . 10
- Riemannian . . . . 123
- stable . . . . 15
- unstable . . . . 15
- with boundary, corners . . . . 22

map . . . . 4
- antipodal . . . . 11
- boundary . . . . 62
- chain . . . . 74, 84
- choice . . . . 170
- coboundary . . . . 113, 184
- coding . . . . 189
- definable . . . . 49
- degeneracy . . . . 214
- diffeomorphism . . . . 10
- differentiable . . . . 13
- extension . . . . 198
- face . . . . 214
- homeomorphism . . . . 4
- kinematic . . . . 76
- local homeomorphism . . . . 161
- restriction . . . . 180
- tame . . . . 49
- transfer . . . . 97

max-flow min-cut theorem 114, 187, 232
Maxwell's equations . . . . 122
Mayer-Vietoris sequence . . . . 88
microfluidics . . . . 38
Minkowski sum . . . . 54
model category . . . . 175
monic . . . . 211
monoid . . . . 233, 241
morphism . . . . 208
- of sheaves . . . . 193

Morse
- complex . . . . 138
- data, local . . . . 142
- homology . . . . 138–140
- index . . . . 136
- inequalities . . . . 139
- polynomial . . . . 150
- theory . . . . 136
  - discrete . . . . 149–151
  - stratified . . . . 141

motion planning . . . . 17, 173
motivic integration . . . . 60

N

Nash equilibrium . . . . 100
natural
- images . . . . 106–108
- isomorphism . . . . 218
- transformation . . . . 217

naturality . . . . 86, 90, 218
navigation field . . . . 17
nerve
- lemma . . . . 31
- of category . . . . 214
- of cover . . . . 31

network coding . . . . 188
nonevasive . . . . 151
normal cycle . . . . 128
nullhomologous . . . . 65
Nyquist-Shannon theorem . . . . 195

O

$O_n$ . . . . 19
$\Omega^p$ . . . . 121
$\Omega_p$ . . . . 126
o-minimal structure . . . . 49
object . . . . 208
open set . . . . 238
opposite category . . . . 213
orientation
- local . . . . 72
- of cell . . . . 67, 91
- of manifold . . . . 24

of object . . . 13, 76, 163
of simplex . . . 66
presheaf . . . 191
sheaf . . . 183, 233
output complex . . . 36

P
$\mathbb{P}^n$ . . . 11
path space . . . 173
periodic orbit . . . 15
discrete . . . 150
persistence
complex . . . 106
level-set- . . . 203
sublevel set- . . . 137
zigzag- . . . 108–109
persistent homology . . . 104–106
phylogenetic tree . . . 32
Poincaré
-Hopf theorem . . . 48
duality . . . 117–118, 139
polynomial . . . 65
point
-set topology . . . 238
cloud . . . 28
critical- . . . 136
poset . . . 241
precosheaf . . . 200
preimage theorem . . . 19
presentation . . . 240
presheaf . . . 190, 225
orientation- . . . 191
primes . . . 238
prism operator . . . 85
product
cap, $\frown$ . . . 132
cup, $\smile$ . . . 125
cartesian . . . 13, 221
topology . . . 239
wedge, $\wedge$ . . . 121, 124
projective
plane . . . 10
space . . . 11
pullback . . . 222

Q
quaternions . . . 163, 168
quotient
group . . . 241
space . . . 16, 239

R
Radon
theorem . . . 98
transform . . . 54
rank-nullity theorem . . . 19
reconfigurable system . . . 39
recurrence equation . . . 182
reduced homology . . . 69
regular cell complex . . . 28
regular value . . . 18
relation
algebraic . . . 240
sensing . . . 51
relative
homology . . . 70
Mayer-Vietoris sequence . . . 89
sheaf cohomology . . . 186
repeller . . . 17, 146
residual . . . 19
restriction map . . . 180
retraction . . . 4, 80, 218
de Rham cohomology . . . 124
Riemannian manifold . . . 123
ring
algebraic . . . 241
cohomology . . . 124
routing . . . 186

S
$\mathbb{S}^n$ . . . 11
$SO_n$ . . . 19, 68
saddle . . . 15
sampling sheaf . . . 196
scissors equivalence . . . 50
second spectral moment . . . 59
section
of fiber bundle . . . 172
of sheaf . . . 181
of tangent bundle . . . 14
sectional category . . . 173
Seifert
conjecture . . . 165
surface . . . 163
semiflow . . . 165
semigroup . . . 241
sensor
cover . . . 92–94
relation . . . 51
support . . . 51
set

boundary of . . . 239
closed . . . 238
closure of . . . 239
compact . . . 239
excursion . . . 59, 136
exit . . . 144
interior of . . . 239
invariant . . . 143
open . . . 238
partially ordered . . . 241
residual . . . 19
simplicial . . . 213

shadow . . . 29

sheaf
ambiguity . . . 196
capacity . . . 233
cellular- . . . 180–183, 224
cohomology
Čech . . . 190
cellular . . . 184
compactly supported . . . 194
relative . . . 186
topological . . . 192
constant- . . . 181, 192
constructible . . . 197
flow . . . 186–188
morphism . . . 193
network coding . . . 189
orientation- . . . 183, 233
sampling- . . . 196
skyscraper- . . . 181
subordinate to cover . . . 190
topological . . . 225

short exact sequence . . . 86

signal
of opportunity . . . 20
processing . . . 54, 141, 195

simplex . . . 26, 209
category . . . 214

simplicial
complex . . . 26
homology . . . 62–64
set . . . 213

singular homology . . . 68

sink . . . 15

skeleton . . . 27

skyscraper sheaf . . . 181

snake lemma . . . 86

social choice . . . 170

source . . . 15

space
base . . . 171
configuration . . . 12, 16
contractible . . . 4
coordination . . . 41
cotangent . . . 112
covering . . . 160
definable . . . 49
Eilenberg-MacLane . . . 169
Grassmannian . . . 11
lens . . . 161
manifold . . . 10
stratified . . . 23
tame . . . 49
tangent . . . 13
total . . . 171
tree . . . 32

Sperner's lemma . . . 96

sphere . . . 10

spike trains . . . 32

spline . . . 201

stability . . . 220

stable manifold . . . 15

stalk . . . 181, 191

state complex . . . 39

stochastic process . . . 58

Stokes' theorem . . . 123

Stone-Tukey theorem . . . 98

strategy
complex . . . 34
space . . . 101

stratified
Morse theory . . . 141
space . . . 23

structure theorem . . . 105

subgroup . . . 240

sublevel set persistence . . . 137

subspace topology . . . 239

surface group . . . 159

surgery . . . 137

symplectic form . . . 147

T

$\mathbb{T}^n$ . . . 13

$\tau$ . . . 99, 149

tame . . . 49

tangent
bundle . . . 14
bundle, unit . . . 171
space . . . 13

target enumeration . . . . . . . . . . . . . . . . . . 51
TDOA . . . . . . . . . . . . . . . . . . . . . . . . . . . . 22
template . . . . . . . . . . . . . . . . . . . . . . . . . 165
terminal . . . . . . . . . . . . . . . . . . . . . . . . . . 212
TOA . . . . . . . . . . . . . . . . . . . . . . . . . . . . . 22
topological complexity . . . . . . . . . . . . . . 174
topology . . . . . . . . . . . . . . . . . . . . . . . . 4, 238
    basis of . . . . . . . . . . . . . . . . . . . . . . 238
    compact-open . . . . . . . . . . . . . . . . . 239
    point-set . . . . . . . . . . . . . . . . . . . . . 238
    product . . . . . . . . . . . . . . . . . . . . . . 239
    quotient . . . . . . . . . . . . . . . . . . 16, 239
    standard . . . . . . . . . . . . . . . . . . . . . 238
    subspace . . . . . . . . . . . . . . . . . . . . . 239
    weak . . . . . . . . . . . . . . . . . . . . . . . . . 27
torus . . . . . . . . . . . . . . . . . . . . . . . . . . . . 10
total space . . . . . . . . . . . . . . . . . . . . . . . 171
transfer map . . . . . . . . . . . . . . . . . . . . . . . 97
transform, integral
    convolution . . . . . . . . . . . . . . . . . . . . 54
    Euler-Bessel . . . . . . . . . . . . . . . . . . 141
    Euler-Fourier . . . . . . . . . . . . . . . . . . 141
    Fourier-Sato . . . . . . . . . . . . . . . . . . . 56
    Radon . . . . . . . . . . . . . . . . . . . . . . . . . 54
transition graph . . . . . . . . . . . . . . . . . . . . 34
transversality . . . . . . . . . . . . . . . . . . . 17–20
    theorem . . . . . . . . . . . . . . . . . . . . . . . 19
travelling wave . . . . . . . . . . . . . . . . . . . 145
tree . . . . . . . . . . . . . . . . . . . . . . . . . . . . . 32
tree space . . . . . . . . . . . . . . . . . . . . . . . . 32
triangulation theorem . . . . . . . . . . . 49, 140
type theory . . . . . . . . . . . . . . . . . . . . . . . 176

U
$\mathfrak{UC}^n$ . . . . . . . . . . . . . . . . . . . . . . . . . . . 16
unimodal category . . . . . . . . . . . . . . . . . 153
unit tangent bundle . . . . . . . . . . . . . . . . 171
universal coefficient theorem . . . . . . . . 116
universal cover . . . . . . . . . . . . . . . . . . . . 161
unknot . . . . . . . . . . . . . . . . . . . . . . . . . . . 163
unstable manifold . . . . . . . . . . . . . . . . . . 15

V
$VR_\epsilon$ . . . . . . . . . . . . . . . . . . . . . . . . . . . . . 28
valuation . . . . . . . . . . . . . . . . . . . . . . . . . 57
van Kampen theorem . . . . . . . . . . . . . . . 160
vector
    bundle . . . . . . . . . . . . . . . . . . . . . . . . 171
vector field . . . . . . . . . . . . . . . . . . . . . . . 14
    discrete . . . . . . . . . . . . . . . . . . . . . . 149
    flow of . . . . . . . . . . . . . . . . . . . . . . . . 14
Verdier duality . . . . . . . . . . . . . . . . . . . . 132
Vietoris
    -Rips complex . . . . . . . . . . . . . . . . . . 28
    mapping theorem . . . . . . . . . . . . . . . 194
von Neumann - Mullins formula . . . . . . 57
vorticity . . . . . . . . . . . . . . . . . . . . . . . . . 123

W
$W^s, W^u$ . . . . . . . . . . . . . . . . . . . . . . . . . . 15
weak topology . . . . . . . . . . . . . . . . . . . . . 27
wedge
    product, $\wedge$ . . . . . . . . . . . . . . . 121, 124
    sum, of spaces . . . . . . . . . . . . . . . . . . 65
Whitehead theorem . . . . . . . . . . . . . . . . 169
Whitney embedding theorem . . . . . . . . . 21
winding number . . . . . . . . . . . . . . . 77, 124
witness complex . . . . . . . . . . . . . . . . . . . 29

Z
zero-section . . . . . . . . . . . . . . . . . . . . . . . 20
zigzag
    cosheaf . . . . . . . . . . . . . . . . . . . . . . . 202
    persistence . . . . . . . . . . . . . . . 108–109

Made in the USA
Lexington, KY
15 April 2015